EINSTEIN AND OPPENHEIMER

SILVAN S. SCHWEBER

# EINSTEIN AND OPPENHEIMER

## THE MEANING OF GENIUS

**HARVARD UNIVERSITY PRESS**
Cambridge, Massachusetts, and London, England
2008

*Library of Congress Cataloging-in-Publication Data*

Schweber, S. S. (Silvan S.)

    Einstein and Oppenheimer : the meaning of genius / Silvan S. Schweber.

       p.   cm.

    Includes bibliographical references and index.

    ISBN 978-0-674-02828-9 (alk. paper)

    1. Einstein, Albert, 1879–1955.   2. Oppenheimer, J. Robert, 1904–1967.   3. Physicists—Intellectual life—20th century.

    4. Physicists—Psychology.   5. Science—History—20th century.

    I. Title.

    QC16.E5S3285   2008

    530.092′2—dc22       2007043108

*For Snait*

*without whom it would not have been done*

*and*

*without whom it could not have been done*

# Contents

# Preface

In the light of the numerous publications connected with the centennial of Albert Einstein's *annus mirabilis* of 1905 and the several biographies of J. Robert Oppenheimer that have appeared in the past few years, one may well ask whether another book on Einstein and Oppenheimer is warranted.

A possible justification for the present volume is that it explores aspects of the lives and personalities of Einstein and of Oppenheimer that have received less attention: their views of individual and collective creativity, their link to Buddhist thought, their metaphysics, and in particular, how they coped with their lives after having climbed to summits that are unreachable to almost everyone else—this last, an aspect of their lives that is put into sharper relief by a comparative study.

One of my aims in the book was to banish the term *genius* when referring to these two extraordinary individuals by emphasizing how they created their science and made use of the cognitive and intellectual resources of their community; how they interacted with their colleagues and friends, and with the communities they were part of; how they molded themselves after they had become "great men"; how they lived with their "greatness"; and how they saw each other's greatness and interacted with one another. I refrained from using the attribution of "genius" because I wanted to stress the resources available to the individual and the role of the community. Calling Einstein a "genius" dwarfs the background against which his work was done—so that only the foreground, he himself, remains.

Both Einstein and Oppenheimer were singular human beings—off-scale as I am wont to say. But I am very conscious of the fact—and emphasize—that context and the vagaries of circumstances were factors in allowing them to become "great." Einstein was born at the right time and was at the right place to be able to make his truly remarkable contributions to the development of physics. Einstein could achieve his particular greatness by arriving on the scene at the particular time that he did: it was precisely the time when the limitations and fissures of "classical" physics had become manifestly evident and that it had been concluded that Newtonian gravitational theory could not account for the anomaly in the precession of the perihelion of Mercury. It was Einstein's creation of general relativity that makes him the extraordinary figure that he is in the pantheon of natural scientists. Any number of other people, individually and collectively, would have come up with the other outstanding contributions he made to the development of physics. Einstein did publish five remarkable papers during 1905, but they were the product of much prior research and rumination by him and by others—Heinrich Hertz, Hendrik Lorentz, Max Planck, and Henri Poincaré, in particular. And though perhaps in a somewhat different form, even general relativity would in time have been formulated by others. The fact is that there are crucial constraints of nature. Moreover, there is something to the "pre-established harmony between mathematics and physics" that Einstein believed in, howsoever that "harmony" is interpreted. A connection between geometry and gravitation would eventually have become established in other hands—but probably only later. In addition, it should also be remembered that Einstein had intense interactions with his friend Marcel Grossman and with David Hilbert in order to assimilate and master the mathematical resources necessary to formulate his final version of general relativity.

I have had the privilege to come into contact with or study the lives of some of the outstanding members of the theoretical physics community, including Paul Dirac, Werner Heisenberg, Wolfgang Pauli, Hans Bethe, Rudolf Peierls, Lev Landau, Eugene Wigner, Edward Teller, Robert Oppenheimer, Freeman Dyson, Richard Feynman, Julian Schwinger, Sin-Itiro Tomonaga, Steven Weinberg, Jeffrey Goldstone, and David Gross. There is a great deal of contingency in what such people accomplish. Much depends on the problems faced by the community, the intellectual resources then available, the freedom—economic, professional, and political—to pursue their researches. It is my contention that some of them would have accomplished many of the things Einstein did—except perhaps for general relativity.

Keep also in mind how different the contexts were in the first two decades of the twentieth century and during the 1930–1970 period regarding the size of the community and the experimental data available. It is with that in mind that I believe that Oppenheimer, given his talents, could have made a singular contribution to the creation of quantum mechanics during the decisive year 1925–1926 had he arrived on the scene at that time. Because of the trench dysentery he contracted in 1921 on a trip to Germany while searching for mineral samples, he had to postpone entering Harvard for a year, and thus came into his own a year too late to make contributions to the genesis of quantum mechanics of the same importance, if not as those of Heisenberg, Erwin Schrödinger, Dirac, and Wolfgang Pauli, then perhaps those of Max Born and Pascual Jordan. Bear also in mind that quantum mechanics was the result of an intense collective and communal effort, while the general theory of relativity is the work of one person.

I believe that there are always people like Einstein about. I believe that Julian Schwinger, Richard Feynman, Murray Gell-Mann, Frank Yang, Steven Weinberg, Gerard 't Hooft, Kenneth Wilson, David Gross, Edward Witten, and Frank Wilczek are such people. The fact is that the range of phenomena they had to deal with was much more extensive and the amount of reliable, accurate data they had to assimilate, comprehend, and explain was much greater than what Einstein had to confront with his gravitational theory. Also, their community was and is much, much larger, and the competition they faced much more intense. Put somewhat factitiously, Einstein's theory of general relativity accounted for two refractory pieces of data: the equality of the inertial and gravitational mass of an object and the advance of the perihelion of Mercury of 43 arcseconds per century. Einstein also made two predictions: the deflection of light passing near the sun and the redshift of light emitted on the surface of a heavy star. Of course, not to be forgotten and most important is the fact that under specified conditions, Newton's gravitational theory is a limiting case of general relativity, and therefore all the successes of that theory can be recovered from general relativity. But compare this with the phenomena and data accounted for by quantum mechanics, by the electroweak theory of Glashow, Salam, and Weinberg, and more generally by the standard model.

Given the size of the community, the selection process that operates in the choice of scientific fields that attracts gifted young people, and the cultural context, perhaps it is not possible for an individual at present to attain the "mythical" greatness status that Newton and Einstein were able to achieve. Should string theory be confirmed, I believe Witten would join their rank.

Schwinger, Feynman, Gell-Mann, Weinberg, and Wilson were lucky to be born at a time when their tradition and their culture at the local level inculcated in them a desire to emulate Einstein and others like him, and the culture and politics at the national level supported and valued their talents and strivings. There are many gifted young people like Lee and Yang in China today. Whether they will follow in Lee and Yang's path depends on the cultural, political, social, and economic values the Chinese government will promulgate, which in turn will determine the opportunities and channels such young people will have in order to develop their talents. Governmental Chinese policy will also determine whether there will be positions that will give such people the freedom to chart their own paths. Lev Landau did what he did because the only channels open to him in the Soviet Union during the 1930s and 1940s were condensed matter and nuclear physics. If born elsewhere, who knows what he might have accomplished. It is this set of underlying suppositions that has molded my exposition in the book.

Oppenheimer's singular standing derives from his directorship of Los Alamos. He helped bring about a victorious end to the last "just war," and in the process the products of the laboratory he directed transformed the world. But this achievement was at a personal price that he never came to terms with, and I should add, at a collective price that humankind has not yet come to terms with. Oppenheimer was aware of the deep transformation that had been wrought by nuclear weapons from the moment of their creation and use. He was never at peace with himself in the various roles he assumed thereafter—as a citizen, as a scientist, as an adviser to the highest echelon of his government, and as an intellectual. Nor was he able to integrate these roles so as to render him effective and durable in the part he played in averting the catastrophes that the use of fission or fusion bombs would bring about. The task he faced after World War II was superhuman.

My intent in putting this book together was only coincidentally to offer further biographical materials regarding the life of Einstein and Oppenheimer. As I have stated, during the past decade there have been a number of excellent studies of Einstein: the biographical works of David Cassidy, Gerald Holton, Albrecht Fölsing, John Stachel, and Palle Yourgrau and the researches of Jean Eisenstadt, Don Howard, John Norton, Jürgen Renn, John Stachel, and others on the genesis of general relativity readily come to mind. In addition, 2005 has generated a rich assortment of more general books on Einstein addressed to a wide readership. Similarly, the impressive biographies of Oppenheimer by Jeremy Bernstein, Kai Bird and Martin Sherwin, David Cassidy, Abraham Pais and Robert Crease, and Charles Thorpe have given us accounts of his life that will be definitive for a long time to come. Also, Priscilla McMillan's careful, thorough, and sensitive depiction of Oppenheimer's "ruin," and the works by Barton Bernstein, Gregg Herken, and others have presented us with as detailed a report as is possible at the present time of the complex story of Oppenheimer's involvement with the creation of nuclear weapons and the factors that led to the revocation of his security clearance.

I would also like to stress that this book should not be seen as an endeavor to compare Oppenheimer and Einstein, and certainly it is not to be interpreted as an attempt to diminish Einstein or elevate Oppenheimer, nor is it to be read as judging them. I have looked at their lives and their interactions in order to better understand them, both individually and in their context. As one of the reviewers of the manuscript indicated, a possible title for the book might have been *Einstein and Oppenheimer: The Scientific and Political Scene of Their Times*.

The book should be read as addressing the question: How did Einstein and Oppenheimer try to remain relevant after they had made their singular contributions? Though subtitled "The Meaning of Genius," the book is really about the meaning of greatness, of individuality, and of community. It does not address the questions: Why has Einstein become the mythic figure that he has? And: Why our fascination with Oppenheimer? Nonetheless, I hope that the book gives some insights into the distinctive character of their individuality to enable giving answers to these questions.

EINSTEIN AND OPPENHEIMER

# Introduction

The real is not given us, but set as a task
(in the manner of a riddle).

—*Albert Einstein (1949a, 681), quoting Immanuel Kant*

On one of their walks together in the early 1950s, Albert Einstein told J. Robert Oppenheimer, "When it has once been given you to do something rather reasonable, forever afterward your work and life are a little strange" (Oppenheimer 1956b, 2). Einstein had not been modest as a young man, but as a man in his seventies he characterized what he had achieved as "something reasonable" rather than as "something great." Both Einstein and Oppenheimer had done "great things"; both were "great"; both transformed our ideas concerning what human beings can be or do. In his essay on Chaim Weizmann, Isaiah Berlin noted that

> Greatness is not a specifically moral attribute. It is not one of the private virtues. It does not belong to the realm of personal relations. A great man need not be morally good, or upright, or kind or sensitive, or delightful, or possess artistic or scientific talents. To call someone a great man is to claim that he has intentionally taken . . . a large step, one far beyond the normal capacities of normal men, in satisfying, or materially affecting, central human interests. (Berlin 1981, 32)

To deserve the attribution of "greatness," Berlin required a thinker or artist to advance a society to an exceptional degree toward some intellectual or aesthetic goal, toward which it was already, in some sense, groping; or alternatively, to change its ways of thinking or feeling to a degree that would *not*, until the task had been performed, have been conceived as being within the powers of a single individual. Also, for Berlin, to be great, a thinker or artist need not have been a "genius."

Similarly, in the realm of action, greatness, for Berlin, demanded that the individual seemed able, "almost alone and single-handed, to transform one form of life into another; or—what in the end comes to the same—permanently and radically alters the outlook and values of a significant body of human beings." For this title to be truly deserved, the transformation that was effected had to have been such that those best qualified to judge considered it to be "antecedently improbable . . . something unlikely to occur without the intervention" of the person who for this very reason deserved to be described as great (Berlin 1981, 32–33).

Einstein surely was "great" by virtue of his extraordinary scientific accomplishments; Oppenheimer was deserving of the description by virtue of what he accomplished as a teacher in Berkeley during the 1930s and as director of the Los Alamos weapons laboratory during World War II. For both Einstein and Oppenheimer, science, and physics in particular, had a special meaning. Both were outstanding physicists, and becoming outstanding physicists was a necessary condition for their becoming "great" in Berlin's sense.

Both Einstein and Oppenheimer became iconic figures at a certain stage of their lives. After the successful verification in 1919 of one of general relativity's predictions—the bending of light in its passage near the sun—Einstein was hailed in the public sphere as a "genius," "the greatest scientist of the world," and "the most revolutionary innovator in physics" since Newton, a "universe maker" (Berlin 1981, 144–145). Myths were created about him.[1] In 1939, on the occasion of Einstein's sixtieth birthday, Oppenheimer could state in a public address that "[Einstein's] name is perhaps more widely known than that of any other living scientist; to many millions of people it has come to stand for science itself, and for all we admire in the way of life and thought of the scientist" (Oppenheimer 1939, 335).[2]

As time went on, in addition to his scientific stature Einstein became acclaimed for his humanity, his humility, and his lack of pretension. He also was admired for his political stance during the Weimar period in Germany, for his fight against fascism during the 1930s, after World War II for his commitment to racial equality, for his stand against McCarthyism, for his campaigning for peace, and for his efforts to eliminate nuclear weapons. He became seen not only as a courageous guardian of democracy and an outspoken foe of militarism, but as the embodiment of reason and of ethical behavior. And his concerns with morality and with religion found resonance in the public sphere. Perhaps no more admiring summary of Einstein's life, character, and accomplishments has been written than the *Biographical Memoir* by John Archibald Wheeler, himself an outstanding physicist who devoted a good part of his life to extending general relativity. This is what Wheeler said:

> Of all the questions with which the great thinkers have occupied themselves in all lands and all centuries, none has ever claimed greater primacy than the origin of the universe, and no contributions to this issue ever made by any man anytime have proved themselves richer in illuminating power than those that Einstein made. (Wheeler 1980b, 97)

Oppenheimer, for his part, was the "boy-wonder" of the American physics community. During the 1930s he created an outstanding school of theoretical physics at the University of California in Berkeley. After 1945, having overseen the making of weaponry that terminated World War II, Oppenheimer quite suddenly rose to public eminence and recognition. This new prominence had not been presaged by events preceding the war. As Isidor Rabi noted in his eulogy for Oppenheimer in 1967:

> [Before World War II] Oppenheimer's reputation and influence were centered around the small and close circle of physicists. As the wartime director of Los Alamos Laboratory, he was bound to receive important public attention, but there were other directors of great laboratories, and other physicists, who shared equal esteem but did not become objects of

such general interest. Oppenheimer after Einstein emerged as the great
charismatic figure of the scientific world. (Rabi 1969, 4)

By 1946, with the end of World War II and with the retirement of the
older American statesmen of science whose leadership had been generally accepted by the physicists, by the press, and by the public at large,

> this mantle naturally fell on the shoulders of Robert Oppenheimer. Al-
> though other eminent scientists exerted strong influences, for example
> Ernest O. Lawrence, Harold Urey, Arthur Compton, Lee DuBridge, and
> James B. Conant [and one should add Rabi himself], Oppenheimer's
> leadership was recognized more universally, both at home and abroad,
> even though he held no high position and was not the recipient of ex-
> traordinary scientific honors. (Rabi 1969, 3–4)

Oppenheimer became identified in the public mind with the awesome instrumental power of science. But in addition, by virtue of his public statements regarding nuclear weapons, he became recognized as someone who
understood the gravity of the crisis humankind faced by virtue of the existence of atomic bombs; as someone who was aware of how difficult it
would be to control their proliferation, and as someone who had thought
deeply about the transformations that were necessary in the world in order to integrate these developments into human life. He also became
recognized as someone struggling to define the new responsibilities of
scientists given their new knowledge and new powers. The creation of nuclear weapons had made warfare "inescapably a civilian as well as a military affair" and had made obsolete the separation and isolation of the
world of science and of the intellect from that of politics and practical affairs. Oppenheimer came to symbolize these changes, and was acknowledged as having the moral courage to act accordingly (Price 1967).

But whereas Einstein in the public mind represented the lonely "genius," Oppenheimer embodied the collective charisma of the physicists
who, after the war, became seen as special, dynamic people who had created a new world full of perils and new hopes; and also as an élite and as
entrepreneurs able to offer solutions to the problems they had helped
bring about.[3]

Yet there is a paradoxical element associated with Einstein and Oppenheimer's rise as iconic members of the scientific community, since after they acquired this position they were for the most part outside it. Though Einstein recognized the revolutionary changes in our understanding of the microscopic world brought about by the advent of quantum mechanics in 1925, he did not accept them. And after 1946 Oppenheimer, though continuing to keep abreast of developments in physics and identifying himself as a physicist, no longer did any research in physics. In 1956 he described himself as "a somewhat old and not so lively professional physicist" (Oppenheimer 1956a, 127). He became a scientific statesman, the director of the Institute for Advanced Study in Princeton, New Jersey, and the person overseeing the very active school of physics there. Although he tirelessly kept up with developments in quantum field theory and high energy physics, he did not set its intellectual agenda or that of the wider theoretical physics community.[4] Thus both Einstein and Oppenheimer were iconic and idiosyncratic at the same time, representatives of the community yet outsiders.

It is of course clear why the physics community should wish to accept Einstein as an iconic figure: Einstein's accomplishments in physics prior to 1927 were so extraordinary that his influence within physics and in many areas of culture in general has extended into the twenty-first century.[5] And there were good reasons in the case of Oppenheimer as well: he had been a first-rate theorist during the 1930s, the person responsible for creating at Berkeley during that decade the outstanding school of theoretical physics in the United States, and the person whose leadership at Los Alamos had enabled scientists, and physicists in particular, to design and assemble weapons that transformed the world. In that new world physicists had acquired unusual importance, status, and influence, had new responsibilities thrust upon them, and had gained some measure of power.[6]

How does one cope with being the kind of public figure that Einstein and Oppenheimer became? Surely, self-knowledge is a prerequisite. Perhaps Oppenheimer never came to a full understanding of himself, and it seems that he never achieved a stable sense of identity. Einstein on the

other hand was confident of who he was and on several occasions stated
so clearly. He never looked "upon ease and happiness as ends in themselves." The trite objects of human efforts—possessions, outward success, luxury—always seemed to him "contemptible . . . [and] without the
kinship of men of like mind, without the occupation with the objective
world, the eternally unattainable in the field of art and scientific endeavor, life would have seemed empty" (Einstein 1954, 9). In his tribute
to Max Planck on the occasion of Planck's sixtieth birthday, Einstein
stated that, like Arthur Schopenhauer, he believed that "one of the
strongest motives that lead men to art and science is escape from everyday life with its painful crudity and hopeless dreariness, from the fetters
of one's own ever shifting desires. A finely tempered nature longs to escape from personal life into the world of objective perception and
thought" (Einstein 1954, 225). As everyday life was *ephemeral,* the *permanent* world of physical nature became Einstein's escape. When in 1945
Herman Broch sent him a copy of his *The Death of Virgil,* Einstein wrote
him: "The book shows me clearly what I fled from when I sold myself,
body and soul, to science—the flight from the 'I' and 'We' to the 'It'."[7]

For Oppenheimer too, science was an escape "from the fetters of
one's own ever shifting desires." I do not know whether by 1932 he had
read some of Einstein's statements regarding his *Weltbilt* (worldview),
and in particular his homage to Max Planck. In a letter to his brother in
March 1932 Oppenheimer declared that

> through discipline, though not through discipline alone, we can achieve
> serenity, and a certain small but precious measure of freedom from the
> accidents of incarnation, and charity, and that detachment which pre-
> serves the world that it renounces. I believe that through discipline we
> learn to preserve what is essential to our happiness in more and more ad-
> verse circumstances, and to abandon with simplicity what would else
> have seemed to us indispensable; that we come a little to see the world
> without the gross distortion of personal desires, and seeing it so accept
> more easily our earthly privation and its earthly horror. . . . [I]n its nature
> discipline involves the subjection of the soul to some perhaps minor end;
> and that end must be real, if the discipline is not to be factitious. There-

fore I think that all things which evoke discipline: study, and our duties to men and to the commonwealth, war, and personal hardship, and even the need for subsistence, ought to be greeted by us with profound gratitude; for only through them can we attain to the least detachment and only so can we know peace. (Smith and Weiner 1980, 155–156)

In coming to his views concerning detachment, Einstein had been influenced by his early readings of Schopenhauer,[8] and in particular by Schopenhauer's insights into Buddhism. Oppenheimer, too, became deeply affected by his studies of Vedic texts at Harvard and at Berkeley, as is reflected in his comments regarding detachment, discipline, and incarnation in the letter to his brother. In Book 3 of the *Bhagavad Gita*, which Oppenheimer was studying with Arthur Ryder at the time he wrote his letter to his brother, Krishna proclaims to Arjuna: "Except by works done through sacrifice, this world is bound by works; therefore do thou, son of Kunti, carry out thy work to that end, free from attachment" (Johnston 1908, 58).[9] One of the aims of this book is to highlight these parallel proclivities regarding Eastern philosophy in the two men. These proclivities are addressed in Chapter 6.

Einstein had taken to heart what Schopenhauer had said about finding happiness in "what one is":

What a person is for himself, what abides with him in his loneliness and isolation, and what no one can give or take away from him, this is obviously more essential for him than everything that he possesses or what he may be in the eyes of others . . . for one's happiness in this life, that which one is, one's personality is absolutely the first and most essential thing. (Schopenhauer 1851, vol. 1, 348–349; quoted in Howard 1997)

If Einstein achieved some measure of personal happiness by being what he was for himself, there was nonetheless a tragic element in his life. In the last twenty-five years of his life, Einstein stopped questioning the validity of his approach in the unification of gravitation and electromagnetism *at the classical level*. He thus became isolated from the theoretical physics community, which was primarily concerned with the applications and extensions of quantum mechanics. Though very much outside the mainstream of contemporary physics, some of Einstein's most important work in general relativity—the proof of the existence of gravitational waves and the proof that the equations of motions of compact sources of gravitation were determined by the gravitational field equations—was carried out in Princeton during the late 1930s with his assistants, Nathan Rosen, Leopold Infeld, and Banesh Hoffman. Similarly, the questions he raised regarding the description of "physical reality" by quantum mechanics in the famous 1935 Einstein, Podolsky, and Rosen (EPR) article have proven to be consequential and seminal. The EPR paper is one of the most cited papers of the past few decades (Einstein et al. 1935).

How Einstein came to write the final version of his paper on gravitational waves[10] in 1936 illustrates the cultural dissonance he experienced in coming to the United States, and also reveals certain facets of his character.

On the heels of the first of his three winter visits to the California Institute of Technology, Einstein started publishing in the *Physical Review,* the leading American physics journal. The first of these communications was a letter written with colleagues at Cal Tech (Einstein et al. 1931) that was submitted to the *Physical Review* on February 26, 1931. In 1935, together with Boris Podolsky and Nathan Rosen, he published in the *Physical Review* the famous EPR paper (Einstein et al. 1935), which argued that quantum mechanics is not a "complete" physical theory and demonstrated that it exhibited strange nonlocal features. That same year there appeared in the *Physical Review* an article with Rosen on the equations of motion of particles in general relativity. In 1936, again with Rosen as coauthor, he published a paper in the *Physical Review* on the two-body problem in general relativity. In late May 1936, Einstein submitted to the *Physical Review* another paper he had written with Rosen with the title "Do Gravitational Waves Exist?" By 1936 most workers in the field believed that, by analogy with the electromagnetic field, gravitational waves must exist, although no experimental evidence existed at the time.[11] But in a letter to Max Born that mentioned the content of his paper with Rosen, Einstein wrote that they had arrived "at the interesting result that gravitational waves *do not* exist.... This shows that the non-linear general relativistic field equations can tell us more or, rather, limit us more than we have believed up to now" (Einstein 2005, 122).

On July 23, 1936, the editor of the *Physical Review,* John Tate, returned the manuscript to Einstein with a critical review by its referee and indicated that he "would be glad to have [Einstein's] reaction to the various comments and criticisms the referee had made."[12] Upon receiving the letter, Einstein wrote back to Tate on July 27:

Dear Sir,

We (Mr. Rosen and I) had sent you our manuscript for publication and had not authorized you to show it to specialists before it is printed. I see no reason to address the—in any case erroneous—comments of your anonymous expert. On the basis of this incident I prefer to publish the paper elsewhere.

Respectfully

*Albert Einstein*

P.S. Mr. Rosen, who has left for the Soviet Union, has authorized me to represent him in this matter.[13]

In his reply to the retraction of the paper, Tate wrote Einstein that he regretted Einstein's decision, but averred that he would not set aside the *Physical Review*'s review procedure, and therefore that he could not accept and publish a paper that the author was unwilling to submit to the Editorial Board before publication. Thereupon, Einstein proceeded to submit the manuscript to the *Journal of the Franklin Institute.*

It had been the case that any paper that Einstein had submitted to the Prussian Academy was automatically published in its *Sitzungberichte.* Similarly, the editors of the *Annalen der Physik* or the *Zeitschrift für Physik* would be the only persons deciding on the acceptance of a paper submitted to their journal. By 1930, every European scientific journal would automatically accept and publish any paper that Einstein had submitted.

The referee for the *Physical Review* was the distinguished Princeton cosmologist Howard P. Robertson, and his criticisms were correct. In the fall of 1936, upon his return to Princeton from the California Institute of Technology, where he had been on sabbatical leave for the academic year 1935–1936, Robertson convinced Leopold Infeld, Einstein's new assistant for the academic year 1936–1937, of a mistake in the paper. When Infeld informed Einstein of his meeting with Robertson and of the result of their discussion, Einstein agreed that he and Rosen had made a mistake. Einstein thereafter drastically edited in proof the manuscript that he had submitted to the *Journal of the Franklin Institute* and added the following note:

> The second part of this paper was considerably altered by me after the departure of Mr. Rosen for Russia since we had originally interpreted our results erroneously. I wish to thank my colleague Professor Robertson for his friendly assistance in the clarification of the original error. I thank also Mr. Hoffman for kind assistance in translation.
>
> *A. Einstein*[14]

Einstein and Rosen's original mistake stemmed from the fact that they had sought to find an exact solution for plane gravitational waves. They found that they could not do so without introducing singularities in the metric that described the waves. Thereupon they turned their finding into a nonexistence proof for solutions representing gravitational waves. As Daniel Dennefink explains in his article on the story behind the writing of the Einstein-Rosen gravitational wave paper, the singularity that Einstein and Rosen had encountered was an apparent singularity introduced by their choice of coordinate system, similar to the singularity one encounters when attempting to find the longitude of the North Pole.[15] In fact, in his referee report Robertson had indicated that the singularity was removed by a change to a cylindrical coordinate system.[16]

Some of the traits that Einstein displayed on this occasion were in evidence at other times.[17] His dismissive assessment of the referee's criticisms was an expression of his self-confidence, but also an indication of his courage—and daring—when he committed himself to devote most of his energies to a given undertaking and, in particular, to the unification of gravitation and electromagnetism. Nothing could sway him from his commitment to the task.[18]

Einstein's reaction to Tate's rejection of the paper in its original form was so strong that he never again submitted a paper to the *Physical Review*. As we shall see, he reacted this same way in other situations as well. Having once been "rejected"—whether it was a paper or the recommendations that he had made—it seems that he would harbor resentment for a long time thereafter.[19]

But there is one other aspect of the gravitational wave paper that should be mentioned. Infeld, in his autobiography, related that on the day after hearing of Robertson's criticism Einstein had been scheduled to give a lecture on his nonexistence proof of gravitational waves. He did not know yet how to rectify his mistake, but he was not fazed as he set out to give a lecture on the incorrectness of his and Rosen's proof. He evidently concluded his lecture by saying: "If you ask me whether there are gravitational waves, I must answer that I do not know. But it is a highly interesting problem" (Infeld 1980, p. 269).

One can point to similarities in the personalitiy and character of Einstein and Oppenheimer. They both were considered very attractive by women. Both had "charisma," though of a different type: Einstein's charismatic qualities were of a personal nature; Oppenheimer's were of a collective nature. Cornelius Lanczos, who knew Einstein from the late 1910s on and became one of his closest friends, wrote of "the magic of his personality" (Lanczos 1974, 12) that imposed itself on almost everybody who came in touch with him and of "the way the name 'Einstein' has assumed a charismatic sound with which only few words can compete" (Lanczos 1974, ix). Eugene Wigner, who knew Einstein from 1920 on, corroborated this assessment and added, "we all not only admired but also liked him [Einstein]" (Wigner 1980, 462). Many people admired Oppenheimer, and said so, but it is difficult to find people who said they "liked" him. He was endowed with charismatic qualities as a teacher and mentor at Berkeley and Cal Tech; this was attested to by all his students of the 1930s. And the charismatic qualities he acquired and displayed at

Los Alamos were important elements in the success of that enterprise (Schweber 2000; Thorpe 2006). But he was distant, and the charisma always expressed itself as "one to many" rather than "one to one."

Both Einstein and Oppenheimer exhibited stoical traits, both could be extremely stubborn and strong-minded, and both were at times self-centered and insensitive to others. Neither one was particularly successful in his personal relationship with his wife or as a parent to his children, though they both were very attached to their children when they were young. And there were truly tragic elements in the lives of their children: One of Einstein's sons eventually became institutionalized because of his schizophrenia and Oppenheimer's daughter committed suicide. Both men faced death simply and with great fortitude. They could talk about their eventual death as straightforwardly as they could discuss a conclusion in physics.

Both Einstein and Oppenheimer were deeply attracted to and interested in philosophy. In fact, in the late 1930s Einstein once told Infeld, "I am really more of a philosopher than a physicist" (Infeld 1941, 258). Einstein's reflections on Hume's analysis of causality, on the role of determinism in Spinoza's writings, on Mach's critical analysis of the long durée development of mechanics, were crucial in his arriving at both the special and the general theory of relativity. For Einstein, as well as for the other outstanding scientists and mathematicians of his professional milieu, the philosophical inquiries by Hume, Kant, Schopenhauer, Hegel, Mach, and the neo-Kantians of the Marburg school were closely connected to their knowledge of the past scientific accomplishments of Newton, Maxwell, Helmholtz, Boltzmann, Hertz, and Planck. And for Einstein in particular, philosophical issues were integrally linked to his knowledge of astronomical observational data and to his assumptions regarding the properties of blackbody radiation, of atomic and molecular sizes, of capillarity, and of the relation between the gravitational and the inertial mass of an object.

Oppenheimer, a product of the American context, reflected the influence of pragmatism and its limitations. He fully embraced Bohr's approach to the integration of philosophy and science but did not, perhaps could not, critically confront Bohr's presuppositions and metaphysics. He accepted Bohr's assessment, in his reply to the EPR paper, that the principle of complementarity was the central idea of the quantum revolution and that it left no escape from "a radical revision of our attitude as regard physical reality" and a fundamental modification of all ideas regarding the absolute character of physical phenomena (Bohr 1935, 702).

If there were similarities in Einstein and Oppenheimer's character, there were also striking differences. Einstein did not see himself as a tragic figure, whereas Oppenheimer most probably did for a time, especially after the revocation of his security clearance and the outcome of his "trial" in 1954. He *was* seen as a tragic figure—a symbol of the fallen hero—by many of his fellow physicists and by liberal intellectuals,[20] and was at times compared to Galileo and to Dreyfus.[21] There were yet other differences in their character: Einstein could poke fun at himself and others, but this element of "lightness" and "fun" was absent in Oppenheimer. Likewise, Einstein's deep connection with young children and empathy for them were traits that were not present in Oppenheimer. There was no element of ostentation in Einstein, whereas Oppenheimer often seemed to be playing a role and thrived on being oracular and "deep."[22] Similarly, anything to do with a cult of his personality was "painful" to Einstein,[23] whereas Oppenheimer did not distance himself from such manifestations.

Perhaps the most striking difference between these two men lies in their self-image. Einstein had a coherent image of his "self," and a cumulative biographical narrative can be attempted in which consistency outweighs rupture. A large part of his appeal as an iconic figure was precisely that he stood for individuality, consistency, and solidity. This consistency in later life was shaped by a singular vision of the world: a belief that an ultimate theory existed from which everything could be deduced. Moreover, after his remarkable accomplishment with general relativity, he came to believe that his vision of an ultimate theory could be implemented by a further geometrization of space-time. The character that he molded for himself reflected this vision, both its unity and its determinism.

Oppenheimer in 1966, after receiving his honorary degree
from Princeton University. (Photograph by William H.
Regan, Los Alamos National Laboratory; Argonne National
Laboratory, courtesy AIP Emilio Segre Visual Archives)

In contrast, it seems that Oppenheimer had no coherent image of his
"self," and ruptures can be said to characterize his biographical narrative. Nor did he believe that a unitary, ultimate theory that would encompass all phenomena was achievable. Though Oppenheimer was
deeply committed to physics, its approach to the understanding of the
world, *and* its communal aspects, physics could not supply the glue that
would give coherence and consistency to his character.

One of the most interesting psychological differences between these
two men will not be addressed, namely, their way of thinking. Much has

been written on the way Einstein thought and on the relation of language to his creativity.[24] To Jacques Hadamard, who had asked Einstein about the kind of signs that emerged in his mind when he was absorbed in scientific discoveries, Einstein replied: "The words of the language, as they are written or spoken, do not seem to play any role in my mechanism of thought." By contrast, Oppenheimer could be said to think by talking. And whereas Oppenheimer's mode of expression tended to be arcane, convoluted, and mellifluous, Einstein's was precise, economical, and direct, the result of his years at the Bern patent office where he had to "explain very briefly . . . why a device will or why it won't work; why the application [for a patent] should be granted or why it should be denied" (Wheeler 1980b, 103).

Einstein had a pictorial and geometric way of addressing scientific problems before translating his insights into mathematics (Miller 1984, 1996). His partiality to geometrical concepts was reinforced by the success of his description of gravity in terms of Riemannian geometry. He regarded electromagnetism as geometrical. In an essay written in 1934, entitled "The Problems of Space, Ether, and the Field in Physics," Einstein stated that "the idea that there exist two *structures* of space independent of each other, the metric-gravitational field and the electromagnetic [is] intolerable to the theoretical spirit. We are prompted to the belief that both sorts of fields must correspond to a *unified structure* of space" 278; my emphasis). For Einstein, electromagnetism was a structure of space, as was gravitation—hence the need for a unified representation of these seemingly disparate set of phenomena.

Oppenheimer, in contrast, was more analytic and formalistic. His pragmatic proclivity disposed him toward modeling and seeking solutions to problems at hand, with the tools at hand.

Since I shall be concerned primarily with events after 1925, it should be kept in mind that the late 1920s and early 1930s were a period of great change for Einstein. First, he found himself differing sharply with Bohr and the rest of the physics community over how fundamental an advance quantum mechanics was, particularly if the Copenhagen interpretation of the formalism was accepted. And after 1933, after Hitler came to power and Einstein resigned his position in Berlin and accepted a professorship at the Institute for Advanced Study in Princeton, Einstein became an "outsider" in his adopted country both as a physicist and as an émigré, never accepting the professional mores or mastering the national language. Anyone who has worked in the Einstein Archives must have been struck by the fact that after 1933 all the drafts of his speeches and the answers to the letters he received were first handwritten in German and then translated into English. What is also striking is his fluency in German, his mastery of that language, and his ability to bend it to his will and convey deep sensitivity, irony, sarcasm, elation, and grief, seemingly effortlessly. It is the transplanted Einstein that is my primary concern here.

Chapter 1 narrates Einstein's fairly extensive involvement in getting the American nuclear weapons program started, as well as his reactions to Hiroshima and Nagasaki, and, after the war, to the threat of global annihilation posed by the proliferation of fission and H-bombs. We thereby get a glimpse into the kinds of associations he was willing to enter in order to show his concerns. Perhaps the most striking revelation for me was the parallelism between Einstein's physical and political reasoning. In dealing with problems, both in science and in politics, he sought an ever more encompassing set of principles that would apply to an ever widening set of experiences: in special relativity, the equivalence of all inertial frames for all physical phenomena and the associated Lorentz covariance of the equations describing the phenomena; in general relativity, the local equivalence of gravitational fields and accelerated frames of reference and the principle of general covariance; in world politics, where the elimination of war was his primary objective, the establishment of a supranational government and a world court.

Similarly, Chapter 2, which deals with Einstein's involvement with the establishment of Brandeis University, not only indicates the importance he attached to his being Jewish, but also reveals how he reacted when what happened turned out to be different from what he had envisioned. It is well known that Einstein was persistent to the point of stubbornness—sticking to his beliefs and the results of his research in spite of pressures against them (as in the case of light quanta until the Compton effect established the validity of his viewpoint), stubbornly rejecting quantum mechanics as a final statement of the dynamics of the microscopic realm, and stubbornly clinging to his program to unify gravitation and general relativity. What Chapter 2 indicates is that Einstein could also be stubborn over more mundane things.

Chapter 3 attempts to sketch a biographical portrait of Oppenheimer by describing his reaction to three crises in his life: the breakdown he suffered in 1925–1926 while a graduate student in J.J. Thompson's laboratory at the Cavendish Laboratory in Cambridge, England, and his determination thereafter to become a theoretical physicist; his giving up doing research in physics in the aftermath of his proposal for the international control of atomic energy, the Acheson-Lilienthal plan, not being accepted by Bernard Baruch, James Byrnes, and Harry Truman; and the revocation of his security clearance and his evolution as a public intellectual. These incidents reveal how complex an individual Oppenheimer was. Oppenheimer was ambitious and charismatic, and as a young man he was arrogant and self-important. He could be extremely charming, but also harsh and inconsiderate.[25] His work in physics during the 1930s, at Los Alamos during World War II, and as governmental adviser in the postwar period gave him a deep sense of connection with communities that had distinctive purposes. But he found it difficult either to conceive an overall creative vision for himself or to devise a compelling objective for the community he belonged to if one had not been formulated at the time he assumed its leadership.

Wolfgang Pauli, for whom Oppenheimer worked as an assistant in 1929, already recognized this trait in him. In the fall of 1928 Oppenheimer spent a few months with Paul Ehrenfest in Leyden as a National Research Council fellow. Ehrenfest then sent him to Zurich to be under Pauli's tutelage because he thought that "for the development of his [Oppenheimer's] great gifts he needs to be knocked into shape [*zurecht-gepprügelt*] RIGHT AWAY AND WITH A LITTLE TENDERNESS." Ehrenfest hoped that Pauli would be able to do so.[26] In February 1929, after Oppenheimer had been with him for a while, Pauli wrote Ehrenfest that

Oppenheimer feels well here in Zurich, that he can work here, and that scientifically much good can still be got out of him. His strength is that he has many ideas and good ones, and, chiefly, a lot of imagination. His weaknesses are that he satisfies himself too often too quickly with poorly based claims; does not answer his own questions, which are often very interesting, because of lack of persistence and thoroughness. . . .

Unfortunately, Oppenheimer has a very bad characteristic: he comes to me with a rather unconditional belief in authority and regards everything that I say (or at least to a very high degree of approximation) as a last and final truth. I know how he came to need outside authorities. They are to solve his problems and answer his questions so that he does not have to do so himself. (This connection naturally is not *consciously* clear to him, but only latent to his unconscious.) But how I am to cure him of this I do not know. (Pauli 1979–2005, vol. 1, 486–487)[27]

This trait in Oppenheimer's character—the need for, and deference to, authority—is evidently what helped General Leslie Groves decide to appoint him as the director of Los Alamos. Murray Kempton, in his analysis of Oppenheimer's behavior during his 1954 Atomic Energy Commission (AEC) hearing, attributed Oppenheimer's "cravenness" to his subservience to authority (Kempton 1994). I would suggest that Oppenheimer's deference to authority should be linked to his attraction to "all things which evoke discipline: study, and our duties to men and to the commonwealth, war, and personal hardship," things which he thought "ought to be greeted by us with profound gratitude; for only through them can we attain to the least detachment and only so can we know peace."[28] But as Michel Foucault noted, "Discipline produces subjected and practiced bodies, 'docile' bodies," and "bodies" can be understood as "minds and bodies." Oppenheimer disciplined both his body and his mind. And as Foucault also observed, "Discipline increases the forces of the body (in economic terms of utility) and diminishes these same forces (in political terms of obedience)" (Foucault 1979, 138). The danger is that a perfectly disciplined body is at the same time a perfectly obedient body, and that a wholly disciplined mind is a mind that defers to authority.

In Chapter 3 I also analyze some of the reasons for Oppenheimer's successes: the demands of physics during the 1930s, the make-up of Los

Alamos, and the challenges of the postwar atomic world. In each of these enterprises he assumed a distinctive role and came to represent a distinct persona—but he could not integrate all these different aspects of his personality into a coherent whole that might serve as a model in the new world he had helped to create.

Oppenheimer was like a great conductor. But a great conductor, to consistently give exceptional, memorable performances, needs a superb orchestra staffed by excellent instrumentalists *and* an outstanding musical score. At Berkeley, Oppenheimer had outstanding students and postdoctoral fellows—among them Melba Phillips, Arnold Nordsieck, Wendell Furry, Willis Lamb, Robert Serber, Philip Morrison, Leonard Schiff, George Volkoff, Joseph Keller, Shuichi Kusaka, Hartland Snyder, Julian Schwinger, Sidney Dancoff, Bernard Peters, Edward Gerjoy, Leslie Foldy, and David Bohm—and a clear agenda: to make sense of the phenomena of nuclear and high energy physics and thereby make Berkeley and American physics outstanding. The same was true at Los Alamos. Its technical staff was made up of the best scientists in the United States and Great Britain, and the task to be accomplished was clear cut: make a uranium and a plutonium bomb. Part of his greatness as director of the Los Alamos Laboratory was his ability to create an atmosphere in which all the staff members—whether they were Nobel laureates or merely undergraduate physics majors—felt they had a voice in determining how to get the task done.

Similarly, in his governmental activities after the war, the people he led and was influenced by on the General Advisory Committee of the AEC (such as James Conant, Enrico Fermi, Isidor Rabi, and Cyril Smith) and those he came into contact with on the various high-level governmental committees he served on (George Marshall, Dean Acheson, Vannevar Bush, David Lilienthal, and William Clayton) were all remarkably able and gifted people who respected one another and could work together. Again the agenda was fairly clear cut: find a *modus vivendi* with the Soviet Union, given the political and military realities that existed in the Soviet Union *and* in the United States. If the musical metaphor is to be stretched further, at the Institute for Advanced Study Oppenheimer

had outstanding individual players who collectively could *not* play coherently as an orchestra, nor could he interpret the score either *to them*
or *with them* so that a great performance would result.

If one can invoke a musical metaphor to help characterize
Oppenheimer—he was an exceptional conductor under certain circumstances, but could not achieve greatness as a composer—it is more difficult to find an appropriate metaphor for Einstein. If we call upon the
musical metaphor, however, we can say that Einstein unquestionably
was *the* Mozart-*like* composer of scientific advances of the twentieth
century.[29] In 1916, Einstein's friend Paul Ehrenfest compared Einstein's
visit to Leyden with a marvelous concert, composed entirely of the
finest music, a concert whose " 'echoes, that go on resounding inside
you afterwards' offer as much satisfaction as the event itself" (Klein
1986, 329). For Ehrenfest, Einstein was one of the "wonders of Nature"
by virtue of his intellect, but also "a marvelous interweaving of simplicity and subtlety, of strength and tenderness, of honesty and humor, of
profundity and serenity (a somewhat melancholy serenity, to be sure)."[30]

Although he longed to escape from personal life, Einstein fashioned
himself into a deeply committed activist in the realm of civil liberties.
He became an energetic rebel against every form of authoritarianism, a
vociferous advocate of a Zionism that would respect the aspirations and
rights of both Jews and Arabs, and an active participant in the efforts to
control nuclear weapons. He raised his voice for justice, for the down-
trodden, for tolerance, and against war in any form.[31] In the public's
mind, he also became a saintly figure.

In 1923, in a letter he wrote to Hendrik Lorentz on the occasion of
Lorentz's seventieth birthday, Einstein revealed what he hoped to become:

> In daily life shame prevents us from declaring our love to those for whom
> we have the greatest admiration. However, your seventieth birthday may
> break this ban. How often when the human affairs surrounding me
> seemed hopelessly sad, have I found deep consolation in your noble and
> outstanding personality. A man like you gives consolation and exaltation
> by his mere existence and example. Furthermore, I am especially glad to

Einstein and Lorentz, 1921. (Photograph by Ehrenfest, courtesy of Museum Boerhaave, Leiden)

have such a close relationship with you, since I can look up to you as my teacher in matters of science. To follow you on your path has been the most important tenor of my life *(Lebensinhalt)*. However, not only in science, but also in your attitude towards the individual and towards human affairs you are and you remain for me a shining but unreachable ideal.[32]

Einstein clearly strove to achieve his goal. He came to be seen as a compassionate and concerned teacher—first to the physics community, then to humankind—giving "exaltation by his mere existence and example."

Niels Bohr was perhaps the person Oppenheimer most admired. Oppenheimer undoubtedly believed that Bohr gave "exaltation by his mere existence and example," and he looked up to Bohr in matters of science and general comportment. Although he certainly thought that Bohr's attitude toward the individual and toward human affairs was "a shining example" for him, Oppenheimer never had the "close relationship" with Bohr that Einstein had with Lorentz. Between August 1963 and May 1964, Oppenheimer gave a series of lectures on "Neils Bohr and Atomic Weapons" at the Brookhaven National Laboratory, at Cal Tech, at Berkeley, and at Los Alamos. The nontechnical part of these lectures was printed in the December 17, 1964 issue of the *New York Review of Books*. The article is a very informative account of Bohr's efforts to prevent an arms race between the Soviet Union and the United States once he had become aware of the likely success of manufacturing atomic bombs while the war was still going on. In his lecture, Oppenheimer painted a deeply admiring portrait of Bohr and implied that it was Bohr's March 1945 memorandum to President Roosevelt, written in the hope of altering Roosevelt and Churchill's stand with respect to sharing nuclear know-how with the Soviets, that prompted Secretary of War Henry Stimson to appoint the Interim Committee to advise him and President Truman on the future of atomic energy, and to help answer the more immediate question of how to use the atomic bombs. Actually, the memorandum never reached Roosevelt but instead landed on Stimson's desk. Oppenheimer served on the technical subcommittee of the Interim Committee and thus became involved with the decisions of where, when, and how to use the bombs. And the discussions he had had with Bohr surely influenced him in his actions on the committee.

In the last paragraph of his article on Bohr, Oppenheimer recalled that in late September 1945, on the day that Stimson was to retire as secretary of war, he attended a cabinet meeting at which he once again advocated "in final and eloquent terms" Bohr's recommendation for a friendly approach to the Soviet Union on the problems of the atom.

Later that day he was to receive a farewell salute from the runway of his plane from every officer in the Washington, D.C. area. For this occasion Stimson evidently felt that he had to have his hair trimmed, and asked Oppenheimer to sit with him while he was in the barber's chair. When it was time to go Stimson said, "Now it is in your hands." Oppenheimer concluded his article with the statement: "Bohr never said that. He did not need to." But the fact that Stimson did say it may have been as important in shaping Oppenheimer's life after the war as his discussions with Bohr at Los Alamos and thereafter.

Einstein and Oppenheimer belonged to two different generations, with the advent of quantum mechanics delineating the intergenerational boundary. Einstein was fortunate to have matured at a time when the theoretical physics community was in its infancy, and when the Newtonian framework was showing its limitations and inadequacies— in particular, the limits of its assumptions regarding space and time and its presupposition that its descriptions were valid at all length scales (i.e., for both macroscopic *and* microscopic phenomena). One might say of Einstein what he said of Newton: "The figure of Newton [read Einstein] has, however, an even greater importance than his genius warrants because destiny placed him at a turning point in the history of the human intellect" (Einstein 1954, 254). By the mid-1920s, when Oppenheimer began his graduate studies, theoretical physics was an established discipline with many more *young* practitioners. By then the division of labor between experimental and theoretical physics had become a fact, and it became the exception for a physicist to be both. I believe that Oppenheimer may have felt that, had he arrived on the scene one year earlier, he might have made a contribution as important as those of Born, Heisenberg, Schrödinger, Dirac, Jordan, and Pauli.

Also, by the end of the 1920s, the center of gravity of physics activities had shifted from Europe to the United States. Physics was done differently in the United States because the institutional framework there was different from the European one. A professor of theoretical physics on the continent usually had his own institute and determined his own individual intellectual agenda. In the United States, theoretical physicists were members of physics departments in which experimental and theoretical activities were carried out under one roof; moreover, these departments were almost always headed by an experimentalist.[33] Thus physics was much more of an *interdependent* experimental and theoretical endeavor in the United States than on the continent, especially in the 1930s with the development of nuclear and cosmic ray physics and their associated technologies: cyclotrons, cloud chambers, and counters. Oppenheimer became aware of this interdependence very early on, both at Cal Tech and at Berkeley. At Berkeley, his close friendship with Ernest Lawrence and his strong connection to Lawrence's Radiation Laboratory with its ever larger cyclotrons, and at Cal Tech his association with the nuclear and cosmic ray physicists in the Kellogg Laboratory and with the astronomers on Mount Wilson, had made clear to him the nature of this interdependence and that of science and technology. Already in the 1930s he would characterize technology as both the instrument and the consequence of science.

If it was natural for Oppenheimer to consider experimental physics and theoretical physics as cooperative and interdependent endeavors, his wide knowledge of the material practices of the other physical and natural sciences made him see them in a similar manner. In 1950, as the editor of a set of articles in *Scientific American* celebrating the accomplishments of the sciences during the first half of the twentieth century, Oppenheimer stressed "the extraordinary diversification and specialization of the several sciences." He noted that they differed from one another "by experimental techniques, by emphasis, by the kind of regularities that research reveals, by almost everything that might be codified as method," and he emphasized that "science [at mid-century] is a cooperative enterprise resting on specialization; *its unity is based on the fullest exploitation and encouragement of diversity*" (my emphasis). Furthermore, "the order that characterizes the relations of one part of science with another *is not primarily an hierarchical order*" (again, my emphasis). Although attempts had been made to sketch out possible hierarchies, Oppenheimer did not believe that these schemes had contributed much either to the growth of science or to its general understanding; certainly they d[id] not describe at all the benign and tolerant symbiosis in which the sciences have flourished and nourished one another. Tolerance, open-mindedness and confidence in the resolution of conflict by further inquiry—these constitute the liberalism of the sciences in their relations to one another. (Oppenheimer 1950, 22)

Einstein, on the other hand, did believe in a hierarchical ordering of the sciences. In an oft-quoted statement from his homage to Planck in 1918, he declared that "The supreme test of the physicist is to arrive at those universal elementary laws from which the cosmos can be built up by pure deduction" (Einstein 1954, 221). Every natural process, including life itself, was to be deducible. Einstein believed that the world could be *reconstructed* from such a reductionist approach—from the bottom up—and that an ahistoric, "final theory" could be apprehended that would explain the observed stabilities in the physical cosmos and in principle could also explain life.

Oppenheimer was skeptical about this thesis. He saw the world, and the knowledge thereof, as interconnected and changing, and he was agnostic as to whether unchanging, ahistorical theories could be apprehended at some truly "fundamental" level.

Oppenheimer and Einstein's differences were manifested in their assessments of quantum mechanics. To the question, "Does quantum mechanics chart the way to find out what holds the world together in its innermost parts?"[34] Einstein consistently gave a negative answer. Einstein never accepted Bohr's interpretation of the quantum mechanical formalism with its notion of complementarity, whereas Oppenheimer became one its leading advocates. In his introductory remarks to the articles he had solicited for the September 1950 *Scientific American* devoted to *The Age of Science: 1900–1950,* Oppenheimer explained complementarity as follows:

In the atomic world it is not possible to describe the atomic system under investigation, in abstraction from the apparatus used for the investigation, by a single unique, objective model. Rather a variety of models, each corresponding to a possible experimental arrangement and all required for a complete description of possible physical experience, stand in a complementary relation to one another, in that the actual realization of any one model excludes the realization of others, yet each is a necessary part of the complete description of experience in the atomic world. (Oppenheimer 1950, 22)[35]

Einstein never agreed with this viewpoint. He believed that a physical description should account for phenomena *independently* of the details of the observational apparatus. He was convinced that the conceptual framework of general relativity showed the way and particularly so after he, Leopold Infeld, and Banesh Hoffman had proved that the nonlinear features of the field equations of general relativity implied that the behavior of the sources of the gravitational field were determined by the gravitational field equations themselves—that the equations of motion of the particles, if these were the sources, were deducible from the gravitational field equations.[36] Perhaps Einstein's most succinct statement regarding the view that quantum mechanics offers as complete a description of individual phenomena as is possible was the remark he made to Bohr: "To believe this is logically possible without contradiction; but it is so contrary to my scientific instinct that I cannot forgo the search for a more complete conception" (in Schilpp 1949, 235).[37]

Oppenheimer, in contrast, believed that quantum mechanics had given deep new insights into the microscopic world, and whatever new theories would be devised to account for the submicroscopic world, they would have to replicate the striking and accurate results of quantum mechanics in the atomic and molecular domains. Like Bohr, Oppenheimer thought that the notion of complementarity could also yield better understanding in other areas of human inquiries, in particular in the study of biological, psychological, and cultural problems. Like Bohr, he believed that "complementarity" had extended and refined the pluralism natural to science and had "added new elements of subtlety to the idea of dialectic." It seemed to Oppenheimer that complementarity offered "a far richer and more adequate general point of view for the comprehension [and integration] of human experience than the misleadingly rigid and unitary philosophies that flowed so naturally from the experiences of Newtonian mechanics." In addition, part of the attraction of complementarity for Oppenheimer was that it allowed him "to emphasize the elements of analogy between the scientific tradition and the great traditions of oriental philosophy, of Lao-tse and of Buddha," traditions that Oppenheimer found very attractive.[38] Chapter 4 describes Oppenheimer's attempts, in his William James lectures at Harvard in 1957, to make complementarity a key to understanding the new post–World War II world.

Bohr thus looms large in the background of my story. During World War II, both Einstein and Oppenheimer were deeply influenced by their interactions with Bohr. Chapters 1 and 3 relate some of these contacts, which were concerned primarily with atomic weapons, their use during the war, and the implications of their existence for postwar relations between the United States and the Soviet Union.

Since Einstein and Oppenheimer belonged to different generations, the stage on which they acted and the freedom and constraints under which they operated made for different presentations of self.[39] The generational disparity helps explain some of the differences in Einstein's and Oppenheimer's views of communities, the differences in their relation to the communities they considered themselves part of, and the disparity in their approach to physics. These matters are taken up in Chapter 6. Here let me make only some brief observations. Einstein was an individualist with regard to intellectual and artistic matters. Because physical theories cannot be *deduced* from the empirical data, Einstein believed that the individual theorist has considerable latitude and freedom of choice in devising the physical concepts necessary to construct the theories that are to account for the regularities found in nature. The "great" scientists of the past were the ones who had been able to give the widest scope to this freedom of conceptualization.[40] More than that, Einstein believed that "great men" shaped history and that advances in the arts, in the humanities, and in science were due to the contributions of outstanding individuals who labored in the solitude of the creative process.[41] Thus he characterized Newton as "this brilliant genius, who determined the course of western thought, research and practice like no one else before or since" (Einstein 1954, 253). And in 1941, in a radio address to the British Association for the Advancement of Science, Einstein asserted:

> The supernational character of scientific concepts and scientific language is due to the fact that they have been set up by the best brains of all countries and all times. In solitude, and yet in co-operative effort as regards the final effect, they created the spiritual tools for the technical revolution which has transformed the life of mankind in the last century. Their system of concepts served as a guide in the bewildering chaos of perceptions so that we learned to grasp general truth from particular observation.[42]

Note Einstein's emphasis on the best minds' work being carried out in "solitude," as well as his somewhat begrudging recognition of their "co-operative" effort—theirs as individuals, not that of the general scientific community—which in the long run overcame the bewildering chaos and confusion that existed before the contributions of these great minds. But concerning more mundane matters, Einstein was deeply aware that his "inner and outer life are based on the labors of other men, living and dead, and that I must exert myself in order to give in the same measure as I have received" (Einstein 1949b, 90). Similarly, in his relation to Judaism Einstein could state:

> The best in man can flourish only when he loses himself in a community. Hence the moral danger of the Jew who has lost touch with his own people and is regarded as a foreigner by the people of his adoption.
>
> The tragedy of the Jews is that they ... lack the support of a community to keep them together. The result is a want of solid foundations in the individual which in its extreme form amounts to moral instability.

Oppenheimer, on the other hand, stressed the creative capacities of collectives, though he came to agree with Einstein to the following extent: "[that when] we work alone trying to get something straight it is right that we be lonely; and I think in the really decisive thoughts that advance a science loneliness is an essential part" (Oppenheimer 1956a, 135). To the extent that "loneliness" implied "individuality," Oppenheimer's stand undoubtedly reflected the fact that after World War II individuals like Julian Schwinger, Richard Feynman, Murray Gell-Mann, and Steven Weinberg could set the agenda of the theoretical high energy physics community, a quite sizable community of very able individuals by the mid-1950s. And of course Einstein's accomplishments gave credence to this position.

But it was also clear to Oppenheimer—as evidenced by the practitioners of high energy physics—that it is the *collective* activities of a scientific community that make advances possible. In addition, he would constantly stress that the successes of the various scientific communities depend on the fact "that as part of our culture, the understanding, the life of the mind, the life of science, in itself, as an end as well as a means, is appreciated, is enjoyed, and is cherished" (Oppenheimer 1956a, 135–136). Great advances could only be made in a culture that nurtured the long-term development of scientific knowledge and that made possible and valued the symbiosis of all knowledge resources–literature, the arts, mathematics, physics, chemistry, biology, the applied sciences, engineering, and the social sciences.

Let me give two examples of the emphasis that Oppenheimer placed on collectivities, one from 1958 and the other from some ten years earlier. In 1958 Oppenheimer and David Lilienthal served on the board of trustees of several foundations that were concerned with arms reduction. At one of these meetings Oppenheimer had expounded some ideas he had had concerning disarmament and had suggested that a few people get together to discuss them. Somewhat later, after the group had gotten together, he told Lilienthal:

> The reason that [meeting] was a unique and fruitful experience was that we began with the purpose of studying, of learning. We worked together as students, in the spirit of being educated by being together. It is part of your genius that you can turn individuals into groups. (Lilienthal 1969, 260)

Similarly, a revealing exchange took place between Oppenheimer and Archibald MacLeish following MacLeish's publication of an article entitled "The Conquest of America" in the August 1949 *Atlantic Monthly*.

Oppenheimer was disturbed by MacLeish's antidote for the anticommunism that was filling the postwar moral vacuum of the United States. MacLeish called for a return to Jeffersonian individualism and for a "redeclaration of the revolution of the individual." Despite his admiration for Jefferson, Oppenheimer wrote MacLeish that "man is both an end and an instrument" and that "culture and society play a profound part in the very definition of human values, human salvation and liberation." Therefore, "what is needed is something far subtler than the emancipation of the individual from society; it involves, with an awareness that the past one hundred and fifty years have rendered progressively more acute, the dependence of man on his fellows." Yet, as we shall see, Oppenheimer could be an individualist when matters of personal ambition were involved; thus he never joined the Federation of Atomic Scientists after the war.

Interestingly, in his letter to MacLeish, Oppenheimer also indicated how an *individual* could be responsible for important advances. He also informed MacLeish that earlier that year he had had a walk with Niels Bohr during which Bohr had given him further details about his notions of "openness" (the notion of free access to information about all aspects of life in every country), which was central in Bohr's ideas regarding the control of atomic energy. The notion of "an open world with common knowledge about social conditions and technical enterprises, including military preparations in every country,"[43] became central in Oppenheimer's assessment of the hopes and perils of the post-Hiroshima world. Bohr's ideas on the subject were reflected in the Acheson-Lilienthal report for the control of atomic energy that Oppenheimer helped draft. In his letter to MacLeish, Oppenheimer went on to expound the relevance of Bohr's views of complementarity as applied outside of physics, and asserted that Bohr had provided "that new insight into the relations of the individual and society without which we can have an effective answer neither to the Communists nor to the antiquarians nor to our confusions."[44]

Bohr's great influence on Oppenheimer was in evidence not only at Los Alamos and thereafter, but had already begun in the 1930s. After Dirac, Heisenberg, Pauli, and Fermi had laid out the basis for quantum electrodynamics in the late 1920s by applying the methods of quantum mechanics to the Maxwell equations describing the electromagnetic field, Oppenheimer was one of the first to investigate its predictions for the level structure of the hydrogen atom, with the electron's motion described relativistically by the single particle, pre-hole-theoretic interpretation of the Dirac equation. Beyond the first approximation, Oppenheimer found infinite—and therefore meaningless—answers. Bohr believed that these divergence difficulties in the formalism would only be overcome by a revolutionary change of the theory, just as quantum mechanics had been a radical transformation of the Newtonian dynamics and its metaphysics. Oppenheimer concurred, particularly when the "conservative" approach he had suggested to one of his students, Sidney Dancoff (essentially the approach that was formulated by Hendrik Kramers, Hans Bethe, Schwinger, and Feynman after the war and referred to as "renormalization theory" thereafter), failed to overcome the difficulties because of a calculational error that Dancoff had made.[45] This episode is detailed in Chapter 3 to indicate how perspicacious Oppenheimer's understanding of quantum field theory had been during the 1930s.

If Oppenheimer agreed with Bohr that a revolutionary change was needed to make sense out of quantum field theory, he never had any doubt that the successes of quantum mechanics at the atomic and molecular level had to be recovered from any future theory. Most theoretical physicists of Oppenheimer's generation thought that Einstein was on the wrong path in his attempt to fuse gravitation and electromagnetism at the *classical level,* and they were skeptical that the nonlinearities of such a unitary theory could recover the probabilistic aspect of the quantum description. Nonetheless, Einstein's efforts nurtured the vision of unification, which had its roots in nineteenth-century German Romanticism. In Chapter 5, I briefly review the kinds of unification that took place in physics during the first two-thirds of the twentieth century and focus on MIT's centennial celebration in 1961 to demonstrate the potency of Einstein's vision that there might be a fundamental theory from which all known theories could be derived. The conference took place at a time when nature was seen to work in terms of four separate and distinct forces: gravity, electromagnetism, the weak interactions (responsible for the radioactivity of nuclei), and the strong interactions (which were to account for the binding of protons and neutrons into nuclei in terms of the interactions of the plethora of mesons then known with one another and with nucleons). The views of Ronald Peierls, Chen Ning Yang, and Richard Feynman on the subject of unification and "final" theories at the MIT colloquium still merit attention. Chapter 5 also presents Oppenheimer's position on these matters.

In that chapter, I also briefly review the development of general relativity, primarily to better understand the context of the critical remarks that Oppenheimer made about Einstein and about general relativity in 1965 on the occasion of the tenth anniversary of Einstein's death.

Although Einstein and Oppenheimer had first met in early 1932, during one of Einstein's stays at the California Institute of Technology, they did not come in close contact until after the autumn of 1947 when they became colleagues at the Institute for Advanced Study in Princeton after Oppenheimer became its director. Einstein had a high regard for Oppenheimer and had only respectful things to say about him in public. Thus on April 13, 1954, in a front page article in the *New York Times,* when James Reston broke the news that the AEC had suspended Oppenheimer's security clearance and that he was to appear before the Gray Board, Einstein issued a simple, unpretentious statement expressing his support: "I admire [Oppenheimer] not only as a scientist but also as a great human being" (Pais 1994, 241). If Einstein knew of some of Oppenheimer's less admirable traits—his occasional intellectual arrogance and his sporadic unkindness—and of some of his weaknesses, he probably would have said, in a Spinozian spirit, "I cannot hate him because he must do what he does,"[46] or he would have quoted Schopenhauer's saying, "A man can do what he wants, but not want what he wants."[47] For his part, Oppenheimer, though deeply admiring of Einstein's achievements, was at times publicly highly critical of him. Chapter 6 gives an account of this facet of their relationship and addresses the question of Oppenheimer's animosity toward Einstein by looking at Oppenheimer's and Einstein's relation to their Jewish roots, their stance on nationalism, and their philosophical commitments.

# CHAPTER ONE

# Albert Einstein and Nuclear Weapons

Through the release of atomic energy, our generation has brought into the world the most revolutionary force since man's discovery of fire. The basic power of the universe cannot be fitted into the outmoded concepts of narrow nationalisms. For there is no secret and there is no defense; there is no possibility of control except through the aroused understanding and insistence of the peoples of the world.

We scientists recognize our inescapable responsibility to carry to our fellow citizens an understanding of the simple facts of atomic energy and its implications for society. In this lies our only security and our only hope— we believe that an informed citizenry will act for life and not death.

—*Albert Einstein*[1]

Quantum mechanics was responsible for a restructuring of the physical sciences that proved as consequential as that brought about by the Scientific Revolution. Just as Newton's physics transformed Western culture, so too did quantum mechanics during the twentieth century. And it is one of the characteristics of that century that the insights and understanding that quantum mechanics provided were soon thereafter marshaled for destructive ends. The design of the first atomic bomb was based on the conceptual tools provided by quantum mechanics: the uranium bomb was never tested before its use on Hiroshima. Similarly, the understanding of the structure and properties of metals and semiconductors given by quantum mechanics made possible the design of the first transistor, and transistors have transformed our world. They led to the chips that have revolutionized our computers and computing, and are revolutionizing our understanding and our means of understanding the biological world.

Though readily conceding its extraordinary success, Einstein never accepted quantum mechanics as *the* theory demarcating the path to a more fundamental understanding of the physical world. Already in December 1926, when the formalism of the new quantum theory had been put together and its statistical interpretation had been advanced by Max Born, Einstein wrote him:

> Quantum mechanics is certainly imposing. But an inner voice tells me that it is not yet the real thing. The theory says a lot, but does not bring us closer to the secret of the "old one." I, at any rate, am convinced that He is not playing at dice. (Einstein 1971, 91)

Einstein never wavered from this belief.[2] His initial criticism had been aimed at the formulation of quantum mechanics that described systems with a finite number of nonrelativistic particles interacting with one another. His objections continued with the further development of quantum theory. Thus, in 1934 he wrote Born: "I am greatly interested in your attempt to attack the quantum problem of the field from a new angle, but I am not exactly convinced. I still believe that the probability interpretation does not represent a practicable possibility for the relativistic generalization, in spite of its great success" (Einstein 1971, 122). Similarly, in an interview in 1945 he stated: "The quantum theory is without a doubt a useful theory, but it does not reach to the bottom of things. I never believed that it constitutes the true conception of nature" (Stern 1945, 240).[3] Einstein did not believe that the probabilistic elements that quantum mechanics had introduced were an irreducible characteristic of nature. Einstein's metaphysics was "realistic." He believed that just as macroscopic bodies always have "objective," quasi-sharply defined positions under any circumstance—that is, "irrespective of the experimental arrangement used to examine the system, or to which the system is being 'subjected"—this was also the case for microscopic entities, such as electrons. The fact that quantum mechanical description in terms of wave functions did not yield this information for microscopic objects rendered it incomplete. Einstein did not accept the fundamental tenet of the Copenhagen interpretation of quantum mechanics that *the state of a system is defined only by the specification of an experimental arrangement.* Pauli called Einstein's position "the idea (or the 'ideal') of the detached observer" (Pauli in Einstein 2005, 214–215).[4]

The period from 1927 to 1933, during which quantum mechanics was established as the dynamics governing the atomic and nuclear domains, was a time of great change for Einstein. By virtue of his stand with respect to quantum mechanics and of his views regarding the unification of gravitation and electrodynamics at the classical level, he became a marginal figure in setting the intellectual agenda of the physics community. The physics community in the United States was then primarily occupied with exploring and understanding the nuclear, atomic, and molecular domains and explaining quantum mechanically the properties of macroscopic systems in terms of the properties and interactions of their microscopic constituents. In addition, when in 1933, fleeing Nazi Germany, Einstein immigrated to the United States, he lost German, his mother tongue, as his means of communication with colleagues. He was unable to express himself with ease in English and thus became even more of an outsider in the physics community. But his stature as the greatest scientist since Newton, together with his modesty, his courage, and his forthrightness, made all his statements important and newsworthy and at times transformed them—even the political ones—into moral pronouncements. Einstein was well aware of his unique position and was always ready to use it to call attention to the dangers facing humankind and to condemn injustices done to individuals. After World War I as the evils of fascism and Nazism became clear, he renounced his pacifism. After World War II as the danger posed by American militarism and Stalinist totalitarianism became more apparent and the threat of a new war mounted, he became ever more vociferous in his defense of civil liberties and of the need to establish a world government. Still, none of the many like-minded individuals who appealed to him for aid could be sure of his support in advance. Einstein always was circumspect about which causes he would support—and how he would do so.[5] Soon after the end of World War II he started refusing "to act the bow-vow again" for fear "that my barking won't be of any use in more blatant cases" (Fölsing 1997, 492).

Einstein's political engagement reflected his belief that scientists should make their voices heard. In February 1939, when the Lincoln Birthday Committee for Democracy and Intellectual Freedom[6] asked him, "How can the scientist insure freedom of research and socially useful application of the fruits of his research?" he responded with a most succinct and incisive statement regarding the scientist as citizen:

> Freedom of research and the socially useful application of its results are dependent upon social factors. This explains why the scientists can exert their influence not as professionals but only as citizens. It further explains why scientists have an obligation to become politically active in the interest of free scientific research. They must have the courage, both as teachers and publicists, to enunciate with clarity their hard won political and economic convictions. Through organization and collective action, they must attempt to protect both themselves and society from any infringement upon the freedom to speak and the freedom to teach, and they must ever keep a vigilant eye in this field. (Nathan and Norden 1968, 283–284)[7]

However, Einstein's political and ideological positions were often bolder and more far reaching than others could support. And so his statements frequently became the pronouncements of a wise, insightful, iconoclastic, idealistic prophet rather than the rallying points for realistic political action. Consequently, his influence on events was highly variable.

After he had formulated the general theory of relativity in 1915, Einstein's research agenda was molded by his conviction that nature could be described by deterministic field equations that do not relate solely to probability amplitudes or potentialities and their changes—as the Copenhagen interpretation of quantum mechanics did—but to the temporal changes of an exterior real world. Underlying his quest for a nonlinear field theory that unified gravitation and electromagnetism was the hope that it would yield a description of nature that would encompass the explanation of microscopic phenomena given by quantum mechanics. Einstein was constant in his belief that this was the right path.[8]

Einstein was also constant in his views about war, the causes of war, and the consequent ills of humankind. Shortly after the outbreak of World War I, a few months after arriving in Berlin, Einstein signed Georg Friedrich Nicolai's statement challenging the October 1914 "Manifesto to the Civilized World" that had been signed by ninety-three leading German intellectuals and university professors (Wolff 2000). The Manifesto had defended German militarism, had sought to justify the German invasion of Belgium, and had ranted against the "shameful spectacle . . . of Russian hordes . . . allied with Mongols and Negroes . . . unleashed against the white race" (Nathan and Norden 1968, 3). The Nicolai statement called for "good Europeans" to join forces and organize a League of Europeans to raise its voice and take action against the war. Given the Germans' overwhelming support of the war, this was a very courageous act. Einstein's commitment to pacifism and his stand against the war, together with his contemporaneous formulation of his general theory of relativity, earned him the respect and admiration of scientists in England, possibly making their elated reaction to the successful 1919 observations of the bending of the light from distant stars in its passage near the sun more fervent.

For Einstein World War II, *qua* war, was quantitatively but not *qualitatively* different from World War I. What had been qualitatively different was the collective inhumanity the Germans demonstrated by their organized murder of six million of his brethren, of six million Jews.[9] For him atomic bombs and hydrogen bombs were a quantitative change in the scale of destructiveness that weapons and war could wreak. In this he differed from almost all the scientists who had been part of the Manhattan Project.

Thus Oppenheimer, in his farewell speech as director of the Los Alamos Laboratory, forcefully stated the opposite viewpoint. At the time, Oppenheimer was undoubtedly the most knowledgeable person concerning the manufacture of nuclear weapons and had thought deeply about the political consequences of their invention. On November 2,

1945, three months after Hiroshima and Nagasaki, he addressed the members of the Association of Los Alamos Scientists on the revolutionary nature of the new atomic age:

> What has happened to us—it is really major, it is so major that I think in some ways one returns to the greatest developments of the twentieth century, to the discovery of relativity, and to the whole development of atomic theory and its interpretation of complementarity, for analogy. These things, forced us to reconsider the relations between science and common sense. They forced on us the recognition that the fact that we were in the habit of talking a certain language and using certain concepts did not necessarily imply that there was anything in the real world to correspond to these. They forced us to be prepared for the inadequacy of the ways in which human beings attempted to deal with reality.[10]

In these sentiments Oppenheimer was echoing Neils Bohr, who had made precisely these points in connection with the description of the atomic realm: the habit of talking the language of classical physics and using classical concepts—such as attributing at a given time a definite position and velocity to microscopic particles such as electrons—made it necessary to introduce a new conceptual framework, epitomized by the uncertainty principle and the complementarity principle, to interpret the phenomena encountered at the microscopic level. Oppenheimer went on to state that the impact of the creation of atomic weapons should be compared to the threat that the development of science had posed to the Christian world during the Renaissance or to the threat that the theory of evolution presented to the "values that men lived by." That the impact of atomic weapons had been so great was due to

1. the extraordinary speed with which things that were on the frontier of science were translated into terms where they affected many living people, and potentially all people;
2. the scientists themselves playing such a large part, not merely in providing the foundations for atomic weapons, but in actually making them; and

3. the fact that the weapons arrived in the world with such a shat-
tering reality and suddenness that there was no opportunity for
the edges to be worn off.

Oppenheimer continued, indicating that he thought that the advent of
atomic bombs—because of the magnitude of the devastation they could
inflict, because they were relatively cheap to make, and because they had
shifted the balance between attack and defense—had made a quantitative change and that

> this quantitative change has all the character of a change in quality, of a
> change in the nature of the world. Atomic bombs have created a new sit-
> uation, so new that there is some danger, even some danger in believing,
> that what we have is a new argument for arrangements, for hopes, that ex-
> isted before this development took place. By that I mean that much as I
> like to hear advocates of a world federation, or advocates of a United Na-
> tions organization, who have been talking of these things for years—
> much as I like to hear them say that here is a new argument, I think they
> are missing the point, because the point is not that atomic weapons con-
> stitute a new argument. There have always been good arguments. The
> point is that atomic weapons constitute also a field, a new field, and a new
> opportunity for realizing pre-conditions. I think when people talk of the
> fact that this is not only a great peril, but a great hope, this is what they
> should mean. . . . [B]ecause it is a threat, because it is a peril, and because
> it has certain special characteristics, . . . there exists the possibility of real-
> izing, of beginning to realize, those changes which are needed if there is
> to be any peace.

Oppenheimer next commented on possible agreements between nations regarding the control of atomic energy. He stressed that no a priori
answer should be given to the question of the nature of such agreements: "that is something that is going to take constant working out."

> It would not hurt to have some concrete proposal—but that even as cru-
> cial a question as that of secrecy—which perplexes scientists and other
> people—*was not a suitable subject for unilateral action*. If atomic energy is to
> be treated as an international problem, as I think it must be, it is to be

treated on the basis of international responsibility and an international common concern. . . .

The first thing I would say about any proposals is that they ought to be regarded as interim proposals, and that whenever they are made it be understood and agreed that within a year or two years—whatever seems a reasonable time—they will be reconsidered and the problems that have arisen, and the new developments which have occurred, will cause a rewriting. (my emphasis)

Oppenheimer suggested that the nations participating in the interim agreement establish a joint atomic energy commission that would have exclusive control over all matters dealing with nuclear energy, and that the production of atomic weapons be banned. These suggestions would later be amplified in his contribution to the Acheson-Lilienthal plan.

Similarly, in his contribution, "The New Weapon: The Turn of the Screw," to *One World or None,* the pamphlet that the Federation of American Scientists had issued in 1946 to inform the general public about atomic bombs, Oppenheimer stressed that "the power of destruction that has come into men's hand has in fact been qualitatively altered by atomic weapons" (Masters and Way 1946, 59).

In contrast to Oppenheimer, Einstein had a more sweeping and radical vision. As far as Einstein was concerned, the control, even the elimination of atomic weapons, did not constitute the solution to how to avert wars, if for no other reason that other technological means such as explosive-laden rockets or biological weapons would be developed. The solution to the problem was not arms reduction or arms limitation but the elimination of war through the establishment of a world government with military powers to enforce the keeping of peace, together with a world court that would arbitrate conflicts between national states and that would issue rulings binding on and abided by the member states. This had been Einstein's belief ever since World War I. In 1932, while still a convinced pacifist, he stated his views as follows:

The greatest obstacle to disarmament has been the inability of most people to appreciate the enormity of the problem. Most objectives are ac-

complished in small steps. Think, for example, of the transition from absolute monarchy to democracy! But we are here concerned with an objective that cannot be attained step by step.

So long as the possibility of war exists, nations will continue to insist
on being as perfectly military prepared as they can, in order to emerge tri-
umphant from the next war. They will also find it unavoidable to educate
the youth people in warlike traditions and cultivating narrow nationalistic vanities. The glorification of the war spirit will proceed as long as
there is reason to believe that situations will arise where that spirit will
need to be invoked for the purpose of waging war. To arm means simply
to approve and prepare for war, not for peace. Hence, disarmament cannot come in small steps, it must come about in one stroke or not at all.

To accomplish so profound a change in the life of nations, a mighty
moral effort and a deliberate departure from deeply ingrained traditions
are required. Anyone who is not prepared to let the fate of his country, in
the event of conflict, depend without qualification upon the decisions of
an international court of arbitration, and who is not prepared to see his
country enter into treaties that provide for such a procedure without any
reservation, is not really resolved to avoid war. This is a case of all or nothing. (Nathan and Norden 1968, 163–164)[11]

Einstein's views did not change after World War II.

Much has been written on the subject of Einstein and nuclear
weapons. I refer the reader to Otto Nathan and Heinz Norden's fairly
comprehensive *Einstein on Peace* as a valuable source for many of Einstein's statements on nuclear weapons,[12] to the articles on the subject by
Paul Doty and by Bernard Feld in the volume edited by Gerald Holton
and Jehuda Elkana that contains the papers presented at the 1979 Einstein Centennial Symposium in Jerusalem, and to the recently published
*Einstein on Politics,* edited by David E. Rowe and Robert Schulmann. Much
material is also available on the Web.[13] My intent here is to review Einstein's involvement at the political level in the development of the
atomic bomb and to point out that, based on the Sachs files in the Einstein Archives in Jerusalem, his involvement was greater than he himself
recalled and depicted later on. I also focus on his activities following
President Truman's directive in early 1951 to develop a hydrogen bomb;
these activities indicate an aspect of Einstein's character that is less well

known: his elitism. They also illustrate the steadfastness—bordering at times on inflexibility—with which he held to his principles.

In this chapter I reexamine Einstein's participation in the development of nuclear weapons and review Einstein's reaction to the use of atomic bombs over Hiroshima and Nagasaki and his subsequent efforts to ban nuclear armament. I also compare the stand he took after the war regarding their proliferation and control with that taken by Niels Bohr and Robert Oppenheimer. Bohr and Einstein had been friends since the early 1920s and had a deep affection and respect for one another.[14] During World War II they had several consequential interactions. The rest of the chapter deals with Einstein's relentless focus on the need for a supranational government, his reaction to the development of hydrogen bombs, and an analysis of the difference in his stand when acting individually and when he was part of a collective effort. In 1954, shortly before his death, Einstein contacted Bohr to join him in initiating what became known as the "Einstein-Russell Manifesto." (This initiative is discussed later in this chapter.)

## Einstein and the Atomic Bomb

The young physicists who invented quantum mechanics—Werner Heisenberg, Paul Dirac, Wolfgang Pauli, and Pascual Jordan—were an extraordinary lot. The economic and academic context at the time dictated that only the very best could hope for a professorial position, and indeed only outstanding young researchers gained entrance into the inner circle around Bohr, Max Born, Einstein, David Hilbert, Max von Laue, Max Planck, or Arnold Sommerfeld. One such young man was Leo Szilard[15]—a member of the group of brilliant young Hungarians that included Michael Polyani, John von Neumann, and Eugene Wigner. They had come to Berlin in the early 1920s to study engineering but soon became captivated by science and mathematics. Szilard was not quite as proficient in mathematics as his two dazzling friends, von Neumann and Wigner, but for his Ph.D. dissertation with von Laue he had submitted a path-breaking analysis of the relation between entropy and information, whose importance became generally recognized only much later.

During the 1930s, Szilard made seminal contributions to nuclear physics. Very soon after the discovery of the neutron in 1932, sensitized by H. G. Wells's *The World Set Free: A Story of Mankind*[16]—a book he read as a youth not long after its publication in 1914 in which the devastation from an atomic bomb had led humankind to renounce warfare—Szilard became obsessed with the possibility of a nuclear chain reaction. The immediate stimulus had been Ernest Rutherford's assertion at a meeting of the British Association for the Advancement of Science that nuclear energy would have no practical application. Szilard was not convinced and, in fact, patented the idea of a nuclear chain reaction. After the discovery of fission by Otto Hahn and Fritz Strassmann in the winter of 1938 and the interpretation of the phenomenon by Otto Frisch and Lise Meitner, he immediately realized that the imagined chain reaction was now a real possibility should a uranium nucleus also emit neutrons when undergoing fission.[17] In early spring 1939 Herbert Anderson, Enrico Fermi, and Szilard, working together at Columbia University, were the first to measure the average number of neutrons released during fission and to establish the viability of a chain reaction. Szilard and Fermi shortly thereafter designed the first nuclear reactor.[18]

After Fermi and Szilard had established that on average 2.5 neutrons were emitted in the fission of uranium 235 (U235),[19] Szilard became deeply troubled about the dangers that an atomic bomb—now a real possibility—would pose if developed by Nazi Germany. Given the possibility of war breaking out soon and the strong likelihood that Nazi Germany was working on the production of nuclear weapons because it had cut off the supply of uranium ores from Czechoslovakia, Szilard believed that it was very important to insure the availability of uranium ores to the United States for its researches on the fission processes. As is well known, Szilard, after conferring with Eugene Wigner and Edward Teller, enlisted Einstein's help. Initially Szilard, Wigner, and Teller had thought of obtaining Einstein's help to have the Queen of Belgium—a personal friend of Einstein—use her influence to prevent the sale of uranium ores from the Belgian Congo to Germany and for the United States to buy the available raw materials. However, the mathematician Oswald Veblen, a colleague of Wigner in Princeton, convinced him that the U.S. government ought to be informed of the implication of Hahn and Strassmann's discovery and of Szilard's findings, that is, of the possibility of creating nuclear weaponry and of constructing nuclear reactors for the propulsion of ships and submarines (Pais 1997). Szilard, who together with Fermi had already approached the U.S. Navy for support and had been rebuffed, was skeptical. However, they agreed that they would submit the letter to be sent to the Queen to the State Department for its approval.

On July 12, 1939, Wigner met Szilard early in the morning at the hotel that was his residence, the King's Crown Hotel, which was located at 420 West 116th Street next to Columbia University in Manhattan. They were to drive to Peconic on Long Island to see Einstein, who was spending his summer vacation there in the house of his friend, Dr. Moore. Wigner and Szilard had known Einstein since the time they were university students in Berlin in the early 1920s. Both had attended Einstein's course on statistical mechanics in 1922,[20] and both had emigrated to the United States following the rise of Hitler in Germany. By 1939 both Wigner and Szilard were highly respected physicists. Wigner was the Thomas D. Jones Professor of Mathematical Physics at Princeton University, and the peripatetic Szilard was a research fellow at Columbia working with Fermi to establish all the experimental facts concerning the fission of U238 and U235 by slow and fast neutrons.

The discovery of the fission of uranium had electrified the nuclear physics community. American nuclear physicists had first heard of it from Niels Bohr on his arrival to the United States in January 1939 and immediately thereafter corroborated Hahn and Strassman's findings. Einstein became aware of nuclear fission in the spring of 1939. Bohr, who was staying at the Institute for Advanced Study for the spring term, had probably mentioned it to him. In any case there had been widespread reports in the press concerning these developments.[21] Thus, in response to the question William L. Laurence had posed to him on the

occasion of his sixtieth birthday in April 1939—whether Einstein believed the recent progress in the release of vast amount of energy from
the uranium atom justified the hope that humankind would be able to
tap the enormous stores of energy known to be locked up within
atoms[22]—Einstein gave the following answer (which was printed in an
article featured on the front page of the *New York Times*):

> The results gained thus far concerning the splitting of the atom do not
> justify the assumption that the atomic energy released in the process
> could be economically utilized. Yet, there can hardly be a physicist with so
> little intellectual curiosity that his interest in this important subject
> could become impaired because of the unfavorable conclusion to be
> drawn from past experimentation. (Nathan and Norden, 1968, 291)

After Szilard and Wigner had found Einstein, and Szilard told him
about his experiments establishing secondary neutron emission in fission and of his calculations indicating the possibility of a chain reaction
in a uranium pile moderated by graphite, Einstein exclaimed: "That
never occurred to me!" (Daran habe ich gar nicht gedacht!). Einstein
readily agreed to have a letter sent under his signature to one of the Belgian cabinet ministers, and he dictated a draft of the letter. It was agreed
that Szilard and Wigner would edit it and send it to Einstein for his signature. Wigner did so and sent the draft of his letter to Szilard.[23]

In the meanwhile, Szilard received a letter from Gustav Stopler, a
friend with whom he had consulted about these matters, indicating that
Stopler had discussed the problems with Dr. Alexander Sachs, a vice-
president of the investment firm Lehman Corporation and that Sachs
wanted to talk to him.[24] In the 1932 presidential campaign, Sachs had
written speeches dealing with the economy for presidential candidate
Franklin Roosevelt, and from 1933 until 1936 he had held a fairly high
position in the National Recovery Administration. Most importantly,
he had direct access to Roosevelt. A day or two after he received Stopler's
letter, Szilard saw Sachs. At their meeting Sachs convinced Szilard that
Einstein's letter should be addressed to Roosevelt to inform him of the
dramatic gains—military and economic—should the fission process be

mastered and successfully controlled to yield bombs and reactors and to alert the president to the danger that Nazi Germany might develop nuclear weapons first.

As Wigner had gone West on vacation, Szilard took it upon himself to alter the previous draft of a letter to be submitted to Roosevelt, with Sachs as a possible intermediary. He mailed the draft to Einstein,[25] who in turn indicated that he would like Szilard to come again to Peconic to revise the letter. This time it was Edward Teller, who was teaching at Columbia that summer, who on August 2 drove Szilard to Long Island to see Einstein. Einstein again dictated a draft, this time to Teller, and asked Szilard to convert it into a shorter and a longer version. Einstein would decide which one to send upon receiving the two versions. As none of the three felt they were experienced enough to decide who would be the best person to transmit the letter to Roosevelt, upon his return to New York Szilard phoned Sachs. Sachs, in turn, suggested three further names as possible couriers to Roosevelt: the financier and elder statesman Bernard Baruch, MIT president Karl T. Compton, and the aviator Charles Lindbergh, with Lindbergh as *"Favorit."*[26] Given Lindbergh's Nazi sympathies, this last suggestion was somewhat odd.[27] Szilard thereafter penned both a lengthier and a shorter version of the letter and sent them to Einstein. Einstein indicated that he preferred the more detailed letter but suggested that Szilard make it somewhat more "straightforward," for "it always gives one pause for thought when a person wants to do something too smartly." Szilard did so and it was this version, dated August 2, that Sachs transmitted to Roosevelt on October 12. Szilard also wrote a technical addendum to the letter, in which he indicated that the research in nuclear physics over the past five years had led to the conclusion that "a nuclear chain reaction could be maintained under certain well defined conditions in a large mass of uranium," even though it "still remains to prove this conclusion by actually setting up such a chain reaction in a large-scale experiment." Szilard also indicated that large quantities of energy could likely be obtained from the process in stationary power plants. As large quantities of uranium ores were necessary to implement the project, he stressed the im-

Einstein and Szilard, 1945. (Time Life Pictures/Getty Images)

portance of securing an adequate supply of uranium from the Belgian
Congo and of denying that source of ores to Germany. Until the time of
this writing, only slow neutrons had been used to induce fission;
whether fast neutrons would also work had not yet been established.
But if fast neutrons could be used, "it would be easy to construct extremely dangerous bombs" and even though "the destructive power of
these bombs can only be roughly estimated, . . . there is no doubt that it
would go far beyond all military conceptions." He also suggested that if
the possibility of a chain reaction were established, in view of its far-
reaching military consequences scientists should refrain from publishing on this subject. Appended to the memorandum was a reprint of
Szilard and Zinn's (1939) paper on the "Instantaneous Emission of Fast
Neutrons in the Interaction of Slow Neutrons with Uranium" that they
had published in the *Physical Review*.

The delay in Einstein's letter reaching Roosevelt[28] was due to Hitler's
invasion of Poland on September 1, the subsequent declaration of war

by France and Great Britain, and the consequent reassessment of U.S. policy in the face of these developments. The content of Einstein's letter is well known, so I will only focus on its main points here.

1. That the recent work of Fermi and Szilard implied that "uranium may be turned into a new and important source of energy in the immediate future"; and furthermore that it appeared almost certain that a chain reaction could be induced in a large mass of uranium. The new phenomenon could also lead to the construction of extremely powerful bombs. The situation called for watchfulness and, if necessary, "for quick action on the part of the Administration."

2. In view of this situation, it would desirable to maintain permanent contact between the administration and the physicists working on chain reactions in the United States. One way of achieving this contact could be to have Roosevelt appoint a trusted person to serve in an unofficial capacity (a) to help speed up experimental work by providing governmental funds or help get such funds from private sources, and (b) to keep the appropriate governmental departments informed of further developments and make recommendations for governmental actions.

The letter made a third point: there were reliable indications that Germany was actively pursuing these lines of research; and furthermore, that Germany had "stopped the sale of uranium from the Czechoslovakian mines which she has taken over."

Upon hearing Sachs out, Roosevelt asked General Edwin M. Watson, his military aide, executive secretary, and "gate-keeper," to act as his liaison. He also instructed Watson to introduce Sachs to Lyman Briggs, the director of the National Bureau of Standards, and to constitute, in the president's behalf, a committee, with Briggs as chairman, of representatives from the armed services to consider the ideas and memoranda submitted by Sachs.[29] Roosevelt also suggested that Sachs stay through the following day to meet Briggs. The choice of Briggs was somewhat unfortunate, however, for he was inarticulate and evidently not a very forceful man.

At their meeting on October 12, Sachs and Briggs agreed to include nongovernmental representatives as well, with Sachs as presidential representative and representatives from the Army and Navy to be selected by Briggs in consultation with Watson. The committee convened on October 21 with Lt. Col. Keith F. Adamson and Commander Gilbert C. Hoover as the representatives for the Army and the Navy, respectively, and Wigner, Teller, Fermi, and Szilard as the nongovernmental representatives. The discussions at the meeting were not friendly. The military representatives maintained that it would take several years before a favorable opinion could be given "as to whether it [the project] was worth being considered by the Armed Services and the Government" and that "those interested in the political-military implications were much too previous [i.e., premature] in converting a mere potential into an actual result of research." Hence they urged that the government should leave this project to the universities, "which anyhow have evinced active interest."[30] Briggs countered their argument by indicating that the world situation and American national interests must be taken into account in "the equations of probabilities." The impact of this particular "mere scientific possibility" on national defense was such that it warranted a different assessment than would be the case normally "of the risk coefficients attached to even remote possibilities."[31]

On November 1, Briggs, on behalf of the committee, submitted a report to Roosevelt that recognized the implied military and naval applications, though it characterized them as only possibilities and recommended that "In view of the fundamental importance of these uranium reactions and their potential military value, we believe that adequate support for a thorough investigation of the subject should be provided."[32] Subsequently, the sum of $6,000 was made available to the Columbia investigators. The report also recommended the enlargement of the committee "to provide for the support and coordination of these investigations in different universities" and suggested that MIT president

Karl Compton, Sachs, Einstein, and George Pegram, the dean of science at Columbia University, be invited to join the committee.

The Sachs correspondence in the Einstein Archives in Jerusalem documents Einstein's considerable involvement in the work of the Briggs Committee that oversaw the atomic energy program until mid-June 1940. At that time the committee was placed under the supervision of the newly organized National Defense Research Committee (NDRC), whose head, Vannevar Bush, reported directly to the president. Both Sachs's and Einstein's connections with governmental atomic energy matters ended with that changeover.

Until June 1940, Sachs kept Einstein informed of all of his governmental activities, as did Szilard and Wigner of their scientific researches.[33] Sachs's and Szilard's reports to Einstein that the scope and pace of the work were unsatisfactory prompted a suggestion that Einstein prepare another review of the situation for submission to Roosevelt. Einstein did write such a letter to Sachs, as drafted by Szilard. In this new letter, Einstein called attention to the intensification of investigations on uranium in Germany, "which research is being carried out in great secrecy,"[34] and emphasized that the German government had assembled a group of physicists and chemists from the Kaiser Wilhelm Institutes for Physics and for Chemistry to work on uranium problems and had placed the group under the leadership of Carl F. von Weizsäcker, a highly respected nuclear physicist and former student of Heisenberg. He added that "Should you think it advisable to relay this information to the President, please consider yourself free to do so." Sachs did transmit Einstein's letter to Roosevelt. In reply, Roosevelt instructed Watson once again to convene a conference that would include the members of the Briggs Committee "at a time convenient for you and Dr. Einstein."[35] Watson thereafter wrote Sachs asking him for suggestions of additional academic scientists who should be invited to the conference and indicating that "perhaps Dr. Einstein would have some suggestions to offer as to the attendance of the other professors."[36] After conferring with Einstein and Szilard, Sachs wrote Watson to recommend that Einstein and Fermi also be invited and suggested that the conference be held on April 29.[37] However, after meeting Einstein in Princeton, Sachs wrote that "it became clear that indisposition on account of a cold and the shyness which makes Dr. Einstein recoil from participating in large groups would prevent his attendance."[38]

Einstein himself wrote to Briggs suggesting that Fermi be invited to the conference. Einstein also indicated his interest in Sachs's suggestion "that the special Advisory Committee submit names of persons to serve as a Board of Trustees for a non-profit organization which, with the approval of the Government Committee, should secure from us governmental or private sources, or both, the necessary funds for carrying out the work."[39] The conference took place on April 27.

In public statements issued after World War II, Einstein conveyed the impression that his only connection to the atomic bomb project was writing his letter of August 1939 to Roosevelt. The facts are different, however. Not only was he involved with the activities of the Briggs Committee until June 1940, but he also carried out some research on diffusion processes that he did not associate with the bomb.[40] In addition, he was involved in two other episodes in the wartime history of the bomb, which occurred, respectively, in December 1943 and during the fall of 1944. Both of these episodes involved Neils Bohr. It will also become apparent that, although Einstein did not know any of the technical details of the enterprise, he was aware that important developments were taking place. If nothing else, he must have realized that most of the nuclear physicists from Princeton University had disappeared. Undoubtedly, whoever had specific technical knowledge about the Manhattan Project would have been very guarded and judicious in conversations with Einstein. Nonetheless, broad inferences could have been gleaned. This must have been the case when Bohr suddenly appeared in Princeton on December 22, 1943, shortly after his escape from occupied Denmark. Bohr had become a member of the British team and was on his way to Los Alamos.

Nothing much is known of the conversations that Bohr had with Einstein in December 1943. In some undated notes from the war, Bohr stated simply that he had tea with Einstein, Herman Weyl, Carl Ludwig

Siegel, Ostwald Veblen, James Alexander, Wolfgang Pauli, and others at the Institute for Advanced Study on December 23, 1943.[41] However, Einstein must have gleaned some hints from Szilard because he evidently told Bohr at the tea that he was glad that Bohr had come to help with the atomic bomb project because the "American Army was making a frightful mess of the uranium work and no doubt he [Bohr] would be a able to put this right" (Gowing 1964). A letter from Wallace Akers to M. W. Perrin on January 27, 1944, contains this comment:

> With regard to his own [Bohr's] moves, he is now satisfied that it is quite impossible for him to go to Princeton, to the Institute of Advanced Studies [sic]. It seems that he had a devastating experience when he first went there, on his arrival here; because Einstein, who is a very old friend of his greeted him in a crowded room with the statement that he (Einstein) was delighted that Bohr had come to America as he had heard from Szilard that the American Army was making a frightful mess of the uranium work and no doubt he (Bohr) would be able to put this right.[42]

When Bohr came to London after his escape from Denmark to Sweden in late September 1943, James Chadwick told him of the vast Manhattan Project and hoped that Bohr would go to Los Alamos and "lend his great weight" to Great Britain's contribution to the atomic bomb project.[43] Bohr immediately realized that the creation of an atomic bomb would bring fundamental changes in the world. Thereafter, working out the implications of atomic weaponry and preventing an arms race between the United States and the USSR after the war became Bohr's primary concern. Already after his first visit to Los Alamos in the spring of 1944, he wrote back to London that the future control of atomic energy would face not only elaborate and difficult technical and administrative problems, but more importantly, the need for nations to make concessions regarding the exchange of sensitive information as well as *openness* about their industrial efforts and their military preparedness. Such concessions were clearly difficult to comprehend by the standards of prewar international relationships. But Bohr was hopeful. He believed that atomic bombs were such dramatic and potent new

weapons that they could pave the way to a form of international relations. Bohr also believed that it would be disastrous if the Soviet Union
were not informed of these developments prior to their use in the war,
and so he was determined to discuss the political problems raised by the
bomb with the proper authorities in Great Britain and the United States
and to persuade them to inform the Soviet Union in order, hopefully, to
avoid a nuclear arms race (Gowing 1964, 1986; Aaserud 1999). Given
the caliber of Soviet physicists, he was sure that once a bomb was exploded there would be no atomic secret. In the spring of 1945, Bohr
wrote a memorandum that looked beyond the wartime situation. He
was sure that very soon ways would be devised to

> simplify the methods of production of the active substances and inten-
> sify their effects to an extent which may permit any nation possessing
> great industrial resources to command powers of destruction surpassing
> all previous imagination. Humanity will therefore be confronted with
> dangers of unprecedented character unless in due time measures can be
> taken to forestall a disastrous competition in such formidable arma-
> ments and to establish an international control of the manufacture and
> use of the powerful materials.[44]

Bohr therefore proceeded to outline the necessary administrative machinery, inspections and controls, and the concomitant openness regarding industrial and military facilities to avoid an arms race and
prevent the proliferation of atomic weapons. While at Los Alamos, Bohr
profoundly influenced Oppenheimer regarding the political implications of the possession and use of atomic bombs and had many discussions with him on how to address these problems.[45] The gist of many of
Bohr's ideas was later found in Oppenheimer's views regarding the international control of atomic energy.

A second episode involving Einstein and Bohr occurred in the fall of
1944. Otto Stern, who was a consultant to the Met Lab in Chicago[46] and
whose friendship with Einstein dated back to their association in
Prague and Zurich, told Einstein about the likely success of the bomb
project. They must also have discussed some of the postwar implications

of the existence of such weapons. These talks with Stern deeply troubled Einstein. With the successful building and operation of the reactors at Hanford in the summer of 1944, the Met Lab, whose responsibility had been the design of these reactors, found itself under much less pressure, and informal talks concerning the use of the bombs and the consequences of their existence were taking place.[47] Szilard in particular had prompted many of these discussions. Stern, who had just been awarded a Nobel Prize for the Stern-Gerlach experiment,[48] must have been privy to some of these discussions.

Otto Stern had another conversation with Einstein in early December 1944. On the day after Stern's visit on December 11, Einstein wrote a letter to Bohr, care of Henryk de Kauffman, the Danish ambassador in Washington, D.C.,[49] in which he referred to atomic bombs euphemistically, calling them weapons with technological means. Reflecting the discussions taking place at the Chicago Met Lab, a very anxious ("recht alarmiert") Stern, in his meeting with Einstein, had predicted that after the war in all countries

> there will be a continuation of the secret arms-race of weapons with technological means, which inevitably will lead to preventive wars [veritable wars of annihilation, worse in the loss of life than the present one]. And since the politicians don't know of these possibilities they are unaware of the magnitude of the threat. Every effort must be made to avert such a development.[50]

Einstein went on to say that he shared Stern's view of the situation, but that in the past he and Stern couldn't envision any joint effort that could help. "But yesterday, when Stern was here again, it seemed to us that there might be a way—even though it might be trifling—that could be successful." As Einstein wrote, the relevant, important countries had quite influential scientists to whom the politicians would listen and who could make them aware of the enormity of the danger.

> [Of such eminent scientists] with international contact there is you [Bohr], A. Compton here, Lindemann in England, Kapitza and Joffe in Russia, etc. The idea is to induce them to jointly bring pressure to bear on

the political leaders in their country, in order to achieve an internationalization of military power—a road that was some considerable time ago rejected as being too adventurous. But this radical step with all its
far-reaching political prerequisites concerning supranational government seems the only alternative to a secret technological arms-race.[51]

Stern and Einstein had agreed to have Einstein apprise Bohr of their
plan and to enlist him in its implementation. In his letter, Einstein ad-
monished Bohr not to immediately say "Impossible," but to wait a few
days until he had gotten used to this strange idea. And if he found
something worthwhile in it, "if only with a 0.001 chance" of success,
Einstein entreated him to get together with Stern (in Pittsburgh) and
James Franck[52] (in Chicago) for joint consideration.

The establishment of an effective supranational government and the
internationalization of military power had been the basic components
of Einstein's solution to the problem of war since the early 1930s. They
were to remain the basic components of Einstein's views of any solution
concerning the threat posed by nuclear weapons. Indeed, a single plane
with a single bomb could cause as much damage and kill as many people as what was formerly possible with a thousand planes, each carrying
a huge tonnage of conventional TNT bombs. But it was only a quantitative difference, which, when extrapolated, could annihilate all life on
the planet. Yet the basic issue for Einstein remained the elimination of
war, any kind of war.

For Bohr, even though he had thought deeply about the consequences of the creation of atomic weapons, his immediate goal in the
face of such weapons during the war was more limited. He wanted
wide-ranging discussions of atomic issues among the United States,
Great Britain, and the Soviet Union before any tests of the bombs were
carried out, and he had urged that the Soviet Union be told about the
progress being made toward building atomic bombs. Bohr had come to
the United States not only to participate in the atomic bomb project,
but also to try to convince Roosevelt of the necessity of such an approach. He had drafted a document prior to meeting Roosevelt in August 1944 that outlined his ultimate aim, namely, to seek

an initiative aiming at forestalling a fateful competition about a formidable weapon, [an initiative that] should serve to uproot any cause of distrust between the powers on whose harmonious collaboration the fate of coming generations will depend. . . . Of course, the responsible statesmen alone can have the insight in the actual possibilities.[53]

Supreme Court Justice Felix Frankfurter, whom Bohr knew well, relayed Bohr's views to Roosevelt. Roosevelt agreed with them and urged Frankfurter to "tell our friends in London that the President was anxious to explore ways for achieving safeguards in relation to X" [the atomic bomb]. This in turn led Bohr to seek a meeting with Churchill.

A few weeks before D-Day, Bohr met Churchill with disastrous consequences: Bohr failed to get his points across to an inattentive, otherwise preoccupied Churchill.[54] Although Bohr thought that a subsequent meeting with Roosevelt went well, Roosevelt had been upset by the fact that Justice Felix Frankfurter evidently had been made privy to atomic secrets that he should not have known about.[55] At their meeting in Hyde Park following the Quebec conference in early September 1944, Churchill convinced Roosevelt that the bomb project should be kept strictly secret until further notice. And they agreed that "inquiries should be made regarding the activities of Professor Bohr and steps should be taken to ensure that he be responsible for no leakage of information, particularly to the Russians."[56]

Subsequent conversations with John Anderson,[57] Bush, Frankfurter, Groves, and Richard Tolman[58] made it clear to Bohr that he had gotten nowhere in his discussions with Churchill and Roosevelt, and that the matter had become "enmeshed in the interstices of American politics and had caused difficulties for, and misunderstandings over, Mr. Justice Frankfurter."[59]

It was in this context that Bohr received Einstein's December 12 letter informing him of his talks with Otto Stern. Upon receiving Einstein's letter, Bohr consulted with Felix Frankfurter,[60] who also considered himself a friend of Einstein. It was Frankfurter who had made possible Bohr's interview with Roosevelt and had facilitated the interview with Churchill. According to Frankfurter, Roosevelt had encouraged him to have Bohr inform the British about Frankfurter's interest in discussing the postwar implications of atomic bombs with Churchill.[61] Frankfurter's help had enabled Bohr to go to England and to have Churchill ultimately give him an interview, following pressure on Churchill by John Anderson, Henry Dale, Frederick Lindemann (Lord Cherwell), and Jan Christiaan Smuts[62] to meet Bohr. The subsequent interview with Roosevelt was arranged by Frankfurter as a natural continuation of Bohr's meeting with Churchill.

Bohr had made reasonable and well-considered suggestions to men who wielded considerable power in Great Britain and the United States—John Anderson, Cherwell, Frankfurter, and Bush—and had been rebuffed. Bohr surely must have thought that Einstein's proposal was unrealistic and had no chance of being taken seriously. Furthermore, Bohr probably feared that Einstein might himself write to Pyotr Kapitza and Abram Joffe[63] and thus jeopardize the secrecy surrounding the Manhattan Project, an action that could be interpreted as treasonable. Moreover, the Soviet scientists that Einstein had suggested for Bohr to contact, Kapitza in particular, could hardly have been more ill-chosen. Earlier that year Kapitza had invited Bohr and his family to settle in Russia, and Bohr had graciously declined! Given Churchill's views of the Soviet Union and of communism, these exchanges between Bohr and Kapitza were the grounds for Churchill's apprehension about Bohr.

On December 22 Bohr rushed to Princeton, and in a lengthy meeting persuaded Einstein to remain silent. Bohr wrote a report of his meeting and handed it to the security officials in Washington. The typed copy of the report is among Bohr's papers. It opens with a statement that he had visited Einstein and had indicated to him that "It would be quite illegitimate and might have the most deplorable consequences if anyone who was brought into confidence about the matter concerned, on his own hands should take steps of the kind suggested." Bohr's note continued:

> Confidentially B [i.e., Bohr] could, however inform X [i.e., Einstein] that the responsible statesmen in America and England were fully aware of the scope of the technical development, and that their attention had been

called to the dangers to world security as well as to the unique opportunity for furthering a harmonious relationship between nations, which the great scientific advance involves. In response X assured B that he quite realized the situation and would not only abstain from any action himself, but would also—without any reference to his confidential conversation with B—impress on the friends with whom he had talked about the matter, the undesirability of all discussions which might complicate the delicate task of the statesmen.[64]

On December 26 Einstein wrote Stern a diplomatic letter stating that "a leaden cloud of secrecy" had descended upon him because of his letter to Bohr, so that he could only inform him that they were not the first to consider the situation. Einstein had the impression that the issues would be given serious attention, and "that the matter will be best served, if for the time being one does not speak of it, and especially that one does not promote it in any way, because at the present moment public attention hinges on this. Such nebulous way of talking is difficult for me, but this time I cannot change it."[65]

It is thus clear that as of December 1944, Einstein knew a fair amount about the development of nuclear weapons and about attempts to influence American and British statesmen regarding nuclear weapons policy. It is against this background that Einstein's meeting with Szilard in March 1945 and Einstein's March 25, 1945 letter to Roosevelt to introduce Szilard should be assessed. Einstein's statement "[that he is writing the letter] in spite of the fact that I do not know the substance of the considerations and recommendations which Dr. Szilard proposes to submit to you" is to be taken literally. He did not know what Szilard had written, and he meant to protect Szilard. He wanted to be explicit about his awareness that the terms of secrecy under which Szilard was working "do not permit him to give me information about his work," and that there had not been any breach of secrecy. "However," Einstein could state that "I understand that he now is greatly concerned about the lack of adequate contact between scientists who are doing this work and those members of your cabinet who are responsible for formulating policy."[66]

That Einstein did know a fair amount is revealed by a statement he made shortly before he retired from the Institute in April 1945. In an interview for an article in the *Contemporary Jewish Record,* Alfred Stern had asked him "whether the disintegration of atoms would not be able to release the tremendous atomic energies for warfare?" Einstein had replied, "Unhappily, such a possibility is not entirely in the Utopian domain. When military art is able to utilize nuclear atomic energies it will not be houses or blocks of houses that will be destroyed in a few seconds—it will be entire cities" (Stern 1945, 246). He said this four months before Hiroshima and Nagasaki, and the article appeared in June 1945!

In retrospect, as significant as Szilard, Einstein, Wigner, Teller, and Sachs's diplomatic and political activities had been until June 1940 when the NDRC took charge of the atomic weapons project, the bomb program in the United States effectively started upon the receipt of the British MAUD report in September 1941.[67] That report detailed the Frisch-Peierls calculations[68] of the critical mass of U235 needed for a uranium bomb and their estimation of the feasibility of separating the U235 isotope from "natural" uranium. Only after studying this document did James Conant, who was in charge of the NDRC atomic energy program, become convinced that an atomic bomb could be produced in time to alter the course of the war. Thereafter, in mid-October 1941, Bush recommended that Roosevelt go ahead with the project to build a bomb. This was a few weeks before Pearl Harbor. The timing was crucial: It would not have been possible to obtain the top priority ranking for the project after Pearl Harbor.

Despite his tenuous connection to the process that culminated at Hiroshima and Nagasaki, Einstein faced criticism after the war for having "participated" in the bomb project. His answers, at times tinged with regret, focused on the perceived threat from Germany and were almost always the same:[69]

I have never done research having any bearing upon the production of the atomic bomb. My sole contribution in this field was that in 1905, I established the relationship between mass and energy, a truth about the physical world of a very general nature, whose possible connection with the military potential was completely foreign to my thoughts. My only contribution with respect to the atomic bomb was that, in 1939, I signed a letter to President Roosevelt in which I called attention to the existing possibility of producing such a bomb and the dangers that the Germans might make use of that possibility. I consider this my duty because there were definite indications that the Germans were working on such a project. (January 23, 1950; Nathan and Norden 1968, 519)

In an answer to the editor of *Kaizo,* Einstein expanded on the last sentence:

Yet I felt impelled to take this step because it seemed probable that the Germans might be working on the same problem with every prospect of success. I saw no alternative but to act as I did, although I have always been a convinced pacifist. (Nathan and Norden 1968, 584)

And when the inconsistency of his commitment to pacifism and his "indirect participation" in the production of an atomic bomb were pointed out to him, Einstein answered:

Your reproach is well taken from the point of view of an absolute, i.e., unconditional pacifist. But in my letter to *Kaizo* I did not say that I was an absolute pacifist, but, rather, that I had always been a convinced pacifist. While I am a convinced pacifist, there are circumstances in which I believe the use of force is appropriate—namely, in the face of an enemy unconditionally bent on destroying me and my people. In all other cases I believe it is wrong and pernicious to use force in settling conflicts among nations. (Nathan and Norden 1968, 585, 587)

In addition to these answers to criticisms, Einstein made the following comment to Linus Pauling during Pauling's visit to Princeton on November 16, 1954: "I made one great mistake in my life—when I signed the letter to President Roosevelt recommending that atomic bombs be made. But there was some justification—the danger the Germans would make them."[70]

The last time Einstein commented on these matters was a few weeks before his death. Max von Laue, one of the few German physicists with whom Einstein remained on friendly terms after the war, had asked him for clarification of his letter to Roosevelt. Here is Einstein's answer:

> My action concerning the atomic bomb and Roosevelt consisted merely in the fact that, because of the danger that Hitler might be the first to have the bomb, I signed a letter to the President which had been drafted by Szilard. Had I known that that fear was not justified, I, no more than Szilard, would not have participated in opening this Pandora's box. For my distrust of governments was not limited to Germany. Unfortunately, I had no share in the warning made against using the bomb against Japan. Credit for this must go to James Franck. If they only had listened to him![71]

As long as he was alive Einstein would answer criticisms leveled against him for having written to Roosevelt. The criticism usually came from people who did not know the details of the events that had prompted Einstein to write his letter. Thus the rebuke tendered in the statements Heisenberg made shortly after Einstein's death in 1955 are of a different nature, coming from a person who must have been aware of the exact extent of Einstein's involvement in the atomic bomb project as well as of the context that prompted his letter to Roosevelt.[72] In a popular article entitled "The Scientific Work of Einstein," Heisenberg (1955b) found it difficult to understand "that Einstein, to whom war was hateful, should have been moved by the infamous practices under Nazism to write a letter to President Roosevelt in 1939, urging that the United States *vigorously* set about the making of atomic bombs ... [which eventually] killed many thousands of women and children" (Heisenberg 1974, 6; my emphasis).

But we should recall that Einstein's letter had stated explicitly that its intent was "a call for watchfulness and, if necessary, quick action."[73] Heisenberg criticized Einstein even more severely in 1974, when he spoke at the "Einstein house" in Ulm, Germany. As in 1955, Heisenberg began with a thoughtful and admiring survey of Einstein's work in physics but "in order not to leave the portrait of Einstein all too incomplete" then added that Einstein had written "three letters to President Roosevelt, and thereby contributed decisively to setting in motion the atom bomb project in the United States. And he also collaborated actively, on occasion, in the work on this project" (Heisenberg 1989, 120)! As we have seen Einstein only wrote two letters to Roosevelt. The second one referred to by Heisenberg was addressed to Sachs and forwarded by Sachs to Roosevelt, and the third was a letter of introduction for Szilard, who did not tell him of the details of why he wanted to see Roosevelt, though he had intimations of his intent in doing so. Surely Heisenberg in 1974 must have known that Einstein had not participated in the scientific or technical work of the project.

## After Hiroshima and Nagasaki

On August 6, 1945, a single plane carrying a uranium-based atomic bomb exploded the bomb over Hiroshima.

Later that day, Helen Dukas, Einstein's secretary, heard the news of the bombing and told Einstein what had happened. His response was *"Oh weh"* a wrenching *cri-de-coeur* that is not readily translated.

The Hiroshima A-bomb released the equivalent of 10 kilotons of TNT and reduced the city to rubble; some 140,000 of its inhabitants died by the end of the year, and over 200,000 died within five years of the bombing. Three days after the Hiroshima explosion a plutonium bomb was detonated over Nagasaki, leveling it and causing over 70,000 deaths by the end of 1945.[74]

In a TV interview after Einstein's death in 1955, Szilard recalled that shortly after the bombs were dropped on Hiroshima and Nagasaki, he had visited Einstein in Princeton. Their conversation turned back to Szilard's visit to Einstein on Long Island in 1939 when they had discussed the letter Einstein might write to Roosevelt. The memory of that fateful meeting had prompted Einstein to say, "The ancient Chinese were right. It is not possible to foresee the results of what you do. The only wise thing to do is to take no action—to take absolutely no action."[75] Einstein was alluding to the Dao precept: "Practice no-action; Attend to do-nothing; and therefore do no harm."

For many of the scientists at Los Alamos, having the actual horror of the damage caused by the atomic bomb dropped on Hiroshima shown and described in gruesome details soon after the bombing was "an unforeseen, powerful revelation." For some it was a life-changing "epiphany" (Wilson 1996). Many of them had come to Los Alamos believing that if an atomic bomb could be made, then the only way one could prevent Nazi Germany from using such weapons, if they acquired them, was for Great Britain and the United States to have them too and to threaten retaliation if Hitler used them. Many scientists worked on the bomb in order for it *not* to be used. They thus wanted a demonstration of the bomb. But they also knew—or came to recognize—that the decision was not theirs to make—and most of them believed then, and continued to believe, that use of the bomb did end the war and did save many American and Japanese lives.

After Nagasaki almost all the scientists connected with Los Alamos, Chicago, and Oak Ridge came to the conclusion that a nuclear war must never be waged. A consensus grew among them that

1. There is no secret.
2. There can be no monopoly.
3. There is no defense.
4. More powerful bombs can be built.
5. There must be international control.[76]

They organized themselves to bring these messages to the citizenry of the country.[77]

A letter to Robert M. Hutchins, the chancellor of the University of Chicago, dated September 10, 1945, contains one of Einstein's first reactions to the Hiroshima and Nagasaki bombings. A few days after the Nagasaki bombing, On August 12, Hutchins stated:

> Up to last Monday, I was opposed to the idea of a world state because I believed no moral basis for it existed—no world conscience or conviction to the world community sufficient to keep it from disintegrating.

I do not think we shall be better off because of the bomb. But the alternatives seem clear. Only through the monopoly of atomic force by a world organization can we hope to abolish war. . . .

All the evidence points to the fact that the use of the atomic bomb was unnecessary; therefore the United States has lost its moral prestige. . . . Perhaps the future is more important than the past.[78]

Einstein agreed. In his letter to Hutchins, he reiterated some of his former views:

As long as nations demand unrestricted sovereignty we shall undoubtedly be faced with . . . wars fought with bigger and technologically more advanced weapons. The most important task of intellectuals is to make this clear to the general public and to emphasize over and over again the need to establish a well organized world government. They must advocate the abolition of armaments and of military secrecy by nations.[79]

The question was what kind of world government. The Allies had met in San Francisco in April and May of 1945 to negotiate the charter of the United Nations, which was ratified in early June 1945. The charter created a security council with permanent seats on it for the United States, the Soviet Union, China, Great Britain, and France, each wielding veto power over council decisions.

In the summer of 1945 Emery Reves, an economist, political writer, and successful publisher whom Einstein had known for a number of years,[80] published a book titled *The Anatomy of Peace* to extensive and very favorable reviews.[81] Reves (1904–1981) was a remarkable man. He was born in a small village in southern Hungary to a Jewish family supported by the father's wood and grain business. His father's original name was Rosenbaum, but the name was changed to Revesz so that he could do business with the Hungarian government. A very talented and accomplished musician, Emery almost became a concert pianist. In 1922 he left Hungary to study at the universities of Berlin and Paris, and eventually he earned a doctorate in political economy at the University of Zurich, writing a dissertation on Walter Rathenau. His two best friends in Zurich were fellow Hungarians John von Neumann and

William Fellner. Fellner later became a distinguished economist. Revesz was fluent in nine languages and, during the Weimar Republic, worked in Berlin as a journalist, an activity he had begun in Zurich by interviewing all the prominent people passing through the city.

In Berlin Revesz conceived the notion "that the world needs an organization that can arrange publications in the press of articles on world issues, articles by Englishmen and Germans in the French Press, by Frenchmen and Germans in the English press, and so on." While in Berlin he founded the Co-Operation Press Service for International Understanding in 1930, the first international media company, which eventually serviced 400 newspapers in 60 countries. He and the Press Service left Berlin for Paris in April 1933 when it became clear that the Nazis were going to arrest him and close the Co-Operation Press Service. Revesz first met Winston Churchill in 1937 and enlisted him as one of the writers for the Press Service. Other contributors included Austen Chamberlain, Clement Attlee, Anthony Eden, Leon Blum, Paul Reynaud, Eduard Benes, and Einstein. Revesz's response to Hitler's bombastic speeches was to disseminate Churchill's replies in the newspapers of Europe the next day, which restored Churchill to public awareness and began his political comeback. Churchill got Revesz a position working for the British Ministry of Information, and in February 1940, Revesz became a British citizen.

In June 1940, just before the Germans occupied Paris, Revesz fled to London, where Churchill put him in charge of propaganda to the United States and other neutral countries. He was wounded in a London air raid in September 1940. In January 1941, on his own initiative, Revesz went to New York to help with the public relations campaign designed to enlist the United States' support of Britain in its battle with Germany. While in New York, he changed his name to Emery Reves and met fashion model Wendy Russell, who became his companion and later his wife. After the war, Reves made Churchill and himself very rich by negotiating the American royalties for Churchill's *The Second World War* and by obtaining for himself all foreign-language rights of the volumes. From the proceeds the Reveses bought La Pausa, the Riviera home near Monte Carlo of Coco Chanel, the founder of the Chanel perfume business. Churchill often stayed there for weeks at a time.

To understand the impact of Reves's *The Anatomy of Peace,* the context of its publication should be described. In early 1945 Harvard historian Crane Brinton, in a book entitled *The United States and Britain,* had tried to demonstrate the impracticality of world government at the time. Reves in his book eloquently defended the opposite view. In August 1945 Reves mailed a copy of his book to Einstein with a statement that he was particularly indebted to him, as he felt "that without [Einstein's] philosophical outlook this book could not have been written."[82] In his letter, Reves further indicated that since its publication the book had "provoked rather extraordinary reactions. Columns and columns are devoted to its discussion in the American press."[83] Supreme Court Justice Owen Roberts read the book "twice in two days" and informed Reves that "it expresses entirely his convictions." Roberts had drafted an open letter warning the American people not to believe that the ratification of the UN Charter would bring peace, but that only the thesis in Reves's *The Anatomy of Peace,* "that peace is law and that war between nations can be stopped only if a legal order is established, standing higher than the present nation states,"[84] must be recognized. Three U.S. senators, including J. William Fulbright, endorsed the letter, and Reves asked Einstein if he would sign it.

Two days after receiving Reves's book Einstein answered him, telling him that he had read his book "carefully and finished it in 24 hours. I agree with you wholeheartedly in every essential point and I admire sincerely the clarity of your exposition of the most important problem of our time. . . . I believe Justice Roberts' action is of the greatest value in the matter of enlightenment of public opinion in this country. I find the text of his open letter excellent and convincing and am gladly willing to sign it."[85] On October 10, 1945, the *New York Times,* the *Washington Post,* and some fifty other newspapers published the "Open Letter to the American People" signed by Einstein and eighteen other leading public figures. The letter contained an explicit endorsement of Reves's *The Anatomy of Peace,* for "it expresses clearly and simply" what the signers had been thinking. The letter urged "American men and women to read the book, to think about its conclusions, to discuss it with neighbors and friends privately and publicly."[86]

Later that year Reves excitedly wrote Churchill:

> Events happening around my book almost daily have forced me to postpone my trip [to London] from week to week. It is sweeping this country in an extraordinary way. It became a textbook in several colleges, sermons are made on it in many churches, students in Harvard, Yale and Columbia began to form groups to spread it and new editions of 10,000 copies disappear in two days. . . . The *Reader's Digest* is now organizing for January and February discussions of *The Anatomy of Peace* in 15,000 American clubs and discussion groups, with three speakers in each place presenting the three parts of the book. This is unprecedented in American publishing history and may have unpredictable effects.[87]

In late summer 1945 Reves forwarded Einstein the statement of a group of Oak Ridge scientists in which they recommended that an international agency, a world security council, be made the sole custodian of nuclear power in the world as well as all scientific and technical knowledge relating to it. This agency, the group recommended, would have the right to complete detailed, periodic inspections of all scientific, technical, industrial, and military installations in the world. In addition to the Oak Ridge statement, Reves included a letter to Einstein in which Reves argued that the recommendations of the scientists were "completely fallacious and may have tragic consequences if the peoples are given to believe that on the basis of these suggestions peace between the nations is possible and an atomic war between nation-states is made impossible." They have

> not thought the political problem through and still abide by old-fashioned internationalism, believing in a league of sovereign-states capable of maintaining peace between its member states. . . . If the nation-states would do what they ought to do, what they "must" do, if nation states would willingly and sincerely permit other nations to control and investigate matters within each other's sovereignty, if sovereign states could be trusted to carry out rules agreed upon, the problem of peace would be a very simple problem indeed. There would be no danger of war, there would be not the slightest danger that the atomic bomb would be used anywhere.

Unfortunately, peace among sovereign powers is a daydream. . . . It can be said with mathematical certainty that if control of atomic energy were to be transferred to the United Nations Security Council, a devastating world war with unlimited use of atomic power would be inevitable. . . .

The question of preventing the use of revolutionary weapons such as the atomic bomb is nothing new. Whenever a new weapon was invented—whether . . . gunpowder, dynamite, the machine gun, the tank, gas, the submarine or the bomber—people were always afraid the new weapon would mean the final destruction of civilization and tried to "outlaw" or "control" it. All these attempts failed because war is essentially a political, a social problem and not a technical one.

There is only one way to prevent an atomic war and that is to prevent War. Once war breaks out, it is certain that every nation will use every conceivable and available weapon to defend themselves and to achieve victory. . . .

Analyzing the wars of history. . . . I think it is possible to isolate the virus of war and to define the one and only condition in human society that produces war. This is the non-integrated coexistence of sovereign powers units in contact. . . . Peace is law. Peace between warring sovereign social units . . . can be achieved only by the integration of these conflicting units into a higher sovereignty. Peace between the nation-states in the twentieth century can be achieved only by the transfer of parts of the people's sovereignty into a higher system of government—legislative, judiciary and executive—regulating the relations of man in the international field; by the creation of a world government having direct relations with the individual citizen.

Attempts to maintain peace between nation-states by a league structure such as the San Francisco Organization in which sovereignty continues to reside in the individual members . . . are pitifully outdated and bound to fail. . . . There is only one way and one way alone to make the United States secure from an attack by atomic bombs. The method is the same that today makes the states of New York and California (non producers of atomic bombs) safe from annihilation by the states of Tennessee and New Mexico (producers of atomic bombs).[88]

No group of people today have such influence on the public as do the nuclear physicists. Their responsibility in making political suggestions is tremendous.[89] (Nathan and Norden 1968, 337–338)

I have quoted from the letter at length because it succinctly summarizes the conclusions of Reves's book, and it encapsulates Einstein's own position after the war.

Erroneously, Einstein had assumed that Oppenheimer was one of the signers of the Oak Ridge document. Thus on September 29, 1945, he forwarded Reves's letter to Oppenheimer and added a letter of his own to him to tell him that while he was "very much pleased with the open language [In Nathan and Norden 1968, "the open language" was translated as "candor of the language."] and the sincerity of the [Oak Ridge] statement," he was at the same time somewhat bewildered by its political recommendations, which he considered "obviously inadequate" to meet the challenges of the new realities.

> The wretched attempts to achieve international security do not recognize at all that . . . the real cause of conflicts is due to the existence of competing sovereign nations-states. . . . The conditions existing in the world today force the individual states, out of fear for their own security, to commit acts which inevitably produce war.
>
> At the present stage of industrialism, with the existing complete economic integration of the world, it is unthinkable that we can achieve peace without a genuine supranational organization to create and enforce law on individuals in their international relations. Without such an over-all solution to give up-to-date expression to the democratic sovereignty of the peoples, all attempts to avoid specific dangers in the international field seem to me illusory.

He then recommended Reves's *The Anatomy of Peace* to Oppenheimer and offered to send him a number of copies for distribution. It is a book, Einstein commented, that "explained the problem as clearly and pertinently as anyone ever has" and even though it was written before the use of atomic bombs on Hiroshima and Nagasaki, "it points out the solution to the problem created by this new weapon."[90]

Shortly after receiving Einstein's letter Oppenheimer answered him and indicated that the statements Einstein attributed to him were not his, and that he had never seen them. He knew of the existence of such views and had "attempted where possible to point out their inadequacies." He was in "complete agreement" with the views expressed by Einstein, and "in general agreement with those expressed . . . by Reves." Oppenheimer explained:

If I say "general agreement "I mean only this: the history of this nation through the Civil War shows how difficult the establishment of a federal authority can be when there are profound differences in the structure and values of the societies it attempts to integrate. I view the problem as more, rather than less, difficult than Mr. Reves suggests.

He also let Einstein know that the advice he, Oppenheimer, had given to the American government had been concerned

with the problem of initiating those negotiations which might establish confidence and form the basis for a real unity. We have, I believe, from the first recognized the essentially political character of this problem, and regarded the development of the atomic bomb as of incidental, but perhaps decisive, importance in two respects:

(1) Focusing more sharply the attention of the public on the dangers of international anarchy (and in particular on the dangers of competitive armament between two all-powerful nations.)
(2) Providing a new and specific point of discussion where agreement might be less difficult to achieve. (Smith and Weiner 1980, 309–310)

The exchange between Einstein and Oppenheimer revealed their fundamentally different approaches and positions regarding the avoidance of war and the control of atomic energy.

Einstein's position regarding war reflected his commitment to what Max Weber called "an ethic of ultimate ends."

The believer in an ethic of ultimate ends feels "responsible" only for seeing to it that the flame of pure intentions is not quenched: for example, the flame of protesting against the injustice of the social order. To rekindle the flame ever anew is the purpose of his quite irrational deeds, judged in view of their possible success. They are acts that can and shall have only exemplary value.[91]

Oppenheimer, on the other hand was committed, in Weber's terminology, to an "ethic of responsibility," "in which case one has to give an account of the foreseeable results of one's action."[92] In his contribution to Masters and Way's *One World or None,* Oppenheimer averred: "The obvious consequence of this . . . participation of scientists [in creating atomic weapons] is a quite new sense of responsibility and concern for what they have done and for what may come of it" (Masters and Way 1946, 56).

Oppenheimer was a pragmatist who searched for stepwise ways to establish trust between nations, and between the United States and the Soviet Union in particular. He hoped that the United Nations (UN), the only realistic and supranational organization acceptable at present to the big powers, would be the first step toward a world government. For his part, Einstein believed that the situation required drastic measures and that "The only salvation for civilization and the human race lies in the creation of a world government, with security of nations founded upon law," and that this had to be accomplished and accepted in one fell swoop.

The UN and its organization, in the mode of operation stipulated by its charter, had in fact been created with only the hope, at best, that it might be the first step to a world government. Furthermore, when its charter was signed in San Francisco on June 6, 1945, almost all of the participants were unaware of what had been accomplished at Oak Ridge, Hanford, and Los Alamos. Hiroshima and Nagasaki of course changed that. At their meeting in Washington in November 1945, Prime Minister Clement Attlee of Great Britain, William Lyon MacKenzie King, the premier of Canada, and President Harry Truman, recognizing the need to address the problem of atomic energy, stated in their "Declaration on Atomic Energy" that they would be willing "to proceed with the exchange of fundamental scientific literature for peaceful ends with any nation that would fully reciprocate . . . [but only after it had become] possible to devise effective reciprocal and enforceable safeguards acceptable to all nations" against the use of atomic energy for destructive purposes. They recommended that the newly founded United Nations Organization address the nuclear problem. At a meeting of foreign ministers in Moscow in December 1945, it was agreed that a United Nations Atomic Energy Commission (UNAEC) be created to consider problems arising from the discovery of atomic energy. As a condition for acceptance, the Soviet demanded, and the United States and Great Britain accepted, that the work of the UNAEC would be directly responsible to the Security Council, where the Soviet Union had veto rights. The first resolution of the first session of the UN General Assembly established UNAEC. During the UNAEC's brief period of existence from 1946 until 1948, the declared aim of both the United States and the Soviet Union was not the prevention of the proliferation of nuclear weapons but their elimination. Thus, how to use nuclear power effectively and safely, prevent the use of atomic energy as a weapon of war, and in the meantime curb the spread of nuclear bombs was the challenge that faced the newly founded UNAEC when the Security Council asked it to report on methods to effect these goals.

In early 1946 James Byrnes, the U.S. secretary of state, appointed a committee headed by Undersecretary of State Dean Acheson to formulate U.S. policy regarding atomic energy and to draft a plan that would be presented to the UNAEC outlining the American position regarding the abolition of nuclear weapons and the control of the peaceful uses of nuclear energy. The committee included Groves, Bush, Conant, and John McCloy. To assist them, an advisory panel was set up with David Lilienthal, then the head of the Tennessee Valley Authority, as chair and Oppenheimer as chief scientific adviser. Oppenheimer became the principal architect of the Acheson-Lilienthal report, which was published in March 1946. Its basic premise, due to Oppenheimer, was that "a system of inspection superimposed on *an otherwise uncontrolled exploitation of atomic energy by national governments will not be an adequate safeguard*" (my emphasis). Hence all nuclear activities should be placed under the supervision of an international agency—the Atomic Development Authority (ADA)—which would have exclusive control of all "dangerous" aspects of atomic energy.

But after Truman and Byrnes appointed the seventy-five-year-old Bernard Baruch as the head of the United States delegation to the UNAEC trouble developed. Baruch formulated his proposal—which eventually won Truman's approval. It embodied the outline of the Acheson-Lilienthal report—but not its spirit. It demanded immediate punishment for violations and insisted that no veto power would exist when levying penalties for violations of the rules of the agreement.

For his part, Einstein approved of the veto-relinquishing feature of the Baruch plan. He made his position clear by giving a qualified approval to I. F. Stone's article "Atomic Pie in the Sky," in which the original Acheson-Lilienthal plan was criticized, characterizing the piece as "justified and of a constructive character."[93] But Einstein added:

> It is to be highly appreciated that official authorities, even from army quarters, have openly recognized that security can be reached only on the basis of world government. On the other hand, the measures proposed for the interim period do not appear adequate to the task of bringing us nearer to the goal of world government and convincing other nations of the loyal intentions of our foreign policy. (Nathan and Norden 1968, 375)

It is interesting to compare Einstein's reaction to that of Bohr. In the spring of 1946 Arthur Compton had a conversation with Bohr during which Bohr told him "with all the earnestness at his command":

> What the world needs most just now is for the United States to take decisive moral leadership. For generations the nations have looked to the United States as a land of hope, of freedom and good will. That view is not gone, but it has been shaken.
>
> Your country won the war with a burst of strength at which we marveled, but with a ruthlessness which made us pause. Your armadas of planes bombarded cities of questionable importance. We viewed with horror the use of your atomic bombs. We saw reason for your actions, but we are wondering what you mean. What indeed is in the heart of the American giant? Are his intention good or ill? Or is he, as some would have us think, merely self-centered in his thoughts?
>
> A decisive moral act by the United States now would answer these questionings. New hope would stir. The people of the world would take courage once more with your leadership to building again the life they want.[94]

For Bohr, as for Oppenheimer, the Baruch plan was a grave setback. The importance of finding a way to accommodate the Soviet Union's participation in the control of atomic energy—but accepting the limitations of what was possible given the political situation in the Soviet Union and the United States—shaped Bohr's opinions and guided his actions. For Einstein, his belief in the idea that a world government was essential for peace was incontrovertible and polarized his stand on the issue of the control of atomic weapons. What changed over time, reflecting the intensification of the Cold War, was the particular form of world government and which nations should constitute its initial membership.

## Einstein on World Government

One of the most complete expositions of Einstein's postwar views concerning world government and disarmament was presented in the November 1945 issue of the *Atlantic Monthly*. The article was based on Einstein's interview with Raymond Gram Swing, the American Broadcasting Company nightly news commentator who had advocated the need of an effective world government. Einstein had written to him in late August 1945 commending him for his stand. Swing thereafter visited him in Princeton and interviewed him to obtain his views on the implications of atomic energy. In the article Einstein stated that the release of atomic energy did not create a new problem, but merely made the necessity of solving the problem of avoidance of war more urgent and more imperative given the destructiveness of atomic bombs. He did not believe that the secret of the bomb should be given to the United Nations Organization, but should be committed to a world government. He thought that

> the United States should immediately announce its readiness to do so. Such a world government should be established by the United States, the Soviet Union and Great Britain, the only three powers which possess great military strength. The three of them should commit to this world government all of their military resources.

The drafting and negotiation of the constitution of the world government, Einstein wrote, should be carried out by one American, one

Briton, and one Russian working together; and since the Soviet Union did not possess an atomic bomb the initial draft should be prepared by the Russian. It would be up to these three individuals "to devise ways for collaboration despite the different structures of their government," structures that Einstein did not believe necessarily had to be changed. The world government would have jurisdiction over all military matters and would have the power to intervene in countries where a minority was oppressing a majority, thereby creating the kind of instability that leads to war.

Einstein indicated that he was aware that some people in the United States favored world government as the ultimate objective but advocated a gradual approach to its establishment. The trouble with this approach, Einstein thought, was that while such steps were being taken, the United States continued to keep the bomb without convincing those who did not have the bomb as to U.S. intentions. This in turn created fear, suspicion, and mistrust, with the relation between rival countries deteriorating to a dangerous level—and thus opened the possibility of war. Furthermore, Einstein asserted that

a great era of atomic science cannot be assured by organizing science in the way large corporations are organized. One can organize the application of a discovery already made, but one cannot organize the discovery itself. . . . However, there can be a kind of organization wherein the scientist is assured freedom and proper conditions of work.

I do not believe that the vast private corporations of the United States are suitable to the needs of the times. [Is it not] strange that, in this country, private corporations are permitted to wield so much power without having to assume commensurate responsibilities? I say this to stress my conviction that the American government must retain control of atomic energy, not because socialism is necessarily desirable but because atomic energy was developed by the government; it would be unthinkable to turn over this property of the people to any individual or group of individuals. As for socialism, unless it is international to the extent of producing a world government which controls all military power, it might lead to wars even more easily than capitalism because it represents an even greater concentration of power.

In concluding his article, Einstein again recommended Emery Reves's *The Anatomy of Peace* as an intelligent, clear, concise formulation of the need for world government.

Interestingly, even though the letter to the editor to the *New York Times* that Einstein had recently signed had carried Truman's remarks about the need of a high court to whose decisions the states that participate in the world organization must abide by, the *Atlantic* article made no reference to such a court. It was, however, included in the article he contributed to *One World or None,* the report on the meaning of the atomic bomb and the threat to world survival it posed. The other authors of the report were (General) Hap Arnold, Hans Bethe, Niels Bohr, Arthur Compton, Edward Condon, Irving Langmuir, Walter Lippmann, Phillip Morrison, Louis Ridenour, Frederic Seitz, Harlow Shapley, Szilard, Wigner, and Harold Urey. The book was reviewed on the front page of both the *New York Times*'s and the *Herald Tribune*'s book review section on March 17, 1946. The *Times*'s review encircled an ominous black and white sketch of Manhattan's skyscrapers being blown away and ships sinking in the Hudson, as a result of an atom bomb having exploded a half mile in the air above the corner of Third Avenue and East Twentieth Street. Based on the contribution to the volume by Philip Morrison, who had been sent to investigate the damage to Hiroshima and Nagasaki shortly after Japan surrendered, the sketch powerfully illustrated the havoc created by the bomb. Morrison's article estimated that were such a bomb to be detonated around noontime, over 300,000 people would be immediately killed and that this same number would be critically injured. All the contributors agreed that there was no defense against an attack by atomic bombs, with Bethe, Seitz, and Urey suggesting that "any one of several determined foreign nations" might produce such a bomb within five years—and might even be ahead of the United States in five years. Those in the know—such as Hap Arnold and Oppenheimer—stressed that they were actually cheap to produce.[95] That there must not be another war was the moral reached by all the contributors, including General Arnold. *How* to avert war was the question, however. Szilard in his article outlined an inspection system that would be effective if every country, including the United States and the Soviet Union, would allow its mines and nuclear installations to be opened to UN agents.

In his contribution Einstein again insisted on the "denationalization of military power" and the establishment of a supranational organization. For him, there was only one way out of the present dangerous situation:

> Institutions must be established which will guarantee that any disputes which may arise between individual nations will be solved on the basis of law and under international jurisdiction. A supranational organization must make it impossible for any country to wage war by being able to employ military forces which that nation alone controls. (Nathan and Norden 1968, 362)

Moreover, the establishment of such institutions could not be accomplished in a piecemeal, gradual, stepwise fashion, for without supranational military security, the forces pushing toward war were irresistible.[96] For, as Einstein stressed:

> Even more disastrous than man's lust for power will be his fear of sudden attack—unless we openly and resolutely set ourselves the task to abandon national military power and accept supranational authority. (Nathan and Norden 1968, 362)

Initially, Einstein was convinced that the three great powers must in one fell swoop accept a supranational form of government. The subsequent recruitment "possibly of all nations into the supranational organization, on an entirely voluntary basis" could be a gradual process. But with the intensification of the Cold War Einstein's position regarding the details of the implementation of the organization changed.

Thus on June 9, 1947, in an interview with William Golden,[97] carried out on behalf of Secretary of State James Byrnes, Einstein told Golden that he believed

(a) that the world is heading for an atomic war. The American monopoly cannot be maintained for more than a few years. Russia will surely develop an atomic bomb. With each side having the atomic bomb the premium on surprise attack is greater than ever before. In case of war when the two sides have the bomb, one or the other will surely use it, either "from nerves or fear if not from policy." The devastation that will result from atomic warfare will be fully up to the popular conception.

(b) That the United Nations has been ineffectual and cannot control the situation.

(c) That the only way of averting an atomic war within a few years is through an effective supra-national World government to which military power will transferred. All countries should be invited to join. However, if Russia does not join, then proceed without her. If the World government is strong Russia will join eventually. Non-members of the world government would be invited to send observers to its council so that they could assure themselves that they weren't being plotted against.

(d) that delegates to the world organization should be elected directly by the citizens of the member nations, and not appointed by the national governments of the member states.

Golden noted that although Einstein didn't insist on it, he indicated that "perhaps the number of votes of each country should be proportional to the number of professional men (or college graduates or some other such standard of intellectual hegemony) rather than the total population. I shall return to the issue of Einstein's views of elitism and democracy later in this chapter.

Golden commented further that

Einstein spoke with deep feeling and almost childlike hope for salvation and without appearing to have thought through the details of his solution. He recognizes that World Government would be difficult to accomplish but no matter how remote the chance, every effort should be made to achieve it since otherwise all will be lost.

It pains him to see the development of a spirit of militarism in the United States which follows from its experience in the last war. The

American people are tending to become like the Germans—not like the Nazis but those of the Kaiser. He says that Americans are beginning to feel that the only way to avoid war is through a Pax Americana, a benevolent world domination by the United States. He pointed out that history shows this to be impossible and the certain precursor of war and grief. There can be no lasting enforced peace. The benevolent despot becomes a tyrant or a weakling; either way the structure crumbles.

The German people have been ruined by their military spirit which stems from Bismarck.[98]

Einstein made these views public in an article entitled "The Military Mentality," published in the summer of 1947 in *The American Scholar*, the quarterly of Phi Beta Kappa, the honor society at American colleges. In this article, Einstein decried the "tendencies" that he was observing in contemporary United States as the Cold War was becoming increasingly frigid.

The[se tendencies] arose when under the influence of the two World Wars and the consequent concentrations of all forces on a military goal, a predominantly military mentality developed, with which the almost sudden victory became even more accentuated. The characteristic feature of this mentality is that people place the importance of what Bertrand Russell so tellingly terms "naked power" far above all other factors which affect the relations between peoples. . . .

It is characteristic of the military mentality that non-human factors (atom-bombs, strategic bases, weapons of all sorts, the possession of raw materials, etc.) are held essential, while the human being, his desires and thoughts—in short, the psychological factors—are considered unimportant and secondary. . . . The individual is degraded to a mere instrument; he becomes "human materiel." The normal ends of human aspirations vanish with such a viewpoint. Instead, the military raises "naked power" as a goal in itself—one of the strangest illusions to which men can succumb.

In our time the military mentality is still more dangerous than formerly because the offensive weapons have become much more powerful than the defensive ones. Therefore it leads by necessity to preventive war. The general insecurity that goes hand in hand with this, results in the sacrifice of the citizen's civil rights to the supposed welfare of the state. Political witch-hunting, controls of all sorts (e.g. control of the teaching and research, of the press, and so forth) appear inevitable, and for this reason do not encounter that popular resistance, which, were it not for the military mentality, would provide a protection. A reappraisal of all values gradually takes place in so far as everything that does not clearly serve the utopian ends is regarded and treated as inferior.

I see no other way out of prevailing conditions than a farseeing, honest, and courageous policy with the aim of establishing security on supranational foundations. Let us hope that men will be found, sufficient in number and moral force, to guide the nation on this path so long as a leading role is imposed on her by external circumstances. (Nathan and Norden 1968, 422–425; Einstein 1954, 32–134)

On a later occasion Einstein commented as follows:

I know very well that a world government may have both good and bad qualities. Nonetheless, it is the only conceivable machinery which can prevent war. I do believe that a world government would be just in all its decisions; but with technology at its present level, even a poor world government is preferable to none, since our first goal must be to avoid total destruction through war. (Nathan and Norden 1968, 468; March 8, 1948)

Moreover,

The problem of peace and security is indeed far more important than the conflict between socialism and capitalism. Man must first ensure his survival; only then can he ask himself what type of existence he prefers.[99]

Upon receiving the Wendell Willkie One World Award, Einstein would state these views succinctly and sharply:

Whenever the belief in the omnipotence of physical force dominates the political life of a nation, this force takes on a life on its own and becomes even stronger than the very men who intended to use it as a tool. The proposed militarization of the nation not only immediately threatens us with war; it will also slowly but surely undermine the democratic spirit and the dignity of the individual in our land.[100]

And after the detonation of Joe 1[101] in late August 1949, Einstein could encapsulate his views in proverb-like fashion:

No peace without disarmament; no disarmament without confidence; no confidence without mutual and effective economic relations.[102] (Nathan and Norden 1968, 517; October 31, 1949)

## Hydrogen Bombs

During the fall of 1949, following the detonation of Joe 1 on August 29, important deliberations took place within government circles to determine what the American response to this perceived threat should be. Although the feasibility of fusion weapons had been considered as early as 1942, there were as yet no known methods to trigger and sustain a fusion process that could translate into an effective and deliverable weapon. At a meeting in late October 1949, the General Advisory Committee to the AEC, which Oppenheimer chaired, recommended unanimously that the United States not embark on a crash program to build an H-bomb, a weapon whose destructive power would be a thousand times greater than the Hiroshima or Nagasaki bombs and could be made larger at will.[103] Nonetheless, President Harry Truman ordered the AEC to do so.[104] The reaction within the scientific community as well as in the population at large was immediate. Some scientists considered the weapon a genocidal one and refused to work on designing it. Others, like Hans Bethe, in addition to refusing to participate in constructing it, wrote and lectured extensively on the threats such a weapon would pose.

Shortly before Truman's decision, Einstein was petitioned to take a stand on the matter. Abraham J. Muste,[105] the secretary of the Fellowship of Reconciliation, a Christian pacifist organization whose activities Einstein had supported in the past, wrote him in January 1950:

Unless you and other leading scientists—but especially yourself—are, in spite of your wishes and protestations, to share in the guilt for that decision, then you must surely throw the full weight of your influence against that decision, no matter what it might cost in reputation or even life to do so. The very least this could mean, it seems to me, would be to say that negotiations with Russia must now be seriously taken up, and that pending these negotiations, and regardless of their outcome, there must be no production of hydrogen bombs, or other super weapons of mass destruction by the United States. It would be necessary to call on your fellow scientists and technicians to join you in making that proposal.

Muste, however, noted that

> mere words of protest or resolutions advocating negotiations with Russia, the establishment of international control, etc. are useless in this situation. The only thing that will count, that may bring the people and our political and military leaders to their senses, would be for you and your leading colleagues to take public action indicating your refusal to have anything more to do with the manufacture of weapons of mass destruction, calling upon all scientists and technicians to take the same position, those now involved to resign their posts.

He concluded his letter with the following statement:

> I trust that you will not think me as importunate. If ever there was a time when it is legitimate for one soul to cry out to another, surely this it![106]

Einstein answered Muste a few days later, telling him that he recognized the sincerity and seriousness of his intentions, but that his request was based on false assumptions. Since he, Einstein, had never participated in any work of a military-technical nature, and except for signing a letter to Roosevelt

> in which attention was drawn to the fact that the possibility existed to make [an atomic] bomb, [and that he] considered it [his] duty [to do so] because there were definite indications that the Germans were working on such a project. It would, therefore, be quite ridiculous if I should make a declaration refusing to participate in armament work.

Einstein also told Muste that he did not believe that Muste's proposal to refrain from making the hydrogen bomb went to the heart of the matter. Those who had the actual power in the United States and in the Soviet Union did not intend to avoid the "cold war." "Both sides," Einstein said, "are using the 'cold war' for their internal political goals—regardless of consequences." In the United States, the process started immediately after Roosevelt's death and the people in power succeeded in deceiving, intimidating, and "fanaticizing" the public. "I see no effective way, how the small group of reasonable and well meaning people could stop this disastrous course of events" since even so-called neutral nations are unwilling to find a way to cooperate with one another on a supranational basis. Einstein concluded his letter by stating that he never shrank from expressing his opinion freely and always considered it his duty to do so. "However, the voice of an individual is powerless against the shouting of the masses—this has always been so."[107]

Muste's reply on January 26 took Einstein to task. He was well aware, he said, that Einstein had not participated in any of the technical aspects of the atomic bomb and that there was no likelihood that he would be asked to work on the H-bomb. Muste explained that he had written because he thought that Einstein believed that "scientific conscience and fidelity to the interests of science" required scientific and technical men and women generally to take a conscientious stand against any involvement in the production of such weapons "whether by direct work or simply by failure to take a public stand on ground of conscience."

Muste went on to ask whether it was not as much Einstein's responsibility and privilege to intervene now with the highest political and military authorities to point out "that it is both unwise and evil, under present conditions, to go for the productions of super atomic bombs" as it had been in 1939, when he advised Roosevelt of the possibility of making atomic bombs. Furthermore, it seemed to Muste that

> quite apart from the immediate political effect of his action or inaction, the individual has the responsibility before God and future generations as well as his own, to witness for what he believes to be right and true, even though he has to stand alone and his is a voice crying in the wilderness.

Einstein did not reply to the letter. Muste, joined by several clergymen, made a further plea in a telegram to Einstein. They urged Einstein to use his great influence to secure the delay of the hydrogen bomb decision "pending thorough public discussion. People must have opportunities to ponder and discuss this life and death issue if the United States is to remain a democratic nation."[108] Einstein replied the same day:

> I received your telegram. But the way you propose seems to me quite ineffective. As long as competitive armament prevails it is not possible to stop the process in one place. The only possible way out is an honest attempt to work for a reasonable agreement with Soviet Russia and beyond this for security on a supra-national basis.[109]

Einstein refused to sign Muste's appeal not because he was reluctant to make his views public, but rather because he thought the approach would prove ineffective given the political and ideological stance, as well as the stature, of the signers of the appeal. His exchange with Muste illustrates that no one could take for granted Einstein's public participation in an appeal that surely resonated with his beliefs.

Einstein did make his assessment of the political situation very clear in remarks he recorded for Eleanor Roosevelt's radio program of February 13, 1950, shortly after Truman's directive. The other speakers on that program were Oppenheimer and Lilienthal, both of whom still held official governmental positions. On that occasion Einstein said:

> The belief that one can achieve security through armaments on a national scale is, in the present state of military technology, a disastrous illusion. In the United States this illusion has been strengthened by the fact that this country was the first country to succeed in producing an atomic bomb. This is why people tended to believe that this country would be able to achieve permanent and decisive military superiority which, it was hoped, would deter any potential enemy and thus bring about the security, so intensely sought by us as well as the rest of the world. The maxim we have followed . . . has been, in short, security through superior force, whatever the cost.
>
> This technological as well as psychological policy has had its inevitable consequences. Every action related to foreign policy is governed by one single consideration: How should we act in order to achieve the utmost superiority over the enemy in the event of war? The answer has been: Outside the United States we must establish military bases at every

possible, strategically important point in the globe as well as arm and strengthen economically our potential allies. And inside the United States, tremendous power is being concentrated in the hands of the military; youth is being militarized; and the loyalty of citizens, particularly civil servants, is carefully supervised by a police force growing more powerful every day. People of independent political thought are harassed. The public is subtly indoctrinated by the radio, press, the schools. Under the pressure of military secrecy, the range of public information is increasingly restricted.

. . . And now the public has been advised that the production of the hydrogen bomb is the new goal. . . . If these efforts should prove successful . . . annihilation of all life on earth will have been brought within the range of what is technically possible. The weird aspect of this development lies in its apparently inexorable character. Each step appears as the inevitable consequence of the one that went before. And at the end, looming ever clearer, lies general annihilation.

Once again Einstein stated his unshakable belief that the way out of this impasse and to peaceful coexistence among all nations was for nations to renounce the policy of violence, not only with respect to weapons of mass destruction but with respect to all armaments. And for the latter to be effective a supranational judicial and executive organization that had the power to settle questions of immediate concern to the security of nations was required. However,

In the last analysis the peaceful coexistence of peoples is primarily dependent upon mutual trust and, only secondarily, upon institutions such as courts of justice and the police. This holds true for nations as well as for individuals. And the basis for trust is a loyal relationship of give-and-take. (Nathan and Norden 1968, 521–522)

Einstein was unswerving in his refusal to sign his name to appeals that he believed would prove to be ineffective. Thus he refused to sign a statement Muste sent to him in September 1950 on the proposed treaty between the United States and Japan[110] telling him that, although he fully agreed with Muste's letter, Muste's effort seemed so hopeless that Einstein couldn't participate. He compared the attempt to "sending a bottle of sugar-water to a chronic alcoholic in the hope to cure him."[111]

Perhaps more disturbing was Einstein's refusal in late March 1954 to sign an appeal to President Eisenhower calling on him to order an immediate suspension of the H-bomb tests scheduled to take place in the Pacific in April. The March 1 Bikini tests had resulted in severe radiation burns to military personnel, to the inhabitants of one of the nearby atolls, and to the crew of the *Lucky Dragon,* the Japanese vessel that had been fishing some 80 miles from the nuclear explosion and had been coated with radioactive dust.[112] Once again, Einstein declined, stating that he appreciated the motive of Muste's action but that he could not participate in the effort.

> Such a small scale enterprise of a few private persons has not the slightest influence on the behavior of people who have already decided and have de facto no freedom to change their attitude. Only powerful political agencies can influence the course of events. I find it not reasonable to do something only to satisfy one's personal urge. Reason alone has no effect even if it speaks convincingly and with the voice of angels.[113]

Einstein was fully aware of the catastrophic consequences of the Bikini tests, as is indicated by the little equation he put on top of his copy of the pamphlet entitled "STOP THE BOMB: An attempt to the Reason of the American People." Muste had sent him the pamphlet, requesting that Einstein sign his appeal to Eisenhower to end testing fusion weapons in the Pacific. The pamphlet described what had happened to the crew of the *Lucky Dragon* and to the fish they had caught, after they had been exposed to radioactive dust from the nuclear explosion. The equation read

A.E.C. = Atomic Extermination Conspiracy[114]

Similarly, perhaps upon reading Admiral Lewis Strauss's statements in which he defended the tests and denied that any of the inhabitants of the Marshall Island atolls, or any of the military personnel, or any of the Japanese fishermen had been exposed to a hazardous level of radiation,

Einstein composed the following aphorism: "The patriotic lie—that un-failing weapon of political scoundrels."[115]

## Individual versus Collective Stands

When Einstein participated in collective political action, he was rather sensitive to who made up the collective, even when he fully agreed with them on the particular issues that had brought them together. Thus, when Einstein refused to affix his signature to appeals that Muste had forwarded to him, he did so not because he disagreed with Muste's position, but rather because he felt the appeal would be ineffective given who the signatories were. Einstein declined to join in other similar instances because he thought the petition ineffective or impracticable. Thus, when two scientists asked him to participate in an appeal to scientists to refuse to work on developing nuclear power because of its possible evil uses, he declined to do so because the answers to two critical questions ("Would any action by a group as small as the one they were contemplating have any decisive influence?" and "Would the physicists and engineers necessarily follow the suggested course of action?") were negative (Nathan and Norden 1968, 455–456). Yet he was willing to sign a letter to warn the American people that the UN was a "tragic illusion unless we are ready to take further steps necessary to organize peace," accept a federal organization and constitution of the world, and create a worldwide legal order. Perhaps he signed this letter because it was drafted shortly after the end of World War II, with the memory and pictures of Hiroshima and Nagasaki still fresh in people's minds, and thus rendered the proposal practicable. But I believe the fact that the co-signers were *eminent* people—Justice Owen Roberts, Senator J. William Fulbright, Albert Lasker, Thomas Mann, Mark van Doren, Garner Cowles, Mortimer J. Adler, and Louis Finkelstein—was also a factor.

Einstein was more willing to affix his name to statements when the co-signers were distinguished scientists, and he was also more ready to participate in collective activities with other scientists. Under those circumstances, he evidently could call on Szilard and others for their assessment of the other participants. Thus at Szilard's urging, in May 1946 Einstein agreed to serve as chairman of the Emergency Committee of Atomic Scientists, whose initial mission was "to advance the use of atomic energy in ways beneficial to mankind," to diffuse knowledge and information about atomic energy, and to promote the general understanding of its consequences. In early 1946 this meant to have the public at large informed of the consensus that had developed on nuclear issues by the members of the Federation of Scientists, namely, to have the public at large recognize that

1. Atomic bombs can now be made cheaply and in large number. They will become more destructive.
2. There is no military defense against the atomic bomb, and none is to be expected.
3. Other nations can rediscover by themselves the secret processes that allowed the United States to manufacture atomic bombs.
4. Preparedness against atomic war is futile, and if attempted will ruin the structure of the American social order.
5. If war breaks out, atomic bombs will be used and they will surely destroy our civilization.
6. There is no solution to this problem except international control of atomic energy, and ultimately the elimination of war.

The Emergency Committee originally consisted of Einstein, Leo Szilard, Hans Bethe, Thorfin Hogness, Harold Urey, and Victor Weisskopf, with Selig Hecht, Philip Morse and Linus Pauling joining later. Because all the members believed that the democratic determination of U.S. policy on atomic energy ultimately rested on the citizenry's understanding of the facts of the matter, the committee took on as one of its principal responsibilities the raising of adequate funds to discharge its educational goals. With Einstein as its prestigious chairman, the committee was able to immediately raise $85,000 when it embarked on a fund-raising campaign to obtain $1 million. Einstein took a fairly active part in the committee's affairs and allowed his name to be used rather freely for fund-raising purposes.

As we shall see in the next section, the issue of the eminence of the co-signers of the Bertrand Russell Manifesto was a factor in both Russell's and Einstein's minds. It will also be recalled that in his interview with Golden, Einstein indicated that, although he didn't insist on it, he thought that representation in a supranational government ought to reflect the number of "educated" persons in the member nations. Scheideler has described this facet of Einstein's views of "mankind" (Scheideler 2003). In his attitude toward the masses, Einstein reflected the elitist outlook of the German professorate, but also that of Schopenhauer, whose essays Einstein read avidly. I will only quote two of Einstein's statements to illustrate his position. The first dates from 1926 and is taken from a letter Einstein wrote to Romain Rolland, who, like Einstein, was a dedicated pacifist during World War I:

> The rude masses are driven by dark passions that dominate both them and the governments that represent them. . . . Those few who do not share in the coarse emotions of the masses and who, unaffected by such passions, cling to the ideals of brotherly love are faced with a more difficult situation. They will be rejected by their fellow men and persecuted like lepers unless they act in a manner inconsistent with their conscience or cravenly conceal their real thoughts and feelings. . . . [However, they form] the community who are immune to the pestilence of hate, who seek to abolish war as the first step towards the moral regeneration of mankind and who view this task as incomparably more important than the special interests of their particular state or nation. (Nathan and Norden 1968, 79–82)

The second is from an article on "Society and Personality" written for an Amsterdam publication in 1934:

> All the valuable achievements, material, spiritual, and moral which we receive from society have been brought about in the course of countless generations by creative individuals. . . . Only the individual can think and thereby create new values for society, nay, even set up new moral standards to which the life of the community conforms. Without creative

personalities able to think and judge independently, the upward development of society is as unthinkable as the development of the individual
personality without the nourishing soil of the community.... The health
of society thus depends quite as much on the independence of the individuals composing it as on their close social cohesion. (Einstein 1954, 13–14)

Einstein's sensitivity to signing documents with others contrasts
sharply with his readiness to take a stand as an individual against the in-
fringement of civil liberties brought about by the "military mentality"
that was taking hold in the United States after World War II. As we have
seen, Einstein and Oppenheimer agreed that a "military mentality" was
becoming widespread in the United States. But whereas Oppenheimer, by
virtue of his role as consultant to the government, chose to remain silent
Einstein was outspoken and passionate in his defense of civil liberties.
Fred Jerome (2002), in his book *The Einstein File*, has given numerous examples of Einstein's courageous stand against the excesses of House and
Senate committees investigating "fellow travelers." He became promi-
nently featured in the news for protesting the "inquisitions" of the House
Un-American Activities Committee and of Senator Joseph McCarthy,
Senator William Jenner, and the Senate Internal Security Subcommittee.

For example, in December 1953, Einstein talked at length with Al
Shadowitz, an electrical engineer working on government contracts,
who had been subpoenaed by Senator McCarthy to appear before his
subcommittee. Shadowitz had helped organize the Federation of Archi-
tects, Engineers, Chemists and Technicians (FAECT) at one of IT&T's
laboratories. Shadowitz, on the advice of Einstein, instead of pleading
the Fifth Amendment as his defense for not answering questions posed
to him, pleaded the First Amendment. On December 16, 1953, the *New
York Times* carried a front-page photo of Shadowitz and an article with
the headline: WITNESS, ON EINSTEIN ADVICE, REFUSES TO SAY IF HE WAS
RED." And a few months earlier, on June 12, 1953, Einstein had made
headlines in the *New York Times* for the advice he had given to William
Frauenglass, a high school English teacher, who had been called before
the Senate Internal Security Subcommittee and had refused to be a cooperative witness even though it meant the loss of his job. Frauenglass

had written Einstein for a letter of support. Einstein complied and agreed to have the letter sent to the *New York Times*. In his letter Einstein advised civil disobedience:

> What ought the minority of intellectuals do against this evil? Frankly, I can only see the revolutionary way of non-cooperation in the sense of Gandhi's. Every intellectual who is called before one of the committees ought to refuse to testify, i.e., he must be prepared for jail and economic ruin, in short for the sacrifice of his personal welfare in the interests of the cultural welfare of his country.[116]

The *New York Times,* in an editorial on June 13, chastised Einstein by interpreting his statement to mean that "Congressional committees have no right to question teachers and scientists or to seek out subversives wherever they can find them." Congressmen have that right, the *Times* asserted. "What is wrong is the way some of them are exercising it. . . . To employ the unnatural and illegal forces of civil disobedience, as Professor Einstein advises, is in this case to attack one evil with another." But of course, Einstein only suggested civil disobedience when congressional committees were abusing their powers; he wasn't questioning the right of Congress to investigate under specified circumstances and specified guidelines.

Moreover, he was prepared to go to jail, face economic ruin, and sacrifice his personal welfare in the interests of the cultural welfare of his adopted country.

## The Einstein-Russell Manifesto

The global reverberations of the Bikini hydrogen bomb tests did not die down. Before 1954 ended, Bertrand Russell stepped up his advocacy of an international authority to own all fissionable raw materials. In Einstein's last days, Russell recruited him for a campaign that helped spur a global movement to ban all testing of nuclear weapons in the atmosphere, the oceans, and space. The movement would have a powerful influence on the Test Ban Treaty of 1963.

Russell's concerns with nuclear weapons had started during World War II when he became aware that work was being done to develop fission bombs. Shortly after the bombing of Hiroshima and Nagasaki, Russell made a speech in the House of Lords in which he warned that nuclear weapons based on the fusion mechanism could be made much more destructive than fission bombs, and that in time the latter would become much cheaper to manufacture. Convinced that the Russians would soon have their own atomic bombs, he recommended that nuclear weapons be placed under international control. Russell had supported the Baruch plan for an International Atomic Development Authority. In fact, he considered the idea of the Soviet Union and other nations possessing nuclear weapons so perilous that in late 1948, when the United States still was the only nuclear power, he advocated that the United States force the Russians to accept nuclear disarmament, even by threatening to go to war against it immediately if necessary.[117] Although during World War I Russell was an absolute pacifist, he later renounced this stand, eventually coming to the position that waging war to establish a world government would be a just and valid cause. The alternative, he thought, was waiting until the Soviets had atomic bombs and choosing between a nuclear war and submission. Russell's anticommunism became more moderate after the death of Joseph Stalin. McCarthyism, the American nuclear policy, and, in particular, the Bikini test in 1954 led Russell slowly but surely to believe that the United States was a greater threat to world peace, and more likely to start a war in which nuclear weapons would be used, than the Soviet Union.[118]

After the war Russell devoted most of his efforts to writing and lecturing about the danger of nuclear weapons, advocating world government, being actively involved in peace negotiations, and engaging in civil disobedience to protest war policies. Like Einstein, he believed that world government was the only alternative to the tragedy and ruin posed by a nuclear war and that a world government was the prerequisite for the prevention of such a war. The biggest obstacle to establishing a world government, he maintained, was the unwillingness of nations to give it enough power to be effective. Yet war was inevitable as long as different sovereign states tried to settle their disagreements by the use of armed force.

On December 23, 1954, Russell gave a radio address over the BBC on "Man's Peril from the Hydrogen Bomb," which evoked a very favorable response in Great Britain. He indicated that he was speaking not as a Briton or a European but as a human being, and he stressed that:

> There lies before us, if we choose, continued progress in happiness, knowledge and wisdom. Shall we instead choose death because we can not forget our quarrels? I appeal as a human being to human beings. Remember your humanity and forget the rest. If you can do so the way lies open to a new Paradise; if you cannot, nothing lies before you but universal death. (Russell 1969, 72)

Following this address, Russell drafted a statement that he asked a number of eminent scientists to sign. On February 11, 1955, preparatory to sending out the statement, he sent the following letter to Einstein:

> In common with every other thinking person, I am profoundly disquieted by the armament race in nuclear weapons. You have on various occasions given expressions to feelings and opinions with which I am in close agreement. I think that eminent men of science ought to do something dramatic to bring home to the public and governments the disasters that may occur. Do you think it would be possible to get, say, six men of the very highest scientific repute, headed by yourself, to make a very solemn statement about the imperative necessity of avoiding war? These men should be so diverse in their politics that any statement signed by all of them would be obviously free from pro-Communist or anti-Communist bias.

Russell considered these points to have exceptional importance:

(1) It is futile to get an agreement banning the H-bomb. It would not be binding in case of war.
(2) It is important not to be sidetracked by the peaceful applications of atomic energy.
(3) Everything which would be said in a statement must be strictly neutral. There must not be any suggestion of seeking advantage

for either side or preferring either side. Everything must be said from the point of view of mankind, not of this or that group.

(4) War may well mean the extinction of life on earth.

(5) Although H-bombs are the immediate concern, they do not exhaust the destructive possibilities of science. Bacteriological warfare may before long pose an equally great threat. War and science can no longer co-exist.

Einstein answered Russell promptly, telling him that he agreed with every word of his letter. He supported the notion of a public declaration "signed by a small number of people whose scientific attainment (scientific in the widest sense) have gained them international stature and whose declarations will not lose any effectiveness on account of their political affiliations. . . . The neutral countries ought to be well represented. It is vital to include Niels Bohr." He also indicated that he would write to some colleagues to obtain names of people in the United States and behind the Iron Curtain who might be willing to sign (Nathan and Norden 1968, 625–626). Russell agreed, and Einstein wrote Bohr asking him to get in touch with Russell. However, Bohr ultimately did not sign the Manifesto. Russell's last letter was dated April 5, 1955, and contained a draft of the Manifesto. A few days before he died, Einstein wrote Russell that he "gladly [is] willing to sign your excellent statement" (Nathan and Norden 1968, 631).

The Russell-Einstein Manifesto[119] was issued in London on July 9, 1955. Its content closely paralleled Russell's BBC Broadcast statement of December 1954. In addition to Russell and Einstein, it was signed by nine other scientists: Max Born (Germany), Percy W. Bridgman (United States), Leopold Infeld (Poland), Frédéric Joliot-Curie (France), Hermann J. Muller (United States), Linus Pauling (United States), Cecil F. Powell (Great Britain), Joseph Rotblat (Geat Britain), and Hideki Yukawa (Japan). The Manifesto stressed that the signers were not speaking "as members of this or that nation, continent, or creed, but as human beings, members of the species Man, whose continued existence is in doubt." It basically was an appeal to undertake the abolition of war through a commitment to arms reduction, the first step of which would be the forsaking of nuclear weaponry. It called on scientists to assume their special social responsibilities and inform the public of the technological threats, particularly the nuclear threats, confronting humanity. Since nuclear weapons threatened the future of humankind, humankind must put aside its differences and address this paramount problem. However, the statement did not offer the renunciation of nuclear and of other modern technological weapons as a solution to the peril; war as an institution must be abolished. It concluded with the following appeal:

> Here, then, is the problem which we present to you, stark and dreadful and inescapable. Shall we put an end to the human race, or shall mankind renounce war? People will not face this alternative because it is so difficult to abolish war. . . .
>
> We appeal, as human beings to human beings: remember your humanity, and forget the rest. If you can do so, the way lies open to a new Paradise; if you cannot, there lies before you the risk of universal death. (Rotblat 1967, 78–79)

Thus, as one of his last public statements, Einstein lent his voice to call for the renunciation of nuclear weapons as a first step toward banning war. The Resolution of the Manifesto succinctly stated what he had struggled for all his life:

> In view of the fact that in any future world war nuclear weapons will certainly be employed, and that such weapons threaten the continued existence of mankind, we urge the Governments of the world to realize, and to acknowledge publicly, that their purpose cannot be furthered by a world war, and we urge them, consequently, to find peaceful means for the settlement of all matters of dispute between them.

For Einstein, one of the peaceful means would be to find ways to establish a world government.

The response to the Manifesto was the creation of the Pugwash Conferences (Rotblat 1967). From the first Pugwash meeting[120] of 1957 to the most recent one in 2004, these conferences have constituted the most sustained attempt by the scientific community to address the threat of nuclear war on an international basis. The Pugwash Conferences, which "bring together, from around the world, influential scholars and public figures concerned with reducing the danger of armed conflict and seeking cooperative solutions for global problems," have been consequential and fruitful. Their importance has been recognized and was rewarded with the Nobel Prize for Peace in 1995.[121] The citation from the Nobel Committee read:

> The Pugwash Conferences are founded in the desire to see all nuclear arms destroyed and ultimately, in a vision of other solutions to international disputes than war. . . . It is the Committee's hope that the award of the Nobel Peace Prize for 1995 to Rotblat[122] and to Pugwash will encourage world leaders to intensify their efforts to rid the world of nuclear weapons.

## Epilogue

There is, to my mind, a striking parallel between Einstein's theorizing in physics, in particular, his struggle to unify gravitation and electromagnetism and geometrize them, his hopes for the ensuing theory, and his views and efforts regarding world government. To understand the parallel, mention must be made of Einstein's belief in the "pre-established harmony between mathematics and physics." Einstein gave the clearest statement of the meaning of this "pre-established harmony" in the Herbert Spencer lecture he delivered in Oxford in 1933. Alluding to David Hilbert's call since the beginning of the century to axiomatize physics, Einstein noted:

> If, then, it is true that the axiomatic basis of theoretical physics cannot be extracted from experience but must be freely invented, can we ever hope to find the right way? Nay, more, has this right way any existence outside our illusions? Can we hope to be guided safely by experience at all when there exist theories (such as classical mechanics) which to a large extent do justice to experience, without getting to the root of the matter? I answer without hesitation that there is, in my opinion, a right way, and that we are capable of finding it. Our experience hitherto justifies us in believing that nature

is the realization of the simplest conceivable mathematical ideas. I am convinced that we can discover by means of pure mathematical constructions the concepts and the laws connecting them with each other, which furnish the key to the understanding of natural phenomena. Experience may suggest the appropriate mathematical concepts, but they most certainly cannot be deduced from it. Experience remains, of course, the sole criterion of the physical utility of a mathematical construction. But the creative principle resides in mathematics. In a certain sense, therefore I hold it true that pure thought can grasp reality, as the ancients dreamed. (Einstein 1954, 274)

Felix Klein, David Hilbert, and Hermann Minkowski—the mathematicians who had dominated Göttingen mathematics in the two decades around the turn of the nineteenth century—had often spoken of the pre-established harmony between mathematics and physics. It was Klein—whose study of Leibniz[123] was an ongoing lifetime project—who had promoted and made central the notion of a pre-established harmony between mathematics and physics in Göttingen. Mathematics in its unity and wholeness mirrored the coherence and wholeness of Leibniz's monads: monads beheld an infinite, coherent, view—an infinite mirror image—of the universe even though self contained and windowless. For Leibniz, the pre-established harmony between monads had its origin in God, but it had been "established," not imposed—in the same way that two-fourths is the same thing as four-eighths, or that different parts in a musical score may yield harmony.

For Hilbert and Minkowski pre-established harmony meant something more specific than it had for Klein. Leo Corry (1998, 2004) has analyzed Hilbert's motivation and intent in axiomatizing mathematics and physics and has convincingly shown that both Hilbert and Minkowski believed that when formulating the mathematical scaffolding of physical theories certain universal principles must be postulated. An example is the principle of relativity, that is, the requirement of covariance of the equations of motion under homogeneous Lorentz transformations. Invariance under space and time translations are likewise such principles—that is, the requirement that the "absolute" position of the origin of the coordinate system cannot be a relevant variable in the description of phenomena. Similarly, the "absolute" time that marks the beginning of an experiment cannot be a relevant variable; only time intervals can enter the description. Time translation invariance will entail the conservation of energy, and space translations the conservation of momentum, if it is postulated that the equations of motion for the closed system are to be obtained from a principle of least action, and that the Lagrangian which describes the system is a scalar (i.e., has the proper covariance properties).

For Hilbert and Minkowski, "general principles"—such as Lorentz covariance—were such that should new empirical data force changes in the particular theory that had been advanced to represent the phenomena, the general principles would remain unchanged and would again impose stringent constraints on any new representation. For them, a belief in the pre-established harmony of mathematics and physics meant belief in the existence of such *universal* principles. Others, in particular Henri Poincaré, had expounded somewhat similar views. Poincaré was committed to a "physics of principles," which demanded that particular (model) theories—what Einstein would later call "constructive theories"—be compatible with the general principles. This was of central importance in his construction of particular representations of physical phenomena. The principle of relativity according to which "the laws of physical phenomena must be the same for a stationary observer as for an observer carried along in a uniform motion of translation; so that we have not and cannot have any means of discerning whether or not we are carried along in such a motion" (Darrigol 1995, 4), which Poincaré stated explicitly in 1904 was one such principle.[124]

Einstein explicitly spoke of his belief in pre-established harmony in his encomium of Planck in 1918—a metaphysics he became committed to *only* after 1915, that is, after arriving at the final version of the field equations of general relativity and his incorporation of Hilbert's approach in terms of an action principle in its formulation.[125] For Einstein, action principles, Lorentz covariance for special relativity, the principle of equivalence,[126] and general covariance for general relativity were examples of the "pre-established harmony."

The role of "principles" was elaborated further by Einstein by explicitly differentiating between what he called principle theories and constructive theories:[127]

> Constructive theories in physics attempt to build up a picture of the more complex phenomena out of the materials of a relatively simple formal scheme from which they start out. . . .
>
> Principle-theories employ the analytic, not the synthetic method. The elements which form their basis and starting point are not hypothetically constructed but empirically discovered ones, general characteristics of natural processes, principles that give rise to mathematically formulated criteria which the separate processes or the theoretical representations of them have to satisfy. (Einstein 1954, 228)

Examples of principle theories are the first and second laws of thermodynamics. They reflect well-confirmed, high-level *empirical* generalizations. Constructive theories provide understanding. However, Einstein believed that progress in understanding was hindered by premature attempts to formulate constructive theories in the absence of the constraints supplied by principle theories. Exploring the constraints and the possibilities offered by principle theories is one way to globally characterize Einstein's greatest works.

Einstein's position regarding a world government and a world court and the particular features of these organizations mirrors the position he enunciated regarding "constructive" theories and "principle" theories. For Einstein, the establishment of a world government and of a world court was the starting point and the basis for any proposal to prevent war. They formed the axiomatic basis for any model delineating particular institutions and specific mechanisms to eliminate wars. Their nonexistence in the past and the concomitant constant occurrence of war served as empirical proof of their necessity. Their stipulation—the charter that would establish them and delineate their broad powers—articulated the analogue of a principle-theory. Einstein's varying, detailed suggestions for their implementation—details about membership, governance, mechanisms for arbitration—were the parallel to constructive theories.

And the same was true regarding his commitment to an "ethic of ultimate ends." It, too, played the analogue of a commitment to a principle theory.

In his struggles to find an ultimate theory and in his unshakable belief in the necessity of a supranational government, Einstein was a visionary, and his was a grand vision; he was a prophet who sought universal features that would help make understandable some aspects of the comprehensibility of the world. And like Isaiah, he yearned for and toiled for that time when "nation shall not lift up sword against nation, neither shall they learn war any more" (Isaiah 2:4).

# CHAPTER TWO

# Albert Einstein and the Founding of Brandeis University

It is not enough for us to play a part as individuals in the
cultural development of the human race, we must also
attempt tasks which only nations as a whole can perform.
Only so can Jews regain social health.

—*Albert Einstein (1954, 182)*

I thank you, even at this late hour, for having helped me
become aware of my Jewish soul.

—*Albert Einstein to Kurt Blumenfeld, March 1955,
less than a month before he died*

John Stachel, in an erudite and sensitive essay (Stachel 2002, 57–84),
chronicled the events shaping Einstein's attitude toward Judaism and
his identification as a Jew. He explored Einstein's mature views on Jews,
Judaism, and Zionism and pointed to the fact that when in 1952 David
Ben-Gurion, then the prime minister of Israel, offered the presidency of
Israel to Einstein to succeed Chaim Weizmann, who had just died, Einstein declined. In his letter to Ben-Gurion expressing his regrets for not
being able to accept, Einstein added: "I am the more distressed because
my relationship to the Jewish people has become my strongest human
bond ever since I became fully aware of our precarious situation [as
Jews] among the nations of the world".[1] Stachel, however, "did not attempt to recount [Einstein's] many activities on behalf of the Jewish
people, individually or collectively" (Stachel 2002, 57). The present
chapter describes one such activity: Einstein's involvement with the
founding of Brandeis University.[2] This episode in Einstein's life gives
further proof that his "relationship with the Jewish people had become
[his] strongest human bond." It again reveals the facet of Einstein's

character we encountered in Chapter 1: his stubbornness and inflexibility, and his insistence on having his way.

As a young boy growing up in Munich, Einstein became aware of anti-Semitism and its pervasiveness in the Bavarian culture. After he graduated from the Swiss Federal Polytechnic (Eidgenössische Technische Hochschule—ETH) and was unable to obtain any academic position, he wrote a letter to Mileva Marić—whom he had promised to marry as soon as he found suitable employment—in which he pointed to "the anti-semitism . . . in the German-speaking countries which is as unpleasant as it is a hindrance" (Einstein and Marić 1992, 39) as the main obstacle for his not getting a job there. He later indicated that he discovered his Jewish identity when he moved to Berlin in 1914:

> I discovered for the first time that I was a Jew, and I owe this discovery more to Gentiles than to Jews. . . . I saw worthy Jews basely caricatured, and the sight made my heart bleed. I saw how schools, comic papers, and innumerable other forces of the Gentile majority undermined the confidence even of the best of my fellow-Jews, and felt that this could not be allowed to continue. (Einstein 1954, 171)

After Germany's defeat in 1918, anti-Semitism became more pronounced and more open in that country, and both the theory of relativity and Einstein personally came under attack from German ultranationalists and proto-Nazis, both in the press and in scientific meetings.

Although he had been aware of the anti-Semitism, Einstein later stated that it was witnessing the plight of the *Ostjuden*[3] who had come to Berlin after World War I and their treatment at the hands of the "Germans of the Jewish faith" that made him identify himself as a secular Jew and to commit himself to the Zionist cause. He strongly supported the establishment of a Jewish national homeland in Palestine, even though he had previously denounced nationalism and would continue to condemn it. He believed Zionism would "give Jews inner security" as well as "independence and inner freedom" (Fölsing 1997, 491). In 1921, he joined Chaim Weizmann on a trip to the United States to raise funds to establish a Jewish university in Jerusalem, having been convinced by Kurt Blumenfeld, a Zionist official, of the importance of creating a Jewish university in Jerusalem. On the way home to Berlin from his trip to Japan in 1922–1923, he stopped in Palestine to inaugurate the Hebrew University of Jerusalem on Mount Scopus. Einstein considered the university as the crucial institution that would revive in modern times the pursuit of truth in the Jewish tradition of learning.

In 1922, on a stopover in Singapore on his trip to Japan, where he had been invited to lecture on the special and general theory of relativity, Einstein raised funds there for the Hebrew University. Addressing the Jewish community in Singapore, he said:

> One may ask—why do we need a Jewish University? Science is international but its success is based on institutions owned by nations. Up to now as individuals we have helped as much as possible in the interest of culture and it would be only fair to ourselves if we now, as a people, add to culture through the medium of our own institutions.

On that same return trip he stopped in Israel and visited Mount Scopus, the site on which the Hebrew University of Jerusalem was to be built, and addressed the audience "from a lectern that had waited for two thousand years" for him. In a statement that he published on the day the Hebrew University was inaugurated in Jerusalem on April 1, 1925, he wrote: "A university is a place where the universality of the human spirit manifests itself," and he expressed the wish that "our University will develop speedily into a great spiritual center which will evoke the respect of cultural mankind world over."[4] And in a letter to Ehrenfest in April 1926 he remarked: "I do believe that in time this endeavor [the Hebrew University] will grow into something splendid; and Jewish Saint that I am my heart rejoices."[5]

After coming to Berlin in 1914, Einstein began to recognize that Jews ought not to press their claim to obtain academic positions, but should instead create positions and, by implication, institutions, to be filled from their own rank—hence his fervent commitment to help establish the Hebrew University of Jerusalem, his strong support of the founding of Brandeis University as a Jewish sponsored University in Waltham, and his insistence that it be and remain in Jewish hands.

Einstein had hoped that the Hebrew University in Jerusalem would be an elite institution devoted to the creation of knowledge for knowledge's sake, but that it would also be responsive to the practical needs of the Jews settling the land. However, American Jewry, which had contributed most of the money toward its founding, wanted it to be an American-style "college." Rabbi Judah Magnes, a champion of the American interest, eventually became the president of the Hebrew University.

In 1928, Magnes, who initially had been responsible only for the finances and for oversight of the administrative staff of the university, had his authority extended to academic matters, including professorial appointments. At this point Einstein, convinced that the principles of academic autonomy had been undermined, resigned. At Weizmann's urging, he did so quietly so as not to harm the university. Einstein, in a letter to Chaim Weizmann in January 1928, following his resignation from the board of Governors of the university, gave an account of how he viewed the history of the events:

> The bad thing about the business was that the good Felix Warburg, thanks to his financial authority ensured that the incapable Magnes was made director of the Institute, a failed American rabbi, who, through his dilettantish enterprises had become uncomfortable to his family in America, who very much hoped to dispatch him honorably to some exotic place. This ambitious and weak person surrounded himself with other morally inferior men, who did not allow any decent person to succeed there. . . . These people managed to poison the atmosphere there totally and to keep the level of the institution low. (Fölsing 1997, 494–495)

Yet, as evidenced by later designating the Hebrew University as the final repository of his papers, he continued to regard the fate of the Hebrew University "as a matter close to his heart" (Fölsing 1997, 495).

Ironically, as we shall see, Einstein had a similar experience when he became involved in the founding of Brandeis University: he exhibited initially strong support for the venture and then experienced a subsequent rupture by virtue of his strong principles, stubbornness, and determination to have his counsel followed.

In the early 1930s, Einstein suffered ever more strident verbal attacks in Germany. Finally, Hitler's accession to power made it clear that Einstein had to leave the country. In 1933, he resigned his post in Berlin and accepted a professorship at the newly founded Institute for Advanced Study in Princeton,[6] but he came to regard the Institute as not particularly hospitable to Jews.[7] Abraham Flexner, director of the Institute during the 1930s, though himself a Jew, distanced himself from this identification. Accusations of latent anti-Semitism in Flexner's appointments to the faculty of the Institute became a factor in the faculty's efforts to have him removed as director in 1939.[8] This reinforced Einstein's belief after the war that a Jewish-sponsored university was a necessity. As he indicated to David Lilienthal in July 1946 when he invited him to join him in his efforts to found Brandeis University as a secular Jewish-sponsored university:

> I am writing ... with regard to a cause which should be of interest to every American Jew. I refer to the establishment of as new University which would be under the influence of reliable Jewish personalities. ...
>
> I have given my name, and should be glad to offer my cooperation, to the new undertaking since I am convinced that selfhelp alone will enable us to free ourselves slowly from a situation which is truly painful for us—to be compelled to knock, often unsuccessfully, at doors which are being opened to us only reluctantly and conditionally. It should be also be borne in mind that, under existing conditions, our young scientific talents have frequently no access to scholarly professions which means that our proudest tradition—the appreciation of productive work—would be faced with slow extinction if we remain as inactive as we have been in the past.[9]

After 1945 the fact that six million Jews had been murdered in Europe during World War II solidified Einstein's support for a Jewish-sponsored institution of learning. While Einstein wanted Brandeis University, as a secular, Jewish-sponsored university, to be staffed by a faculty chosen on the basis of merit, regardless of sex, race, politics, or creed, he above all wanted the university to be in "Jewish hands." He wanted this requirement to be understood, openly stated, and implemented from the start. Thus it was that he vehemently objected both to having "a Gentile as the main speaker"[10] in behalf of the project at one of the first fund-raising dinners and to having Cardinal Spellman give the benediction. These objections led to his break with the more ecumenically minded Rabbi Israel Goldstein, who had initiated the project.

## Israel Goldstein

The demise of Middlesex University, a small veterinary and medical school located in Waltham, Massachusetts, on the western outskirts of Boston, triggered the founding of Brandeis University. By the end of World War II, Middlesex was no longer a viable institution. Its medical school, the only one in the country without a quota system for the admission of Jews and other minorities, had closed its doors in 1943 after it lost its accreditation, and it was clear that a similar fate would soon befall its veterinary school. In the fall of 1945 Joseph Cheskis, the dean of Humanities at Middlesex, realizing that the collapse of Middlesex was imminent, informed Ruggles Smith, the stepson of the founder of Middlesex University and its current president, of the American Jewish community's long-standing aspiration to establish a nonsectarian university. Cheskis suggested to Smith that he get in touch with Joseph Schlossberg, the secretary-treasurer of the Amalgamated Clothing Workers Union and a member of the New York Board of Higher Education. Schlossberg, in turn, advised Smith to write to Rabbi Israel Goldstein, the energetic spiritual leader of Temple B'nai Yeshurun, a large, well-to-do Conservative congregation on West 88th Street in Manhattan, who as early as the 1930s had explored the feasibility of establishing a secular university sponsored by the American Jewish community. This idea had also been advanced during the first third of the century, most notably by Rabbi Louis Newman in 1923.[11] The canonical model for such a university was that of Johns Hopkins and Harvard: a liberal arts college to which were attached a graduate school and some professional schools.

Within a week of receiving letters from Smith and Cheskis in early January 1946, Goldstein visited the Waltham campus and became convinced that it could be the site of the Jewish-sponsored liberal arts college he had envisaged. Moreover, the 100-acre campus could easily accommodate the reopening of a medical and veterinary school later on. Indeed, part of the attraction of Middlesex was that its charter allowed the university to grant B.A. and B.S. degrees, as well as doctorates of medicine and of veterinary medicine. Thus in time, the university would be able to alleviate the difficult conditions faced by Jewish students stemming from the discriminatory practices of medical and veterinary schools.

The first person Goldstein turned to after his visit to Waltham was his friend Julius Silver, a prominent New York attorney, who at the time was vice-president and general counsel of the Polaroid Corporation of Cambridge, Massachusetts.[12] A second visit to the campus with Silver and further discussions with Smith—who had the legal right to dispose of Middlesex University—led to an agreement to have Goldstein and his associates become a majority on the Board of Trustees of Middlesex University. Goldstein, aware that to be successful the project must acquire national prominence and have the backing of a broad spectrum of the Jewish community, then called on Albert Einstein in Princeton to tell him of his plans and to obtain his support. Both Einstein and Goldstein were ardent Zionists; the two had shared the podium at several rallies for the establishment of a Jewish national home in Palestine.[13]

Einstein immediately agreed with Goldstein that his objective was important. At the time Einstein was deeply distressed by the plight of Jewish scholars and scientists who, because of discriminatory practices, were finding it extremely difficult to obtain faculty appointments in American colleges and universities. Einstein's primary concern was that the university Goldstein was planning must be "first-class and free from non-academic control." On January 21, 1946, Einstein wrote Goldstein:

I would approve very much the creation of a Jewish College or University provided that it is sufficiently made sure that the Board and administration will remain permanently in reliable Jewish hands. I am convinced that such an institution will attract our best young Jewish people and not less our young scientists and learned men in all fields. Such an institution, provided it is of a high standard, will improve our situation a good deal and will satisfy a real need. As is well known, under present circumstances, many of our gifted youth see themselves denied the cultural and professional education they are longing for.

I would do anything in my power to help in the creation and guidance of such an institute. It would always be near my heart.

Very sincerely yours

*A. Einstein*[14]


During a subsequent visit Goldstein inquired whether Einstein would allow the university to be named after him. Einstein gracefully refused because he was of the opinion that the university should be named for "a great Jew who was also a great American,"[15] but he did consent to have the fund-raising vehicle for the project named the Albert Einstein Foundation for Higher Learning, Inc.[16]

As he himself was not going to be actively involved with the running of the Foundation, Einstein wrote Goldstein in early March 1946 asking him to get in touch with "one of his nearest friends," Dr. Otto Nathan, who, he was convinced "can in many respect be of valuable help in the upbuilding of the institution."[17] Nathan, a life-long pacifist, was an economist with socialist convictions who became a close friend of Einstein. He had come to the United States in the early 1930s and became an American citizen in 1939. He had taught at Princeton University, and at the time Einstein was writing to Goldstein Nathan held a nontenured appointment at New York University.[18]

Since the institution would not be called Einstein University, Goldstein then recommended that the university be named after Justice Louis Brandeis and proceeded to obtain the permission of Brandeis's daughter to do so. Goldstein also recognized that as the campus was located in the Boston area it was important to involve the Greater Boston Jewish community. At Silver's suggestion, George Alpert, a prominent Boston lawyer active in Jewish philanthropy, and the president of the Boston and Maine Railroad, was invited to join the Board of Trustees of the new institution, and he readily accepted.

In a letter to potential donors that Goldstein had drafted, Goldstein stated that "Our purpose is to make a contribution as a Jewish group to American education by supporting a university which in student body and faculty shall be open to all races and creeds, with merit as the only criterion for admission." In the draft Goldstein also outlined his plans for the university. His aim was to admit students to a College of Liberal Arts in October 1947 and to strengthen and improve the School of Veterinary Medicine that was still operating on a limited scale. He intended to reopen the Medical School only when adequate resources for a "first-class Medical School" had been secured.

By the beginning of April 1946, enough had been accomplished by Goldstein for the *Boston Traveler* to comment in an editorial that one of the genuine deficiencies in local education would be remedied when Dr. Israel Goldstein and his associates took over the physical plant of Middlesex University and made it an institution of the first rank.

> At one step Middlesex has shaken off the shackles of the past and entered upon a period of high promise. It is incumbent on the general public as well as the world of scholarship to know and evaluate fully the fact that Middlesex hereafter will be in the main stream of the world's intellectual tradition and that its future graduates will be full-fledged and fully honored members of the ancient company of scholars.[19]

Einstein became deeply committed to the success of the contemplated university for which he had "assumed responsibility." But he also became frustrated by the "somewhat scanty information" that he was receiving from Goldstein about "the new college projects." In mid-April Einstein complained that Goldstein had not as yet gotten together with Otto Nathan.[20] For his part Nathan saw the school as a place where he might obtain a permanent position.

## Rabbinic Connections

By late April 1946, Goldstein could write Rabbi Stephen S. Wise[21] that at a recent meeting of the New York Board of Jewish Ministers: "the project was endorsed and the reception given to my presentation was warm and cordial. A number of Rabbis asked for how they can be helpful; and one of them, who is in a position to do a good deal, undertook to raise a substantial sum."[22]

Wise—the spiritual leader of the Free Synagogue in New York and arguably the best known and most influential Jewish clergyman in the United States—was friends with both Franklin Delano Roosevelt and Felix Frankfurter and therefore had access to the corridors of power in Washington, D.C. He was also a good friend of both Albert Einstein[23] and Otto Nathan, having made their acquaintance shortly after Einstein joined the Institute for Advanced Study in Princeton. Wise was deeply committed to the view that all Jews were members of *Klal Yisroel,* the community of Israel, and that each member of the *Klal* was responsible for the well-being of the others. Ever since hearing and meeting Theodor Herzl at the second Zionist Congress in Basel in 1898, Wise had been a committed Zionist and had played an important role in organizing American support for the establishment of a Jewish national homeland in Palestine. Wise's views regarding Judaism and Zionism thus resonated with those of Einstein (Stachel 2002), though Wise was clearly more ritualistically and religiously inclined than Einstein. Wise was liberal in his outlook and had supported the labor movement in its efforts to have the right to unionize and obtain better working conditions. He thus had earned the respect of both Einstein and Nathan for his courageous political stand, and in particular that of Nathan who was a politically active Socialist. Nathan, like many other liberals during the 1930s and 1940s, sympathized with the Soviet experiment, and admired the Soviet Union for its opposition to Nazi Germany and for its crucial contributions to the Allied victory. In the immediate postwar years, he believed that a peaceful accommodation could be reached between communist Russia and capitalist United States.

In a recent conversation with Goldstein, Wise had discussed the university project and had been friendly and helpful. Thus, Goldstein in his letter to Wise also asked him whether he would consent to have his name added to the list of sponsors on the letter he was sending out to elicit financial contributions. At the time, the list of sponsors included the following dignitaries: Senators Joseph M. Ball, H. M. Kilgore, Brien McMahon, Wayne Morse, Albert T. Thomas, and Robert F. Wagner; Representative John W. McCormack; Governor Maurice Tobin (Massachusetts); Mayors Fiorello LaGuardia and William O'Dwyer (New York City); Boston Archbishop Richard Cushing; President William Green (American Federation of Labor); university presidents Karl T. Compton (MIT), Paul F. Douglass (The American University), Bryn J. Hovde (New School for Social Research), Daniel Marsh (Boston University), J. E. Newcomb (University of Virginia), and Eduard C. Lindeman (New York School for Social Work); Albert Einstein; and Alvin Johnson.[24]

Although Goldstein was keen on enlisting Wise in the project, Wise was reluctant to accept his invitation. In the middle of May, Wise wrote Otto Nathan:

> I am writing specifically to ask whether you have gotten in touch with Israel Goldstein. I myself have decided not to touch the thing until I have your judgment. In confidence, I may say that Goldstein is a tremendous public relations person. He conceives a newspaper heading to be the surest title to immortality. I think he is in earnest about this. The question is, what is "this" to be? Has he invited you to see him? . . . I want your help, and I think the Great Man in Princeton may need your protection.[25]

Goldstein and Wise had enjoyed warm personal relations until fairly recently.[26] But Zionist politics had made Wise change his opinion of Goldstein. In the negotiations during 1946 that eventually led to the establishment of a Jewish national state in Palestine, Goldstein had sided with Abba Hillel Silver and Ben-Gurion—contra Wise—and had supported their more militant position as against Weizmann's more moderate and patient stand. Weizmann's defeat at the 22nd Zionist Congress in December 1946 triggered the resignation of Stephen Wise from the American Section of the Jewish Agency Executive Committee and his retirement from leadership in the Zionist movement. He thereafter branded the Zionist Organization of America, in which Goldstein remained active, a "collection of personal hatreds, rancours and private ambitions."[27]

Wise's talks with Otto Nathan had likely exacerbated matters. Thus, on the heels of a discussion with Nathan, Wise became convinced "that the Great Man cannot afford to permit his name to be used unless he gets certain guarantees." Wise spelled out these guarantees in a letter to Nathan in early June 1946:

> The most worthwhile guarantee would be the delegating of power to you with a little help from I.G. [Israel Goldstein] and myself to name an organizing academic committee. If I.G. is not willing to do that,—and you will find him rather difficult and dilatory,—then I warn both you and A.E. against going along with him.[28]

Goldstein did agree and an academic committee was set up, but Nathan was not on it. On July 1, 1946, Einstein wrote Goldstein that he was "seriously perturbed about the preparation of the academic institutions of the College." Evidently, after discussions with Einstein, Nathan had prepared a fairly detailed plan of procedure for the academic organization of the college and had transmitted it to Ralph Lazrus, a member of the Board. Lazrus was a wealthy industrialist who had ties with the Benrus Watch Company and with the Allied Department Stores. He was liberal in his political outlook. Lazrus had given Nathan to understand that the plan would be discussed at the next meeting of the Board. However, in a letter to Wise on June 25 Goldstein informed him that an advisory committee had already been appointed to look into these matters and "is beginning to give some thought to the selection of the faculty."[29] When Einstein became aware of this, he wrote Goldstein and reminded him that when he allowed his name to be used he took it for granted that "no important step would be taken concerning organization without my consent." He therefore asked that the Board "promptly decide that a body of outstanding, independent, and objective men [be] charged with the selection of an Acting Academic Head—the real organizer of the University" and of an Advisory Board to advise the acting academic head. And Einstein "of course" expected to be consulted about the composition of that body. Further-more, since he was unable to attend the business meetings, he informed Goldstein that "I have asked my friend, Dr. Nathan to act as my representative and I should appreciate if you and the Board would act accordingly. Only under such circumstances shall I be able to continue lending my name to the project."[30] Upon receiving Einstein's letter Goldstein promptly wrote Nathan, inviting him to be present at a July 8 meeting at which the academic phase of the project would be discussed. And somewhat disingenuously, in the light of his letter to Wise in which he had indicated that the Academic Committee was "beginning to give some thought to the selection of the faculty," Goldstein added: "Up to this point, the Board of Directors of the Foundation has not had any discussions regarding the question of faculty, as we waited until we would see encouraging indication of financial results," which they now had.

Einstein's July 1 letter to Goldstein was written against a background of exchanges between Einstein and Wise. On June 26 Einstein had written Wise that he was concerned about the fact that Nathan had not been able to find a suitable position since he had returned from France where he had taught in the Armed Forces University. He indicated that Nathan "must be quite worried since he has his parents to take care of too." Einstein inquired of Wise whether he had any ideas regarding "what could be done to help him in the present situation? [Nathan] has not spoken to me about this matter and he would probably not like it if he knew of this letter. But having myself no connections and informations I have to ask you as a man of experience and of good heart."[31]

Wise understood Einstein's letter as a plea to have Nathan be given a position in the nascent Brandeis University. In his reply to Einstein on June 28, Wise stated:

> You wrote to me in confidence. I answer you in the same spirit. As a friend, I say to you ought not tie yourself up with the Foundation bearing your name and the Jewish University unless there be some completely trustworthy person, like our friend Otto Nathan, standing at the side of Dr. Israel Goldstein, to give him the benefit of his own wise judgment and

your judgment *and thus ensure for him at once a place in relation to the proposed university.* (my emphasis)

If you would write me a line to that effect—- that you wish to associate yourself more closely with the university, but you must have someone at the side of Dr. Goldstein whom you can trust, I think I would be able to do the rest. Will you not be good enough to send me word as promptly as you can.[32]

Einstein answered Wise the next day and indicated to him that he had "completely" misunderstood him. "I am at the present, not worried about Otto Nathan's relationship to the future Jewish University, but about his immediate future." In that same letter, he sketched his thoughts regarding "the question of organization and selection of the teaching-staff of the university and the general plan for the initial period of the institute." He agreed with Nathan that the solution of this problem was to have at the helm a man who would give the "cause" his "whole time." This man should fulfill the following conditions:

1. He must be a reliable Jew.
2. He must be acquainted with American University-institutions and must have understanding for educational problems and scholarship.
3. He should have experience in organization and *"Menschenkennt-nis"* (judgment of human nature).
4. He must be willing to deal and consult with us.

Einstein added that in discussions with Nathan neither of them could come up with a suitable candidate. Although he had thought of Nathan himself in this connection, Einstein doubted "that he will have enough authority in the eyes of the people who would have to consent to the choice." Nor did he believe that Goldstein had any ambition to undertake this job himself. Moreover, "I would not give my consent and cooperation if he would try to take over the mission and I feel pretty sure that also the very reasonable Mr. Lazrus would feel the same way."[33]

Although Goldstein was aware of the existing tensions and of Nathan's pressure to be involved in the academic planning, he clearly hoped that in time these issues would resolve themselves. At the end of July 1946, Goldstein informed Einstein that at the recent "small dinner" in June that was attended by some fifty people at the Waldorf-Astoria in New York over $250,000 had been raised. Thus "a good beginning had been made toward the realization of your and our dream," and he felt that a point had been reached "when we can begin to think about organizing a faculty."[34] The time had therefore come to consider the composition of the Academic Advisory Committee that would be responsible for selecting "an outstanding man who will be able to give all his time to the organization of the University and who would have to prepare the bylaws and regulations for the School and to appoint the original faculty." The suggestion from the Board of Directors of the Foundation was that Ralph Lazrus, Otto Nathan as Einstein's representative, Paul Klapper, the president of Queens College, and David Lilienthal, the chairman of the Tennessee Valley Authority, be members of the Academic Advisory Committee. Einstein drafted the letter to Lilienthal inviting him to be on the committee. Although he had not met him personally, Einstein informed him that he was writing because the project had reached the point that:

> It is now our most serious concern to do anything in our power that the new School be developed with the highest intellectual and moral standards. We have hence decided to appoint a committee of about five personalities with experience in academic affairs . . . [that] will be responsible for the development of the academic institutions of the School. It will be its foremost job to select an outstanding man who will be able to give all his time to the organization of the University.[35]

In view of the success of the "small" June dinner at the Waldorf, the Board made plans for a major fund-raising dinner, "one that will be a deciding factor in the entire ultimate success of the project." Its date was set for October 27, 1946, and it was to be held at the Hotel Pierre in New York. Given the name of the contemplated university, the Board decided that the main speakers at the dinner should be eminent persons who had known Brandeis, and they recommended that Supreme Court

Justices Frankfurter and Robert H. Jackson be asked. But Einstein balked at the suggestion of Jackson:

> I must confess that I cannot approve the choice of a Gentile as main speaker in behalf of our project. This is a Jewish cause and we have to advocate it ourselves. It is therefore, out of the question that I can sign a letter to Justice Jackson, even if a majority of us is in favor of the action.[36]

Einstein, however, finally relented and did write to Jackson, indicating that "Your own presence at this function and—if I may presume—an address by you in eulogy of the late Justice Brandeis, would contribute immeasurably to the success of the meeting, and assure the opening of Brandeis University under auspices worthy of the name of that great American."[37] In an identical letter to Justice Frankfurter, the conclusion of this last sentence read "under auspices worthy of the name of that great American and outstanding Jew."[38]

Although on the surface Einstein appeared as deeply committed to the project as ever, swayed by Wise's resentment of Goldstein, by Goldstein's reluctance to allow Nathan to have too great an influence at this stage of the developments, and by Nathan's eagerness to have a hand in the affairs of the new university, relations between Einstein and Goldstein deteriorated further and became more strained. An invitation by Goldstein to Cardinal Spellman[39] to deliver the invocation at the October 27 fund-raising dinner was one of the two straws that broke the camel's back. Einstein also took exception to the fact that Goldstein had discussed with Dr. Abram L. Sachar[40] "the possibility of his appointment as chancellor and organizer of the University-faculty without the authorization or even knowledge of the Advisory-Committee." In an angry letter to Goldstein in early September 1946, Einstein noted that:

> Those two facts represent two new breaches of confidence from your side. I have decided, therefore, not to cooperate any longer with you and I will have to make it clear that from now on I cannot take any more responsibility for any of your acts concerning the planned university.
>
>     I also cannot permit that my name is used for fund-raising in behalf of an enterprise in which you play an important part. Finally I must request

that my name be removed entirely from the name of your foundation and I expect to be notified as soon as this has been done.

Einstein added that copies of his letter would be mailed to the members of the Board and to the Academic Advisory Committee.[41]

After it became clear to Goldstein that Einstein would not alter his position as long as he was associated with the project, Goldstein resigned as chairman of the Board of Directors of the Albert Einstein Foundation and as chairman of the Board of Trustees of Brandeis University. He appreciated the fact that Einstein's participation in the project was indispensable to its success, but that his own role could be assumed by others. Nonetheless, in a lengthy letter to Einstein, Goldstein explained and defended his actions. He had, in fact, done nothing wrong as far as his contact with Dr. Sachar was concerned, nor anything unusual in the matter of the fund-raising dinner.[42] At its meeting on September 16, 1946, the Board of Directors of the Albert Einstein Foundation accepted Goldstein's resignation and sent Lazrus, Alpert, and Abraham Wexler to meet with Einstein "for the purpose of conveying to him the action of the Board."[43]

When informed of Goldstein's pending resignation, Nathan contacted Alpert, the likely new chairman of the Board of Trustees of the University. After their meeting Nathan wrote Einstein that he had found Alpert "Ein sehr angenehmer Mann" (a very agreeable gentleman).[44] Einstein therefore gave his blessing to have Alpert become chairman of the Board of Trustees. Stephen Wise, however, did not agree with Einstein and Nathan's assessment of Alpert. In November 1946, after having attended some of the Board meetings, Wise wrote Nathan:

It is now too late to talk about it, for the deed has been done. I wonder whether you and Professor Einstein have wisely suggested Alpert for a key place. He is not big enough a man to be President of the Board of Trustees. I think Lazarus [sic] is both finer and wiser. Remember my prediction: you will not be able to work with him long. He will become odious in time, due—or undue—to Professor Einstein.[45]

The vacancy in the chairmanship of the Board of Directors of the Albert Einstein Foundation created by Goldstein's resignation was filled on September 30, 1946, when Ralph Lazrus became the new chairman.

## The Harold Laski Episode

With Goldstein's resignation, matters seemed to have been resolved to Einstein's satisfaction.[46] He therefore wrote to the various people he had notified that he was severing his ties with the Brandeis Board that "work for the university-project . . . had made considerable progress" and that he was now convinced "that we shall be able to overcome the difficulties involved in such a new enterprise" and asked them to rejoin the enterprise. Writing to Sachar in late October 1946 in connection with a Hillel scholarship for a student, Einstein informed him that "Goldstein is not anymore connected with the University project; he resigned on my initiative. . . . I will do whatever I can to help to realize the project and I trust that it will be possible."[47]

One of the things Einstein "did" was to make Otto Nathan a member of the Board of the Albert Einstein Foundation.[48] With Goldstein's departure Nathan was able to assume a more active role. He and Lazrus took charge of the academic component of the project. Later, Lazrus was blamed for devoting too much of his time to these activities and not enough of his efforts to fund-raising, the primary responsibility of the chairman of the Foundation. Actually, financial matters were going well. At the end of October 1946, Boris Young, the director of the Albert Einstein Foundation Inc., informed Einstein that as of September 1 over $350,000 had been pledged and that the October 27 dinner was likewise a great financial success. Concrete plans for the academic mission of the university were therefore in order. On November 9, 1946, Otto Nathan submitted to the Board "An Outline of Policy for Brandeis University," which was to serve as a basis for discussion. Nathan envisaged that initially the university would consist of a liberal arts college with an enrollment of at least 1,000 students. As far as the curriculum was concerned, he favored

a minimum of compulsory courses. . . . Emphasis should be placed on in-
terdepartmental courses to break down the artificial rigidities of departments and fields of teaching. Independent work of students should be
encouraged in every way possible. . . . Wherever possible, the seminar
method of teaching should be encouraged. . . . The question of the honor
system [as practiced at] Swarthmore [College] should be studied. . . .
In the appointments to the faculty, great emphasis should be placed on
the teaching ability of the applicants (and not on the number of their
publications).

Nathan, wanting to make the college a "living democracy," recognized
that this "would require very different relationships among students,
faculty and college administration very different from those in most existing colleges." He wanted the students to have a voice in the administration of the college and to be treated as "free and adult human beings.
No compulsion to attend classes. There should be fewer examinations
than is customary. . . . No supervision of their private lives. . . . No 'per-
mits' for weekend absences or for 'late hours.' "[49]

Nathan was aware of the experiments being carried out at Swarthmore, Antioch, Sarah Lawrence, Black Mountain College, and the newly
founded Roosevelt College in Chicago, as well as the curricular changes
being introduced at Harvard and Columbia. But probably the model
that most closely approximated what he had in mind was Hutchins University of Chicago with its college and separate teaching faculty for the
college. In order to make recommendations for both the college and the
graduate school component of the university, Nathan visited some of
the leading American universities and in December 1946 went on an
overseas trip to Great Britain "to investigate the methods of instruction
and administration policies. He conferred with many important educators, including Professor Laski, and encountered friendly reactions to
the plan for Brandeis University wherever he went."[50]

Nathan's friend Harold Laski was a prominent, outspoken, and artic-
ulate Socialist and a distinguished political scientist, who taught at the
London School of Economics.[51] In July 1945 Laski had been elected
chairman of the national executive committee of the Labour Party, and

his position became even more prominent in the summer of 1945 after Labour won an overwhelming majority in the British general election. In the middle of June 1945, during the election campaign, Laski made a speech in Newark-on-Trent, a market town in Nottinghamshire in the East Midlands region of England, in support of one of the Labour candidates. At the end of his presentation, he was questioned by a fairly well-known Conservative journalist who had probably been encouraged by Tory interests to ask Laski provocative questions. On the day after Laski's speech a letter appeared in the *Nottingham Journal* signed by one H. C. C. Carlton, a Conservative member of the local county council, in which Carlton alleged that during his speech in Newark, when enumerating the reforms he would like to see enacted, Laski had declared that "if we cannot have [these reforms] by fair means we shall use violence to obtain them." The letter also said that when challenged by a member of the audience who claimed he was "inviting revolution from the platform," Laski had replied: "if we cannot get reforms we desire we shall not hesitate to use violence, even if it means revolution" (Eastwood 1977, 140). Upon reading the letter, Laski immediately issued the following statement:

> I am going to take out a writ for libel against the man who wrote it and against anybody else who reproduces this letter. My answer at the meeting was entirely different. What I said was: it was very much better to make changes in time of war when men were ready for great changes than to wait for the urgency to disappear through victory, and then to find that there was no consent to change what the workers felt an intolerable burden. That was the way a society drifted to violence. We had it in our power to do by consent what other nations have done by violence. (Eastwood 1977, 141)

The incident got further prominence when the *Daily Express,* a Beaverbrook, Tory national newspaper, featured the story under the headline; "Laski unleashes another general but as yet unpublished election broadside: socialism even if it means violence." Then on June 20 the *Newark Advertiser* carried the story of Laski's Newark market place speech as recorded in shorthand by one of their reporters. According to the *Newark Advertiser,* Laski had been asked why he had openly advocated "revolution by violence" in earlier speeches. His answer, according to the *Newark Advertiser,* was:

> If Labour could not obtain what it needed by general consent, "we shall have to use violence even if it means revolution." . . . Great changes are so urgent in this country, that if they were not made by consent they would have to be made by violence. . . . When a situation in any society became intolerable . . . it did not become possible to prevent what was not given by generosity being taken by the organized will of the people.

On June 20 Laski brought a libel suit against the *Nottingham Guardian,* the *Daily Express,* and the *Evening Standard,* all based on the publication and reproduction of Carlton's letter, and on June 22 he sued the *Newark Advertiser* and its editor, C. E. Palby.

The trial lasted five days and was presided over by a very conservative judge, Lord Goddard. During the trial, which opened in late November 1946, it emerged that Laski had in fact never made the statement "we shall have to use violence even if it means revolution." That statement had been introduced into the *Newark Advertiser* article after the reporter had seen Carlton's letter. And in answers to questions posed to him by his lawyer, Laski made clear that he could not be a member of the Labour Party if he advocated revolution by violence, since by its constitution any member of the Labour Party is committed to the acceptance of constitutional democracy. Furthermore, he had been a critic of communism ever since 1920, had frequently and extensively criticized in his writings both communist theory and communist strategy, and had recently opposed the admittance of the Communist Party into membership of the Labour Party.[52] The defense, however, pointed to Laski's repeated advocacy of revolution in his writings and his belief that "the time is ripe for revolution." Laski explained that what he had meant by the latter statement was that "The time is ripe for great changes." In his charge to the special jury, Goddard put the matter thus:

> [You have to decide] whether these speeches . . . , remembering the audience to which they are addressed, would be an incitement to violence or

revolution, whether, it is preaching revolution and violence as part of a political creed and urging the people to adopt it. [Or is Laski] putting forward views which you or other people may abominate and hate, but which he is at liberty to express and has the right to express [if he is stating them not as a matter of incitement and not as a matter of advocacy but as a matter of argument]. (Eastwood 1977)

The jury, after deliberating for 40 minutes, found the article in the *Newark Advertiser* a fair and accurate report of a public meeting, and Laski thus lost his suit. Having incurred expenses amounting to some $40,000 in connection with the trial, he became greatly distressed. However, with the help of friends and contributions from the Labour Party, Laski was able to cover three/fourths of the costs. Max Lerner, Otto Nathan, and others committed themselves to raising the balance in the United States.[53]

Nathan's visit with Laski made a deep impression on Nathan. Ever since Goldstein's resignation he had thought a great deal about Brandeis's academic future.[54] As the American Jewish community's financial support of the institution seemed assured, the crucial question had become: Who would assume the academic leadership of the institution? After his visit with Laski, Nathan came to believe that Laski would be an excellent choice to be the president of Brandeis. An agreement had existed between Einstein and Goldstein stipulating that Einstein would have a major say in recommending and selecting the president. It was also understood that one of the principal functions of the Academic Advisory Committee—of which Nathan and Lazrus were members—was to make recommendations of suitable candidates to the Board. But evidently in the spring of 1947 Nathan convinced Einstein that he should ask the Board that he—Einstein—rather than the Academic Advisory Committee make the recommendations.

## Denouement

On March 30, 1947, Alpert visited Einstein in Princeton. In a letter he later wrote to Ralph Lazrus, Einstein described that meeting:

I discussed with him at great length the difficult problem of selecting a president for Brandeis University. I told him that you [S. R. Lazrus] and Mr. Nathan had suggested that the Board of Trustees be asked to delegate the authority for selection of the President to me. I mentioned to Mr. Alpert that, in case the Board would take such action, I was considering to inquire of Professor Laski whether he might be willing to come over here to help us in organizing the University. Mr. Alpert not only did not object to this suggestion, but approved of it.[55]

Alpert also told Einstein that he would bring the matter up at the next Board meeting for its approval. What happened next is somewhat uncertain for there is no explicit record of the events. It is not clear whether Alpert, given his own conservative inclinations and the tenor of the times, upon learning more about Laski had second thoughts about his candidacy, or whether he became apprehensive about delegating to Einstein—and therefore to Nathan and Lazrus—the authority to pick Brandeis's first president. Most likely, he had strong reservations about both issues. As seemingly there was no quorum[56] at the April 14 meeting of the Board of Directors of the Foundation where Einstein's request was to come up, de facto there was no meeting of the Board, "and no resolution appeared in any minutes authorizing or requesting [Einstein] to select a president for the University."[57] What is known is that at the April 14 meeting Nathan reported that "several difficulties had been encountered in the attempt to form such a committee and since it was deemed imperative to insure that the eventual President's qualification be such as to guarantee the implementation of our purposes, it was proposed to authorize Professor Einstein to make the selection of President." Bluestein, one of the directors, then pointed out that, while approval of Einstein's choice was the prerogative of the Board, such authorization would be equivalent to agreement, in advance, to accept whatever choice Einstein made. Nathan then stated that "Prof. Einstein had been reluctant to assume that responsibility but that he had finally agreed to this procedure and was currently engaged in the search for a suitable President."[58]

Given Alpert's positive response at their March 30 meeting, Einstein had written Laski to ask whether he would be prepared to consider an invitation to become Brandeis University's first president—possibly for only two or three years were he reluctant to leave Great Britain permanently. In his letter Einstein added:

> The University will be in Jewish hands, but we are determined to develop it into an institution which is enlivened by a free, modern spirit, which emphasizes, above all, independent scholarship and research and which does not know of discrimination for or against anybody because of sex, color, creed, national origin or political opinion. All decisions about educational policies, about the organization of teaching and research will be in the hands of the family.
>
> The Board of Trustees has delegated to me the authority of selecting the first President of the University. This man would have the challenging task to help us in determining the basic foundations of the University and to select and organize the initial faculty upon whom so much depends. We all feel that among all living Jews you are the one man who, accepting the great challenge, would be most likely to succeed. . . .
>
> I am writing, therefore, to ask you whether you would be prepared to consider such an invitation. . . . You would oblige me by treating this inquiry confidentially.[59]

Laski answered him promptly. On April 25 he wrote Einstein in his diminutive handwriting:

> Few things in my eyes have done more honour or given me greater pleasure, than your most generous suggestion. But, with my deep respect, I fear that I must decline it. First, I am sure that I am not the right person for the post. I have no financial capacity. I like teaching, and not administration. My roots are firmly fixed, after 27 years in the University of London, in this country. To these reasons I must add that my wife would not now wish to live in the United States, and apart from being morally bound to fight here, as best I can, for socialism, I want to spend such leisure as I can find in writing a book it has been my ambition to write since I was a student. I am sorry to decline any offer which comes from you, and from my good friend Otto Nathan. But I am confident this is the right decision. I am not fit for a post which demands the special qualities of a University-President.
>
> I hope very much that I may, nevertheless, have the privilege to being connected, in some loosely continuous way, with the new institution in a

teaching capacity. I hope I need not say that in that realm I would seek, very gladly, to serve it with all the energy in my power.[60]

Thus by early May Einstein and Nathan knew that Laski would not accept an offer to be president of Brandeis. When this information was made available to the Board is not clear.[61] But it was quite clear to both Einstein and Nathan that the Laski matter was closed.

Einstein's letter to Laski upset some members of the Board since he had not been authorized to write it. On May 11 Alpert and Silver went to Princeton to see Einstein. In his letter to Lazrus reporting on the meeting, Silver indicated that Einstein had expressed surprise when he learned that because of the absence of a quorum de facto no meeting of the Board of Directors of the Foundation had been held, and therefore that no resolution appeared in any minutes authorizing or requesting him to pick a president for the university. Nor had Einstein been made aware of the fact that those directors who had been informally consulted by Alpert had not been informed that Harold Laski was receiving consideration as an active candidate. Einstein expressed the opinion that it was definitely "unfair" not to have communicated this information to him. But Silver went on to say that:

> Professor Einstein readily agreed that selection of the President should be made either at the recommendation of an educational committee or on the nomination of other qualified persons, but subject to full disclosure and approval by the Board of Directors. . . . This removes the chief obstacle to continued progress.[62]

However, no further progress was to be made. Alpert had evidently determined to use the Laski invitation to marginalize Lazrus and Nathan, and thereby also minimize Einstein's say in shaping the university. As the available written documentation makes clear, by mid-April Nathan was playing an ever more active role in determining academic policy. At the time, consideration was being given to hiring a part-time provost to coordinate academic matters until a president was selected, and Max Grossman, a professor of journalism at Boston University, was being considered for this post. Nathan interviewed him, and after their meeting Grossman wrote Nathan to "abandon all other undertakings temporarily and have yourself appointed as 'convener' of the university. This will enable you to do whatever organizational work which is necessary until others are selected to perform tasks specifically." More specifically, Grossman proposed to Nathan that he "ought to be dean of faculty and professor and head of the department of economics." Should his deanship not be compatible with whoever was appointed as president, he certainly would be happy as a professor of economics.[63]

That relations between Alpert and Nathan were strained is evident from the fact that during March 1947 Alpert kept pressing Nathan to set up a Committee on Education made up of prominent educators whose responsibility would be to make recommendations to the Board of suitable candidates for the presidency—evidently to no avail.[64]

Relations between Alpert and Nathan and Lazrus had deteriorated to the breaking point by early May. On May 16 Einstein wrote Lazrus that he had asked Professor Otto Nathan to act as his representative at the meeting of the Board of Trustees that was to be held on May 19.[65] The details of that meeting are scant.

> The President [Lazrus] . . . reported that he was in receipt of a letter from Professor Einstein in which the latter indicated that Dr. Nathan was authorized to act as Professor Einstein's representative at the meeting. The President then called upon Dr. Nathan to make a statement on behalf of Professor Einstein and himself, and added that such statement also represent[ed] the position of the President.
>
> Dr. Nathan stated in substance that disagreements had arisen with regard to the selection of a President of the University and to the policies of the University. Accordingly, Professor Einstein desired to withdraw his name from the Foundation; that the Foundation be liquidated and all of its funds be transferred to the University. Dr. Nathan stated that Professor Einstein, Mr. Lazrus and he would like to help the University in every way.[66]

Thereafter Nathan and Lazrus offered to tender their resignations as soon as the name of the Foundation was changed and an understanding was reached that no public statement would be made.[67] However, relations between Alpert and Lazrus became so rancorous that on June 22 Lazrus did release a statement stating that Einstein was resigning and withdrawing the use of his name because Alpert and the Board were trying "to bring down the educational standards of the university." Having been privy to the increasingly contentious exchanges between Alpert and Lazrus, Einstein took pen to hand and on June 20 wrote Susan Brandeis Gilbert that controversies over the educational policies and academic organization of the university had developed between him and his friends and the Board. As these disputes appeared unbridgeable, he informed her that "I have decided to withdraw my support from the project and therefore to have my name eliminated from the fund-raising Foundation."[68]

Alpert of course denied Lazrus's public statement and got Susan Brandeis to issue a statement asserting that Lazrus and Nathan had "arrogated to themselves the shaping of academic policy, the selection of a president and other important educational functions which could rightfully be performed only by the Board of Trustees."[69] Alpert in turn charged that Lazrus and Nathan had "surreptitiously" made overtures to a "thoroughly unacceptable choice" as president and thus were trying to give the school a "radical, political orientation."

> To establish a Jewish-sponsored University and to place at its head a man utterly alien to American principles of democracy, tarred with the Communist brush, would have condemned the University from the start. I made it perfectly clear to Mr. Lazrus then and later that on the issue of Americanism I cannot compromise.[70]

Alpert was using Iron Curtain rhetoric and McCarthyite tactics to justify his assuming control of the university's affairs. An article by William Zukerman in the July 4, 1947 issue of a small Jewish magazine, *The American Hebrew*, stated the case forcefully:

> Mr. Alpert's statement is not only utterly tactless and irrelevant to the issue, but also untrue and vicious. It is the statement of a narrow partisan reactionary politician behooving a member of the un-American Committee, not a President of a University named after the late Justice Brandeis.

Zukerman identified the man whom Alpert had described in the "entire" American press as "alien to the principles of American democracy" as Harold Laski, "one of the greatest teachers of our age, a man who brought up a generation of youth [from all over the world that] has been the vanguard of humanity's struggle for democracy and social justice." He stated that Laski had been the target of reactionary forces in England and in the United States "chiefly because he was a pioneer of the newer and wider interpretation of democracy from the political to the economic field." Zukerman pointed to John Dewey, Roosevelt, Wendell Wilkie, and Brandeis as people who shared Laski's "alien principle of democracy. . . . Professor Laski is the living incarnation in England of the spirit of the New Deal movement in the United States. He can be said to be its intellectual father even as the late President Roosevelt was its political founder." Moreover, for Zukerman, "It would have been a beautiful gesture on the part of a Jewish institution of learning to register its faith in the Rooseveltian interpretation of American democracy in a period of reaction by naming Professor Laski its President."

Zukerman was willing to concede that the university trustees had "good and weighty" reasons for opposing the nomination of Professor Laski. From the point of view of fund-raising, a philosopher of New Dealism might not be the ideal choice in Truman's United States, and the trustees certainly had the right to oppose the nomination on grounds of expediency. But why, Zukerman asked,

> should the trustees have gone out of their way to denounce Laski's principles as "alien to American way of life and democracy"? Who has made this obscure Board the judges of what is native and alien to American democracy? . . . And why was it necessary to raise altogether the question of Americanism in this connection and to waive [sic] the patriotic American flag? Did they not realize that branding Dr. Einstein's choice as "alien to American way of life, and tarnished with the Communist brush," they . . . practically accused Professor Einstein and his colleagues of un-Americanism?

Zukerman's conclusion was that "the tactless and reactionary manner" in which the trustees chose to justify their opposition demonstrated that they were unsuited to be trustees of a Jewish-sponsored university, "and especially one named after Justice Brandeis."[71]

Upon reading Alpert's comments, Einstein drafted the following statement, which he sent to Stephen Wise:

> The press statement, which Mr. George Alpert and another member of the Board of Trustees of the Brandeis University released on the occasion of the withdrawal of myself and of my friends, Professor Otto Nathan and S. Ralph Lazrus, have surprised, even shocked me.
>
> It was I who suggested that the name of an eminent British scholar and educator, Professor Harold J. Laski of the University of London, be considered in connection with the Presidency of the Brandeis University. Mr. Alpert, who now makes an untruthful charge of radicalism against my two associates, in no wise objected to this suggestion, made to him by me in my own home.
>
> The press releases have convinced me anew that it was none too early for us to sever a connection from which no good was to be expected for the community. I should like to state specifically that my associates, Professor Nathan and Mr. Lazrus, and I have always been and have always acted in complete harmony. I feel deep gratitude particularly to my old friend, Professor Otto Nathan, who with great devotion and complete selflessness gave time and effort to a cause which the three of us alike considered good and most urgent.[72]

The edited version Einstein signed on June 24 was somewhat more restrained. It read:

> The press statements which Mr. George Alpert and another member of the Board of Trustees of Brandeis University released on the occasion of the withdrawal of myself and of my friends, Professor Otto Nathan and Mr. S. Ralph Lazrus, have convinced me that it was none too early for us to sever a connection from which no good was to be expected from the community. My associates and myself had very reluctantly come to the conclusion that the type of academic institution in which we have been interested could not be accomplished under the existing circumstances and the present leadership.[73]

Its second paragraph was the same as the last paragraph in the statement Einstein had sent Wise except that in the last sentence, "worthy and most urgent" had been substituted for "good and most urgent."[74]

It was Alpert's version of the "facts," however, that became the official history.[75] Most newspaper accounts reported the events as follows:[76] Nathan was to form a committee to advise the Board of the Einstein Foundation on the selection of a president. However, Nathan and Lazrus informed Alpert that there would be no Educational Advisory Committee and suggested instead that the Board of Trustees designate Professor Einstein to select a president. Alpert then informed them that this was contrary to the original agreement and contrary to all academic procedures but that he was willing to hear their proposal through. When they recommended Harold Laski as president, Alpert was ready to concede that Laski was brilliant and that he might even be a great educator. But Laski was controversial in his political views since he was "an international socialist of record" who had lost a libel suit against a British newspaper that had called him a communist. Alpert was unwavering in his view that the person heading the institution be an American. The matter came to the Board, which backed Alpert, whereupon Einstein, Nathan, and Lazrus resigned "with good will" and assured Alpert that they would make no statements that would jeopardize the project. However, Lazrus did subsequently release a statement stating that Einstein was resigning and withdrawing the use of his name because Alpert and the Board were trying "to bring down the educational standards of the university," a statement that Alpert in turn denied.

In addition to other inaccuracies, there is a glaring one in this version of the story: the issue of Laski never came in front of the Board! When the Board met on May 19, Laski was no longer a candidate! One can only conclude that the real issue had been who was to control the affairs of the university.[77]

The subsequent course of events has an ironic twist: in 1948 Abram Sachar was installed as the first president of Brandeis University.[78] Sachar had been Goldstein's choice for the presidency, and Goldstein's informal inquiry of him in the summer of 1946 as to whether he might be interested in the position had led to the first rift between Einstein

and the Brandeis Board of Directors. Also, history was to repeat itself:
A struggle between Sachar and Alpert over the issue of who was to control the future of the university eventually led to Alpert's removal as
chairman of the Board of Trustees.

Relations between Einstein and the university remained frigid after
the 1947 break. Einstein's bearing in the matter gives a revealing account of one facet of his character. Sachar's attempt to mend fences in
1948 was rebuffed with Einstein's refusal to see him.[79] In the fall of
1950, Sidney Shalett was writing an article for the *Saturday Evening Post*
on the founding of Brandeis University in which Einstein was to be
mentioned in connection with its early history and in which Rudolf
Kayser, Einstein's son-in-law, was reported to be joining Brandeis's faculty in 1951. Shalett wrote Einstein to corroborate some facts, and in
his reply to Shalett, Einstein asked him not to mention him in connection with the early history of Brandeis, adding:

> For your information I should like to add that I have nothing at all to do
> with Dr. Rudolf Kayser's negotiations with Brandeis University. A possi-
> ble appointment of Dr. Kayser to the faculty of Brandeis will of course, in
> no way affect my attitude towards the University.[80]

In March 1952 Sachar again wrote Einstein in order to be able to visit
him and tell him, "without any obligation to you, something about the
development of the school."[81] Sachar had hoped that Einstein would
draw a distinction between "the one or two people who are, after all, a
temporary part of the life of an institution and the institution itself."
But Einstein answered him as follows:

> I was somewhat astonished by your letter of March 25th. The most con-
> cise answer to it has been formulated hundred years ago by Schopen-
> hauer who said: "Erlittene Unbill vergessen, heisst müsahm erworbenes
> Geld zum Fenster hinauswerfen".[82]

> With kind regards,
> Sincerely,
> *Albert Einstein.*[83]

When later that year a young Indian scholar asked Einstein for his assistance in obtaining a position at Brandeis, Einstein answered that events of the past had made it impossible for him to get in touch with Brandeis University, "directly or indirectly."

> I was connected with Brandeis University at the time of its foundation. It happened, however, that a few of the Trustees behaved quite dishonestly against me and my nearest friends. Therefore I had to sever completely my connections with the institution. This happened before Dr. Sachar was connected with the enterprise so that he is not directly involved. But after he has, so to speak, inherited the "Tabu" it is impossible for me to approach him.[84]

And in 1953, when Sachar wanted Brandeis to give Einstein an honorary degree, Einstein rejected the offer and sent him the following letter:

> Dear Dr. Sacher [*sic*]
>
> It is embarrassing not to be able to repay friendly behavior in kind. However in this case I cannot help it. What transpired in the preparatory phase of Brandeis University certainly was not based on misunderstandings and can no longer be compensated for. Thus I cannot accept your offer of an honorary doctorate.
>
> I would not want to have this matter discussed, for this would be harmful to the University, and I shall tell only my closest friend, who has the right to be informed, about it.
>
> With kind regards,
>
> A. Einstein[85]

Einstein's last exchange with Sachar occurred in January 1954. Sachar had once again tried to see him. Einstein answered him that:

> If you would be simply a private person who has written delightful books, I would gladly accept your kind offer to visit me. Under the prevailing circumstances, however, it is not possible for me to do so. As you are informed about the relevant past events you will easily understand.[86]

It should be noted that Einstein's interaction with Brandeis and Sachar after 1947 was passive, whereas his animosity toward Alpert was active.

Upon learning that Alpert had been appointed honorary chairman of the Albert Einstein College of Medicine campaign, Einstein wrote Nathaniel Goldstein that his confidence in Alpert had been completely destroyed when he dealt with him in the early days of Brandeis and that this had led to his withdrawal from that institution. And he concluded his letter by asserting: "I must tell you that I would never have permitted the use of my name in connection with the College of Medicine had I known that Mr. George Alpert would be asked to play an important role in its development."[87]

## Epilogue

When reviewing this early history of Brandeis University, I was struck by the fact that Israel Goldstein's role seems largely to have been forgotten, overwhelmed by Abram Sachar's subsequent accomplishments in building the institution. But as Sachar himself acknowledged in his letter to Goldstein upon being installed as the first president of Brandeis in 1948: "You are really the 'father' of Brandeis University. You put endless energy and devotion into the building of the concept and the corralling of its first support."[88]

Another point that should be made here is that the sentiments that motivated Einstein's strong endorsement of Goldstein's project were the same as those that inspired the Jewish members of the faculty to join the university when it opened its doors to its first class of 100 freshmen in the fall of 1948. A parallel can be drawn between Einstein's commitment to help create a Jewish national home in Palestine and the commitment of the many secular Jews who comprised the bulk of Brandeis's initial faculty to help build a Jewish-sponsored institution of higher learning.

Einstein—that cosmopolitan, secular Jew who abhorred nationalism, and German nationalism in particular, and who believed primarily in the creativity of *individuals*—in his response to the hatred and the xenophobia that he had encountered in German academic circles during the Weimar era and to the unmistakable vulnerability of the

Jewish community even in "civilized" Germany, was to make the following assertion:

> The best in man can flourish only when he loses himself in a community. Hence the moral danger of the Jew who has lost touch with his own people and is regarded as a foreigner by the people of his adoption.
>
> The tragedy of the Jews is that they . . . lack the support of a community to keep them together. The result is a want of solid foundations in the individual which in its extreme form amounts to moral instability.

And in 1933 after he had left Germany, Einstein declared:

> It is not enough for us to play a part as individuals in the cultural development of the human race, we must also attempt tasks which only nations as a whole can perform. Only so can Jews regain social health.

Consequently:

> Palestine is not primarily a place of refuge for the Jews of Eastern Europe, but the embodiment of the re-awakening of the corporate spirit of the entire Jewish nation.

Einstein also stressed that the "Jewish" homeland he advocated would be a place where three archetypal Jewish ideals would flourish: "the pursuit of knowledge for its own sake; an almost fanatical love of justice; [and] a desire for personal independence."[89]

Einstein's justification for the establishment of a Jewish state can be recast to make manifest why the vision of a Jewish-sponsored university resonated so deeply with the secular Jews who joined the rank of its initial faculty. The statement would then read as follows:

> It is not enough for us to play a part as individuals in the cultural development of the United States, we must also attempt tasks only its entire Jewish community can perform. Only so can Jews regain social health.

Consequently:

> Brandeis will not primarily be a place of refuge for Jewish scholars who have been discriminated against in the elite American universities, but—now

after Auschwitz—it will be the embodiment of the re-awakening of the corporate spirit of the entire American Jewish community.[90]

And the three ideals that Einstein hoped the Jewish state would nurture
were precisely the ones that made a Jewish-sponsored, secular university
also attractive to non- Jewish scholars.

For the most part these ideals have been cultivated at Brandeis. Moreover, Einstein's demand that the university be "first class" and meet the
highest academic standards was heeded from the very beginning. The
faculty that taught the first class to graduate from Brandeis in 1952 was
indeed outstanding.[91] Among them were, in the fine arts: Leonard Bernstein, Erwin Bodky, Arthur Fine, Harold Shapero, Mitch Siporin, Louis
Kronenberger, and Lee Strasberg; in the humanities: Ludwig Lewin-
sohn, Albert Guerard, Milton Hindus, Nahum Glatzer, and Simon Raw-
idowicz; in history, the social sciences, and psychology: Max Lerner,
Frank Manuel, Lewis Coser, Merril Peterson, Leonard Levy, Marie Boas,
and Abraham Maslow; and in the sciences: Saul Cohen, Sidney Golden,
Oscar Goldman, and Albert Kellner.

# CHAPTER THREE

# J. Robert Oppenheimer: Proteus Unbound

> The kind of person I admire most would be one who becomes extraordinarily good at doing lots of things but still maintains a tear-stained countenance.
>
> —*J. Robert Oppenheimer, 1926*[1]

In Greek mythology Proteus was the old, prophetic, uncommunicative shepherd of the seas' flocks who knew all things—past, present, and future. Those who wished to consult him first had to surprise him and bind him during his noonday slumber. Even when caught he would try to escape by assuming all sorts of shapes; but if caught he would then tell all he knew. Protean thus came to mean variable, versatile, taking many forms.

In his Jefferson Lecture of the Humanities in 1973, Erik Erikson called Thomas Jefferson a *Protean* man, meaning that Jefferson was a many-sided man of universal stature; a man of many gifts, competent in each; a man of many appearances, yet centered in a true identity. Erikson noted that the appellation *Protean* could also designate a man of many disguises; a man of chameleon-like adaptation to passing scenes; a man of essential elusiveness. However, such designations in a man of exceptional stature must be seen in relation to the new identity emerging in his time. As a self-made man, a person with a Protean personality would have the ability to make many things of the self, in both a semi-deliberate and defiant fashion (Erikson 1974, 51–52).

Like Jefferson, Oppenheimer can aptly be characterized as a *Protean* man, with both designations applying. But in contrast to Jefferson, whom Oppenheimer admired greatly, Oppenheimer was a man of many appearances. George Kennan, who became acquainted with Oppenheimer after the war[2] and became a personal friend and his colleague at the Institute for Advanced Study in 1951, perceptively described Oppenheimer as

> in some ways very young, in others very old; part scientist, part poet; sometimes proud, sometimes humble; in some ways formidably competent in practical matters, in other ways woefully helpless: . . . a bundle of marvelous contradictions. . . . His mind was one of wholly exceptional power, subtlety, and speed of reaction. . . . The shattering quickness and critical power of his own mind made him . . . impatient of the ponderous, the obvious, and the platitudinous, in the discourse of others. But underneath this edgy impatience there lay one of the most sentimental of natures, an enormous thirst for friendship and affection, and a touching belief . . . in what he thought should be the fraternity of advanced scholarship . . . [a belief that] intellectual friendship was the deepest and finest form of friendship among men; and his attitude towards those whose intellectual qualities he most admired . . . was one of deep, humble devotion and solicitude. (Kennan 1972, 18)

Oppenheimer made many things of himself: creative physicist and influential teacher in an era that revolutionized the physical sciences, charismatic administrator of a wartime project that altered the course of world history, prominent adviser to the highest echelon of American policy makers in the postwar period, vocal intellectual alerting the public to the changed character of the world they were living in and of its dangers. In each of these roles he became the personification of what others should aspire to becoming. Yet for all these accomplishments he could not fashion a sense of identity for himself. A sense of identity means a sense of being at one with oneself as one grows and develops; and it means, at the same time, a sense of affinity with a community's sense of being at one with its future as well as its history—or mythology (Erikson 1974, 27). For a time his work in physics during the 1930s, his directorship of Los Alamos during the 1940s, and his role as governmental adviser in the immediate postwar period did give him a deep sense of connection with communities that had a distinctive purpose. But he found it difficult to integrate these disparate roles—perhaps because he found it difficult to conceive an overall creative vision for himself or to devise a compelling objective for the community he belonged to if one had not been formulated at the time he assumed its leadership; and perhaps also because each of these activities also had been connected with a deep crisis—a deep rupture.

This chapter highlights three such fissures. The first concerns Oppenheimer's determination to become a theoretical physicist; the second, his giving up doing research in physics; and the third, the revocation of his security clearance and his becoming a public intellectual. All three have a bearing on his public persona.

## The Early Years

Robert Oppenheimer was born in New York on April 22, 1904, into a prosperous, emancipated Jewish family of German descent.[3] His father, Julius, was a successful businessman who had come to the United States in 1888 when he was seventeen to work in the cloth-importing business of relatives. His mother, Ella Friedman, was a painter of almost professional standards, whose family had migrated from Germany to Baltimore in the 1840s. Both parents had a taste for the arts, especially for music. A younger brother died shortly after birth, when Robert was four years old, and his brother Frank was born in 1912. The Oppenheimers were well off and lived in a large eleventh-floor apartment overlooking the Hudson River. They also maintained a summer home in Bay Shore on Long Island. Looking back on his growing up, Oppenheimer declared that "My life as a child did not prepare me in any way for the fact that there are cruel and bitter things" (Cassidy 2005, 18).

Robert's lifelong friends, Paul Horgan and Jeffries Wyman, who frequently visited his home during the 1920s, described the household as very handsome—several van Goghs adorned the walls—very formal, and elegant, but somewhat sad and melancholic. The tone was set by the mother, whose crippled right arm was always encased in a gray silk glove. Horgan described her as highly neurotic, highly attenuated emotionally, a mournful person. John Edsall, another of Robert's friends during the 1920s, had the impression that Bob, as Robert was called by his friends until the late 1920s, was strongly attached to his mother but not to his father.

During Robert's youth, the Oppenheimers were active members of the Ethical Culture Society, which Felix Adler had founded in 1876. It was therefore natural for the young Robert to be sent to the Ethical Culture School, a school with high academic standards and liberal educational ideas. He entered its second grade in September 1911 and graduated its high school in February 1921. For the precocious, dazzling, but insecure young Robert, the school was an ideal place to nurture his differences without causing him to feel like an outsider. One of his schoolmates remembered him at age fifteen, "as still a little boy, . . . very frail, . . . very shy, and very brilliant" (Smith and Weiner 1980, 6–7). While he eventually outgrew his shyness, the insecurity and deep unhappiness remained. Paul Horgan, with whom he became good friends during a trip to New Mexico in the summer of 1922 before entering Harvard, recalled that "Robert had bouts of melancholy, deep, deep, depressions as a youngster" (Smith and Weiner 1980, 6).

Oppenheimer entered Harvard in 1922 a year after graduating from the Ethical Culture School; the delay was caused by a bad case of dysentery he had contracted on a trip to Europe during the summer of 1921, necessitating a full year for recovery. During the summer of 1922, he went on a journey to the Southwest with his English teacher at the Ethical Culture School, Herbert Winslow Smith, whom his parents had hired to accompany him. Oppenheimer asked Smith to introduce him as his brother on the trip, presumably to conceal the fact that he was Jewish, but Smith refused. It was on this trip that Oppenheimer fell in love with New Mexico.

As a freshman Oppenheimer intended to become a mining engineer. He majored in chemistry, but took and audited many other courses.

In 2005 the record of the courses Oppenheimer took at Harvard became available.[4] The transcript only indicates the courses he took for credit and excludes those he audited. His transcript also lists the average grade for the courses he took at the Ethical Culture School. While he obtained grades of 99 and 98 for his high school algebra and plane geometry courses and a 95½ in plane trigonometry, he only received an 85½ in solid geometry. In Greek, Latin, and German he obtained the grades of 92, 88, and 91, respectively, but only an average grade of 65 for the three English courses he had taken.

What stands out in his Harvard college record is the number of courses Oppenheimer took in any given year: he took 12 courses a year, instead of the normal 8. This allowed him to graduate in three years. His academic performance during his first two years at Harvard was so stellar that he was elected to Phi Beta Kappa, the national honor society, at the end of his a sophomore year and was made a John Harvard Fellow, a "very high academic distinction," for his third year at Harvard. He majored in chemistry, obtaining A's in all his chemistry courses.[5]

The variety of courses Oppenheimer took and the range of subjects he familiarized himself with as an undergraduate reflect his remarkable capacities to absorb and integrate new information and fresh knowledge. But the path he had chosen to follow was not smooth. In a letter to Herbert Smith, his former teacher at the Ethical Culture School, the nineteen-year-old sophomore gave indications of the inner turmoil he was experiencing:

Aside from the activities exposed in last week's disgusting note, I labor, and write innumerable theses, notes, poems, stories, and junk; I go to the math lib and read and to the Phil lib and divide my time between Meinherr Russell and the contemplation of a most beautiful and lovely lady who is writing a thesis on Spinoza—charmingly ironic, at that, don't you think? I make stenches in three different labs, listen to Allard gossip about Racine, serve tea and talk learnedly to a few lost souls, go off for the weekend to distill the low grade energey [sic] into laughter and exhaustion, read Greek, commit faux pas, search my desk for letters, and wish I were dead. (Smith and Weiner 1980, 54)

A good friend of Oppenheimer described the diffident young Robert at Harvard as somewhat precious, narcissistic, arrogant, and completely deaf to music. But even though he worked like a demon, he found time to cultivate his amity with Paul Horgan and with Francis Fergusson, the latter a classmate from the Ethical Culture School who had also gone to Harvard, and to form new lasting friendships—with John Edsall, Jeffries Wyman, Frederick Bernheim, and others. There was, however, no time for dating women.

Oppenheimer's talents were so numerous and varied that he could have been anything he chose. His friends actually thought that he would become a humanist. Oppenheimer's contact with Percy Bridgman and with Edwin Kemble, however, convinced him that he wanted to be a physicist. Physics offered him a safe, bounded arena in which he excelled, into which he could focus his energies, and in which he didn't have to confront his personal demons. The study and practice of physics required intense discipline.[6] Moreover, physics presented problems he could be *passionate* about, and their solutions indicated how the world was put together.

During his senior year at Harvard, Oppenheimer spent many hours each week in Bridgman's laboratory working on the problem of the effect of pressure on metallic conduction. That experience should have made it clear to him that, even though he was acquiring a deep appreciation and understanding of experimental practices, his genre, whatever it was, was not experimental science. Nevertheless, upon graduating from Harvard, he went off to Cambridge University where he would have a disastrous experience working at the Cavendish.

Oppenheimer aspired to become an experimental physicist but had no great competence in the laboratory, thereby raising doubts in the minds of his teachers at Harvard as to his future prospects. Bridgman, in his letter of recommendation to Ernest Rutherford at the Cavendish, acknowledged that Oppenheimer's weakness was on the experimental side, explaining that he had an analytical, rather than a physical, mind and that he was not at home in the manipulations of the laboratory. Nonetheless, everyone at Harvard regarded Oppenheimer as an exceptional person.

In his letter Bridgman referred to Oppenheimer's perfectly prodigious power of assimilation. Though Bridgman felt unsure whether Oppenheimer would ever make a contribution of any real character, Bridgman indicated that "if he does make good at all, I believe that he will be a very unusual success" (Smith and Weiner 1980, 77).

The year following his graduation from Harvard was one of deep emotional crisis for Oppenheimer. He went to the Cavendish hoping to work with Rutherford, but instead came under the tutelage of J. J. Thomson. "The business in the laboratory was really quite a sham. . . . I was living in a miserable hole," he reminisced to Thomas Kuhn (Kuhn 1963, 2). He couldn't get his experiment on the scattering of electrons by thin metallic foils to measure the electrical conductivity in thin metallic films to work, as he had great difficulty preparing the thin metal targets, and Thomson, by now an old man, was not very helpful. Even though Oppenheimer was terribly excited by the new developments in physics, his failure as an experimentalist raised doubts about the appropriateness of a career in physics. Francis Fergusson and John Edsall noted how acutely troubled Robert was.[7]

During a visit to Fergusson in Paris in the late fall of 1925, Oppenheimer revealed his despair over his inept performance in the laboratory and confided about unsatisfactory sexual ventures. Some of the factors involved in bringing about the crisis in the winter of 1925 have been brought to light by Bird and Sherwin (2005, 41–55). Oppenheimer's frustration in his work at the Cavendish, his general unhappiness with the Cambridge culture, and the cooling of his friendship with some of his Harvard classmates because they were getting married were catalysts in the breakdown. He became deeply depressed and jealous of the success of some of the people around him, in particular Patrick Blackett, a young experimental physicist at the Cavendish some three years his senior who had become his tutor and something of a mentor to him (Nye 2004). Sometime in the fall of 1925 he actually left a "poisoned apple" on Blackett's desk, an apple laced with some chemical, possibly cyanide, that might well have caused Blackett great harm. Fortunately, his deed was discovered, and Blackett did not eat the apple. But Oppenheimer was hauled before the university authorities and nearly expelled. Only the intervention of his parents and the promise that he would seek psychiatric help prevented his expulsion.

Another element in causing Oppenheimer's breakdown may have been confrontations with his sexual identity. Questions of sexual polarity may have arisen; certainly questions about sexual adequacy did.[8] These issues are of relevance only in so far as they may point to another aspect of this exceedingly complex man, a feature that made cohesiveness and integrity of self more difficult.

In mid-March 1926, together with three friends, Frederick Bernheim, John Edsall, and Jeffries Wyman, Oppenheimer went for a vacation to Corsica. A few days before their intended return to England, Oppenheimer announced that he had to leave immediately because he "had left a poisoned apple on Blackett's desk" and he had to go back to see what had happened. But the incident had taken place in the fall of 1925 and not the spring of 1926. The crisis seemed to have abated with the help of some psychoanalysis and Oppenheimer's own insights into his problems. It was overcome by the end of the summer. Many years later Oppenheimer told his Berkeley colleague and friend, Haakon Chevalier, that reading Marcel Proust's *A la Recherche du Temps Perdu* in Corsica was "one of the great experiences in his life," and that the book helped him overcome his depression. He had memorized some passages from it and could quote to Chevalier a particular one that had evidently resonated with him:

> Perhaps she would not have considered evil to be so rare, so extraordinary, so estranging a state, to which it was so restful to emigrate, had she been able to discern in herself, as in everyone, that indifference to the sufferings one causes, an indifference which, whatever other names one may give it, is the terrible and permanent form of cruelty.[9] (Chevalier 1965, 34–35)

Although he never experienced a crisis of similar proportion as the one in Cambridge, Oppenheimer's emotional balance would always be delicate. As Isador Rabi later put it: "In Oppenheimer the element of earthiness was feeble" (Rabi, in Oppenheimer 1967, 3). An urge to test the limits of safe conduct never left him. He would sail boats in terrible weather, drive cars recklessly, explore the Rockies inadequately prepared—not to mention exhibt a nonconventional demeanor in social situations. He never learned "normal, healthy way[s] to be a bastard." Behind a facade of charm, wit, arrogance, and on occasion insensitivity and cruelty lurked deep insecurities and, in particular, doubts about his creativity.

## Becoming a Physicist: Oppenheimer and His School

After he recovered from his crisis, Oppenheimer determined to become a theorist. John Edsall, who was working toward a Ph.D. in biochemistry at St. John's College at the time, remembered Oppenheimer talking to him at length about the papers of Heisenberg, Dirac, and Schrödinger and the meaning of these advances. The year 1925 was the beginning of an exciting and momentous period in theoretical physics. Heisenberg published his first paper on matrix mechanics that year, which was soon followed by Born and Jordan's paper, which indicated that Heisenberg in his quantum mechanical treatment of the anharmonic oscillator had represented the position and momentum of the oscillator by matrices, and showed that these matrices obeyed the commutation rule: $[q, p] = ih/2\pi$ with $h$ Planck's constant. These two papers in turn were rapidly followed by those of Dirac and by the paper of Born, Heisenberg, and Jordan, which formulated quantum mechanics in terms of observables (such as the positions and momenta of the particles involved, their energy, angular momentum, etc.) that—in contrast to their representation in classical mechanics—were operators that satisfied certain commutation rules, but whose equations of motion were structurally the same as the classical ones. In particular, the operators representing the positions of the particles, $q_i^{(n)}$ for the $i$th component of the $n$th particle ($i = 1, 2, 3$), and those representing their momenta, $p_j^{(n)}$, satisfy the following commutation rules:

$$[q^{(n)}_i, p^{(m)}_j] = ih/2\pi\, \delta_{nm}\, \delta_{ij}$$

$$[q^{(m)}_i, q^{(n)}_j] = 0;\ [p^{(m)}_i, p^{(n)}_j] = 0$$

In early 1926 Schrödinger's papers on wave mechanics began to appear. The result was a period of intense activity, during which all atomic phenomena were subjected to an analysis based on the new mechanics. Oppenheimer's abilities and phenomenal quickness in grasping new ideas were so outstanding that in May 1926 his first paper—on the quantum mechanical explanation of the frequencies and intensities of molecular band spectra—was submitted for publication. A second paper—dealing with the continuous spectrum of the hydrogen atom and with the question of how to normalize the wave functions in that case—was finished by July 1926. The two papers acknowledge help from Ralph Fowler and Paul Dirac. They indicate that by mid-1926 Oppenheimer had not only mastered all the mathematical and technical demands of the new methods, but had gone on to carve out an area of his own: problems involving the continuous spectrum.

When Max Born visited Cambridge in the spring of 1926, he invited Oppenheimer to continue his studies at Göttingen and Oppenheimer accepted. At Göttingen he became a member of the intellectual community around Born and blossomed. His close friendships with Paul Dirac and Isador Rabi date from those days. There Oppenheimer continued working on the description of the nonbound states of simple systems in the new wave mechanics. He was one of the first theorists to work on the quantum mechanical description of X-ray emission by atoms and of scattering phenomena—all problems involving continuum wave functions that could not be tackled with the old quantum theory. With Born he wrote an important paper that laid the foundation for the quantum mechanical treatment of molecules, a problem he had already addressed in Cambridge in his paper on the quantum theory of the vibrational and rotational degrees of freedom of molecules. In that paper Born and Oppenheimer formulated a quantum mechanical approach to describing physical phenomena that has become ever more prominent. Born and Oppenheimer recognized that in molecules the lighter electrons move much faster than the heavier nuclei. To describe the nuclear motion they therefore essentially integrated out the (high-frequency) electronic motion and obtained an approximate, effective, wave mechanical description of the nuclear vibrations (Born and Oppenheimer 1927).

Oppenheimer obtained his Ph.D. in the spring of 1927 after less than a year's stay in Göttingen. A glimpse into the young, self-important Oppenheimer is revealed by his behavior in the seminars Born chaired. Oppenheimer evidently felt free to contradict any speaker at the seminar, explain why he was wrong, and walk up to the blackboard and write out a corrected proof. His conduct led the other members of the seminar to send Born a note written in ornamental letters on a large sheet of parchment, asking him to stop Oppenheimer's disruptions or they would no longer attend the seminar. Born, too shy to confront Oppenheimer directly, left their note on his desk for Oppenheimer to see. Oppenheimer saw the note and read it and thereafter restrained himself. Though he deeply admired Oppenheimer's brilliance, Born was relieved when Oppenheimer left Göttingen. He wrote Ehrenfest:

> Oppenheimer who was with me for a long time is now with you. I should like to know what you think of him. Your judgment will not be influenced by the fact I openly admit that I never suffered as much with anybody as with him. He is doubtless very gifted but without mental discipline. He is outwardly modest but inwardly very arrogant. Through his manner to know everything better and to continue any idea you give him, he has paralyzed all of us for three quarters of a year. I can breathe again since he is gone and find courage to work. My young people have the same experience. Do not let yourself keep him for any length of time. Stop! You are supposed to give me your opinion. Perhaps I just got very nervous. (Greenspan 2005, 146)

The note says as much about Born as it does about Oppenheimer. Born was intimidated by his brilliant students and assistants, and would suffer a nervous breakdown in the winter of 1928–1929. The opposite of what Born had said of Oppenheimer was probably also true: "He [Oppenheimer] is outwardly arrogant but inwardly very modest," and if not "modest," then insecure, and seeking approval and admiration.

It was at Göttingen that the deep friendship between Oppenheimer and Dirac was cemented. Dirac came to Göttingen in February 1927 and lived in the same pension as Oppenheimer. They spent a lot of time with one another discussing physics and went on long walks together on weekends. Clearly, Dirac was not fazed by Oppenheimer's demeanor at the seminars and appreciated the quickness and incisiveness of Oppenheimer's mind, though he could not understand Oppenheimer's pleasure in and devotion to poetry. For his part, in his interview with Thomas Kuhn, Oppenheimer indicated that "the most exciting time I had in Göttingen and perhaps the most exciting time of my life was when Dirac arrived and gave me the proofs of his paper on the quantum theory of radiation."

Oppenheimer remained in Europe in 1929, on a fellowship from the International Education Board of the Rockefeller Foundation, spending a semester with Paul Ehrenfest in Leyden and with Pauli in Zurich. He continued working on problems involving the continuous spectrum and became involved with Pauli and Heisenberg's investigations of the quantization of relativistic field theories. He spent the following year as a National Research Fellow at Harvard and at the California Institute of Technology. At Cal Tech, discussions with Robert Millikan and Charles C. Lauritsen, who had just observed the extraction of electrons from metal surfaces by very strong electric fields, led to an extension of his previous treatment of the ionization of hydrogen atoms by electric fields. Oppenheimer's theory of field emission was the first example of the phenomenon of barrier penetration in wave mechanics and antedated Condon and Gurney's and Gamow's explanation of $\alpha$-decay in radioactive nuclei. It was recognized as an important elucidation of the difference between the classical and the quantum mechanical description of the motion of microscopic particles.[10] Oppenheimer's work on cold field emission exhibited a feature that was to become very prominent in his work: his close collaboration with his experimental colleagues.

While at Cal Tech, Oppenheimer visited Berkeley and decided that he would like to go there because it was a desert. "There was no theoretical physics and I thought it would be nice to start something." But he also recognized the danger of being isolated and therefore kept a connection with Cal Tech, which he thought was "a place where I would be checked if I got too far off base." He accepted appointments at Berkeley and at Cal Tech that were to start after he had spent another year in Europe as a postdoctoral fellow (Kuhn 1963, 9–10).

Theoretical physics, as a profession and a vocation, came late to the United States. It became institutionalized there in the 1920s after the advent of quantum mechanics in departments primarily staffed by experimentalists and chaired by an experimental physicist (Schweber 1988). Oppenheimer came into his own in such a department at Berkeley, a department in which Ernest Lawrence's Radiation Laboratory played a very important role (Heilbron and Seidel 1989). Its activities became the focus of much of Oppenheimer's theoretical interests and investigations during the 1930s.

Lawrence and Oppenheimer were close friends until the end of that decade (Davis 1968). Oppenheimer learned much from having been intimately connected with all the developments at Lawrence's Radiation Laboratory during the 1930s. He saw how Lawrence assembled the teams of scientists and engineers that designed and built ever larger cyclotrons. He saw him keep regularly in touch with all the staff members, and noted his presence at key moments in the operation of the cyclotron and at crucial phases of experiments.

When Oppenheimer first came to Berkeley and started giving graduate courses, his lectures left students bewildered by the level of their exposition. But after a few years all his courses—particularly his lectures on quantum mechanics and on electromagnetic theory—became models of clarity, emphasizing utility, but also conveying to the students the beauty of the subject matter. In his eulogy of Oppenheimer in 1966, Hans Bethe noted: "Probably the most important ingredient [Oppenheimer] brought to his teaching was his exquisite taste. He always knew what were the important problems, as shown by his choice of subjects. He truly lived with these problems, struggling for a solution, and he communicated his problems to his group" (Bethe 1968, 396). It should also be noted that Oppenheimer was not only the center of his students' intellectual world, but he was also at the center of their social world. In his eulogy, Serber described Oppenheimer's interaction with his students:

> He met the group [which in the mid-thirties consisted of a dozen graduate students and about half a dozen postdoctoral fellows] once a day in his office. A little before the appointed time the members straggled in and disposed themselves on the tables and about the walls. Oppie came in and discussed with one after another the status of the student's research problem while others listened and offered comments. All were exposed to a broad range of topics. Oppenheimer was interested in everything; one subject after another was introduced and coexisted with all the others. In an afternoon they might discuss electrodynamics, cosmic rays, astrophysics, and nuclear physics. (Serber, in Oppenheimer 1967)[11]

This form of cooperative investigation that made use of the collective knowledge of the group became the characteristic mode of distributed inquiry in research groups—but only much later. This mode of doing research is based on the assumption that creativity is the result of a cooperative enterprise, that it is the result of communal activities. Perhaps the most succinct depiction of the characteristic element of this collective, cooperative activity was given by Oppenheimer himself: "What we don't understand we explain to each other" (Bethe 1967, 1080).

Arnold Sommerfeld's seminar in Munich had initiated this transformation in theoretical physics. Whereas the usual practice at German universities was to invite outside speakers to give lectures on their recent researches, Sommerfeld essentially excluded outside speakers (except for his former students) and had his graduate students, postdoctoral fellows, and assistants as well as the faculty members of the seminar make presentations. These presentations would be semiformal, with frequent interruptions and helpful give and take.

Oppenheimer had early on recognized the transformation that had taken place in theoretical physics, and in physics more generally, since the advent of quantum mechanics: physics had become much more of a cooperative enterprise. The era when single individuals like Bohr and Einstein could transform all of physics was seemingly over. There were also many more young people doing theoretical physics. The change could readily be seen in the different character of physics conferences. At the first Solvay Congress of 1911, the thirty-two-year-old Einstein was the youngest of the invited participants.[12] Again it was primarily older, well-established physicists who were invited to the second one in 1921. By the third congress, in 1927, it was young physicists— Heisenberg, Pauli, Dirac—who set the agenda and were active partici-pants. But it was in the purely theoretical physics conferences—in Copenhagen, at Bohr's Institute in the late 1920s and during the 1930s, and in the United States at the Washington Conferences on Theoretical Physics during the 1930s—that the transformation was clearly evident.

During the 1930s, Oppenheimer was arguably one of the most imagi-native and courageous theoretical physicists working in quantum field theory and high energy physics, the foundational frontiers of the disci-pline. Thus after Dirac had obtained his famous equation describing rela-tivistic spin half particles, such as electrons, he proposed that all the negative energy states it exhibited were filled except for empty states, "holes," which then described positively charged particles, which he iden-tified with protons. Oppenheimer quickly pointed out that in such a representation the hydrogen atom would be unstable and, moreover, that the positively charged particle associated with a hole had to have the same mass as electrons. As Peierls suggested in his entry on Oppenheimer in the *Dictionary of Scientific Biography:* "[Oppenheimer] practically predicted the positron three years before its discovery by Carl Anderson." He and his students kept abreast of all developments in cosmic ray physics, atomic physics, and nuclear physics, and made important contributions to the explanations of the new phenomena encountered in these fields. For ex-ample, when cosmic ray experiments indicated serious discrepancies with theory, Oppenheimer considered the possibility that this might indicate a breakdown of quantum electrodynamics. When the discovery of the mesotron seemingly resolved the difficulty, he and his students studied the properties this particle had to have to account for the experimental data. Similarly, he and his students became deeply involved in explaining the data that Ernest Lawrence's cyclotron was producing.

A word ought to be added on the kinds of problems that Oppenheimer and his students addressed. In order to obtain academic positions, it was necessary for his students to be useful and relevant to the experimental colleagues they would acquire in the departments they would join. This meant that his students had to be good calculators, be able to obtain "numbers" that could be compared with experimental data, and be familiar with the intricacies of obtaining reliable and accurate data in experiments, and thus understand the details of the experimental set-up. These criteria were always met. But this meant that some constraints had to be imposed on the questions that were asked.

Oppenheimer represented a new type of theoretical physicist, who by virtue of his mastery of the novel features that emerged from the synthesis of quantum mechanics and special relativity could not only account for some of the puzzling experimental data generated by the newly built accelerators and by cosmic ray experiments, but could also predict new ontologies: positrons and mesotrons.

Oppenheimer, this Protean personality, had fashioned himself, semi-deliberately, into this new kind of theoretical physicist.[13] And as asserted by Bethe in his eulogy, in his capacity as teacher and as mentor to a generation of American theoretical physicists, "J. Robert Oppenheimer did more than any other man to make American theoretical physics great" (Bethe 1967).

The irony was that during the 1930s Oppenheimer was ahead of his time. Experiments were not accurate enough to corroborate his suspicions that the Dirac equation could not account for the spectrum of the hydrogen atom. Similarly, cosmic ray experiments were difficult to analyze and could not pinpoint the masses nor the spin of the observed mesotrons. Astrophysical observations were such that claiming that a stellar object might be a neutron star would at best be considered an interesting suggestion, and the state of theoretical astrophysics was such that claiming the possible existence of black holes would certainly have been considered a wild conjecture. The fact that many of his most insightful and daring productions could not be experimentally verified prevented him from becoming convinced that he was as creative a physicist as the founders of quantum mechanics; not as creative as Dirac, Heisenberg, Schrödinger, and Pauli, but perhaps as creative as Born. He may have felt that the illness that had prevented him from entering Harvard in 1921[14] had forced him to arrive at the Cavendish and Göttingen a year too late to make singular contributions to the initial development of quantum mechanics.

A mistake in the calculations by one of his students, Sidney Dancoff, in the late 1930s, robbed Oppenheimer of the opportunity to overcome the divergence difficulties that one encounters in the lowest orders of perturbation theory in quantum electrodynamics and to eliminate them by a procedure that later would be called mass and charge renormalization— when in fact he was clearly aware of the physics involved and of the steps to take. To substantiate this assertion, let me turn to the lecture Oppenheimer gave in Philadelphia on September 20, 1940, at the conference that celebrated the bicentennial anniversary of the founding of the University of Pennsylvania. The title of his lecture was "The Mesotron and the Quantum Theory of Fields." Oppenheimer characterized his contribution as "rather negative," for it was the purpose of his presentation to explain why meson theory, and more generally, the quantum theory of fields, "has failed so completely to deepen our understanding of nuclear forces and processes," the subject that the other speakers at the conference—Enrico Fermi, Gregory Breit, Isidor Rabi, Eugene Wigner, and John van Vleck—had reported on. Yet, in spite of this failure it seemed to Oppenheimer that the quantum theory of fields was "still a subject worth reporting on" (Oppenheimer 1941).

Oppenheimer then proceeded to indicate that the infinities encountered in quantum electrodynamics originate in the fact that the electric and magnetic field become operators in the quantum version of the field theory. Thus measurements of the electric field in a little region of space around point x at time t *in identically prepared systems* will yield different results because the electric field has quantum fluctuations. Bohr and Rosenfeld, and Oppenheimer's student, Philip Morrison, who had studied these fluctuations, had shown that the mean square deviation of the electric field averaged over a spherical region of radius *a* centered on the point x diverges as

$$\left\langle 0 \left| \int_a d^3 x \ E^2(\chi) \right| 0 \right\rangle \rightarrow \frac{1}{a^4}$$

as $a \rightarrow 0$. Since the fields at neighboring points that are spacelike with respect to each other will fluctuate independently of one another, the spatial derivative of the electric field with respect to x is ill-defined, and similarly for the time derivative. Oppenheimer thus insisted that "the expressions usually written down for the interactions between fields are meaningless." But, he observed, "it is just these interaction terms, these couplings, used in a special restricted way, that ha[ve] given us most of the successful predictions of the quantum theory: the formulae for the scattering of radiation, its emission and absorption, for pair production, for the Compton effect, for Bremstrahlung." To explain why this is so, Oppenheimer indicated that it came about because one restricts the validity of the theory with its standard coupling to those situations where the coupling energy is small compared to the characteristic energies of the field involved, that is, of the masses of the particles involved. His students, Edwin Uehling and Robert Serber, had previously shown that a proper identification of what corresponded to the observed charge of an electron (when the fluctuations of the electron-positron field are taken into account to lowest order) removed one the infinities encountered in quantum electrodynamics. Oppenheimer had asked Dancoff to investigate whether an identification of the change in the mass of an electron due to the fluctuations in the electromagnetic field and a redefinition of the mass parameter of the electron in the field equations would remove another of the divergences encountered in quantum electrodynamics. In fact, a finite answer would have been obtained for all processes to order $e^2$ had Dancoff not made a mistake.[15]

During the 1930s physics fully engaged Oppenheimer's scientific efforts, and during that period he made outstanding contributions to its development.

The 1930s were tumultuous times: the Great Depression; the Nazi takeover of Germany; Hitler's pathological anti-Semitism and expulsion of all civil servants of Jewish descent from governmental positions and the ensuing migration of German scholars to Great Britain, the

United States, and elsewhere; the Spanish Civil War; the Stalinist purges in the Soviet Union. These events eroded the safe haven that physics had provided Oppenheimer. His involvement with Jean Tatlock[16] introduced him to political action and led him to actively support the cause of the Loyalist Spanish government. In 1940 Oppenheimer married Kitty Puening Harrison, who had been married to a communist labor organizer, Joe Dallett, who had gone to fight in Spain and was killed there.

After the outbreak of World War II in September 1939 and particularly after the fall of France in June 1940, Oppenheimer came to believe that Western civilization was in grave danger. After France, one of the bastions of Western civilization, fell, Oppenheimer committed himself to seeing that Britain and the United States would not fall as well. He publicly stated: "We have to defend western values against the Nazis. And because of the Molotov–von Ribbentrop pact we can have no truck with the Communists."

After the discovery of fission by Otto Hahn and Fritz Strassman[17] was announced in January 1939, Oppenheimer immediately recognized its import. Luis Alvarez, a bright young experimental physicist who had come to work in Lawrence's Radiation Laboratory in 1935, recalled that after hearing about it he tried to duplicate Hahn and Strassmann's finding as everyone at Berkeley had found it hard to believe. He "tracked down Oppenheimer . . . [who] instantly pronounced the reaction impossible and proceeded to prove mathematically to everyone in the room that someone must have made a mistake." The next day Alvarez and his colleague, Ken Green, demonstrated the reaction.

> I invited Robert over to see the very small natural alpha-particle pulses on our oscilloscope and the tall spiking fission pulses, twenty times larger. In less than fifteen minutes he not only agreed that the reaction was authentic but also speculated that in the process extra neutrons would boil off that could be used to split more uranium atoms and thereby generate power or make bombs. It was amazing to see how rapidly his mind worked, and he came to the right conclusions. His response demonstrated the scientific ethic at its best. When we proved his previous posi-

tion was untenable, he accepted the evidence with good grace, and without looking back he immediately turned to examining where the new knowledge might lead. (Alvarez 1987, 75–76; see also Rhodes 1988, 274)

The concept of a fission bomb as well as its feasibility was first analyzed by Otto Frisch and Rudolf Peierls and their associates in England.[18] It became the focus of a thorough study by theorists working with Oppenheimer in Berkeley during the summer of 1942. Oppenheimer directed this workshop, which was concerned with the theoretical design of an atomic bomb and the estimation of its efficiency. The results obtained by this study group bolstered the conclusion that Peierls and Frisch had obtained, refined their estimate of the amount of U235 necessary, and gave a better estimate of the efficiency of such a weapon. When the opportunity to work on the uranium project had presented itself, Oppenheimer had eagerly accepted. He worked on Lawrence's project, whose goal was the electromagnetic separation of U235 from U238 in naturally occurring uranium ores. Early in 1942 Oppenheimer was assigned responsibility for investigating fast neutron fission and for the design of an atomic bomb. When Leslie Groves was put in charge of the atomic bomb project, after Roosevelt had given his approval to Vannevar Bush and James Conant, he not only accepted Oppenheimer's proposal that all atomic bomb activities be concentrated in a single laboratory as well as his recommendation that it be located on a high mesa in Los Alamos, New Mexico, but also selected Oppenheimer to lead the project.

Groves had visited Berkeley in October 1942 to acquaint himself with what was going on in Lawrence's Radiation Laboratory using cyclotrons to separate the uranium 235 and 238 isotopes. He was also considering whether he should put Lawrence in charge of all the work connected with the atomic bomb. While in Berkeley, Oppenheimer deeply impressed Groves during his briefing of the results of the summer study program that he had overseen. Groves picked Oppenheimer as the director of what would become the Los Alamos Laboratory—despite Oppenheimer's vulnerability stemming from his left-wing views and associations during the 1930s—because he had recognized Oppenheimer's unique scientific and technical capabilities and the respect and authority that was accorded him by his colleagues. Groves evidently also thought that he could control Oppenheimer, but that Lawrence by virtue of self-confidence, self-assuredness, and a Nobel Prize would be more assertive and more difficult to manage. He therefore offered the directorship to Oppenheimer.

This was a bold and consequential decision as Oppenheimer had no administrative experience and had never overseen any experimental project. But Groves was a remarkable man who had the uncanny ability to recognize in Oppenheimer what was needed to successfully direct the project: that he was an exceptional physicist who had a mastery not only of the theoretical components of the enterprise but also of its experimental and technical facets; that he commended the respect of the physicists who would work under him; and that he loved his country and was a deeply loyal citizen despite what his security dossier stated. Groves also saw clearly that Oppenheimer would always defer to his authority; this perhaps was the most important factor in arriving at his decision.

In the spring of 1943, the Los Alamos Laboratory was established to develop and produce an atomic bomb, with Oppenheimer as its director.

## Los Alamos

The reminiscences of many of the physicists who participated in the wartime project at Los Alamos give the sense that they looked back upon that experience as if Los Alamos had been a utopia. They had believed that they were in a frantic race to save the Western democracies from the possibility of Nazi Germany obtaining such a weapon first since work on such developments had started there two years earlier. They knew they were involved in an enterprise, which, if successful, would change the course of human affairs.[19] And after the Trinity test, the first nuclear explosion in history, they hoped that atomic bombs would secure a lasting peace. The explosion took place on July 16, 1945, at Alamogordo in the Jornada del Muerto, a ninety-mile stretch of bleak desert in central New Mexico.[20] Oppenheimer had provided the name for the test. He later vaguely remembered having in mind the John Donne poem that begins "Batter my heart, three-person'd God." But it has also been suggested that he did so having in mind the divine Hindu trinity of Brahma (the Creator), Vishnu (the Preserver), and Shiva (the Destroyer).

Los Alamos was unique in its enormous concentration of first-rate people who constantly gave proof of what could be accomplished by their working together on very circumscribed goals. It was, in fact, a collaboration of unparalleled intensity, a cooperative task undertaken by outstanding people into which everyone threw themselves completely and single-mindedly, and to which everyone gave their ideas, experience, and energy fully, freely, and selflessly. The intensity resulted in the total effort being much greater than the sum of its parts. And everyone shared in the credit.

Although not everyone shared the mood of exultation that had permeated Los Alamos, which Bethe and Rabi had described in their eulogies for Oppenheimer, after the Trinity test it became clear that, without Oppenheimer's masterly direction, Los Alamos might not have produced atomic bombs in time to be used on Japan. This implied that the credit he received was justifiable, but also that he bore a greater responsibility for the consequences of the creation of these weapons—and consequently, possibly a greater burden of guilt.

Though isolated—and perhaps because of its isolation—Los Alamos created that rare situation in the lives of individuals and communities when they feel in touch with much more than themselves. During the few years spent there, many of them—and in particular many of the physicists—felt whole. Indeed, an atmosphere of wholeness permeated the entire enterprise, transmuting it into a kind of magic and enshrining it in the minds of those who had been there. Oppenheimer—who was largely responsible for creating this sense of wholeness and maintaining it until the project was successfully completed—personified the integration of the multifaceted aspects of the enterprise: the theoretical and the experimental, the mundane and the idealistic, the individual, the community, and the nation.

Thorpe, in his masterly "sociological biography" of Oppenheimer (Thorpe 2006), has made clear the dynamics involved in the construction of the complex organization of Los Alamos and the *simultaneous* molding of Oppenheimer's role and authority as its charismatic director. It was a recursive process. The organizational order, the assignment of authority, Oppenheimer's charismatic role and identity were *emergent* properties of the social and professional interactions of scientists, technicians, military personnel, and all the other people who had been brought together to accomplish the military mission of building an atomic bomb and of Oppenheimer's interaction with them. Thorpe highlights Oppenheimer and Groves's complementary roles at Los Alamos and the nature of their relationship and interactions. Groves was de facto the person in charge of the project; all authority devolved from him. But despite his own authoritarian, intimidating, military managerial style and his commitment to compartmentalization and to stringent restrictions on the flow of information, Groves supported Oppenheimer's mode of directorship with its absence of coercion, its reliance on consensus, its attempts to defend academic norms and openness as well as to create the maximum amount of collegial equality among all the scientists, engineers, and technicians consistent with the mission-directed and hierarchical structure of the laboratory's organization. Groves did so because he understood that the civilian staff's allegiance to Oppenheimer accommodated them to working on the bomb and to his (Groves's) own authority. Oppenheimer emerged as the accepted and much-admired leader of the atomic bomb project because of his ability to master and keep in focus all aspects of the enterprise. This earned him the admiration and respect of the division and group leaders under him and that of Groves. This mastery allowed him to pull together discussions and bring cohesion to decision making. Oppenheimer had an unparalleled ability to sum up opposing views in ways that seemed to resolve conflicts. It was Oppenheimer's "synthetic knowledge, together with his perceived moral qualities, [that] allowed him to

Oppenheimer and Groves, Los Alamos Scientific Laboratory. Left to right: Oppenheimer; Groves; Dr. Robert Gordon Sproul, president of the University of California; Commodore W. S. Parson, associate directory of the laboratory. (Los Alamos Scientific Laboratory, courtesy AIP Emilio Segre Visual Archives)

reconcile conflicting parties, and made him the 'natural' spokesman for an underlying, though not yet realized, consensus" (Thorpe 2006, 113).

Thus it is a mistake to attribute the success of Los Alamos solely to Oppenheimer. Just as giving the answer "conductor" to the question, "What makes for a great musical ensemble?"[21] is committing the leader attribution error, the same is true for Los Alamos. The standard answer in the case of orchestras reflects the conventional view that the leader's behavior and style shape the team's processes so as to yield an off-scale team performance. The psychologist Richard Hackman, who has studied the performance of many different kinds of teams—emergency room nursing and medical staffs, musical ensembles, athletic teams, airline cockpit crews—has come to the conclusion that to answer the question "What makes for a great ensemble?" requires focusing on *the conditions that support effective team performance; that is, it demands carefully analyzing the enabling conditions* and *not constraining the answer to state causalities* (Allmendinger et al. 1994; Hackman 1990). In Hackman's view, Los Alamos was a success because the following conditions were satisfied:

1. Its various divisions were populated by individuals who possessed an impressive mastery of the requisite technical skills. The laboratory was divided into seven divisions, each of which had definite tasks to perform. Each division was responsible for producing assessable results for which its members had collective responsibility.
2. There was a compelling purpose to the enterprise.
3. The divisions operated in a well-defined organizational context. Furthermore, the structure of the divisions and of the groups within them was enabling: the process of carrying out the tasks enhanced the capability of the members to work together interdependently. Moreover, the group experience contributed to the growth and personal well-being of its members.
4. A very supportive organizational context existed: essentially unlimited resources were channeled into the project. In addition, the ability to requisition the needed materials and manpower—by virtue of General Leslie R. Groves being in charge of the operation—made it possible to maintain a very demanding schedule and tempo for the project.

It is only because all these conditions were in place that Oppenheimer's leadership could eventually be so effective—"eventually," because at the beginning Oppenheimer was unaware of the complexities the operation entailed. Nor did those who knew him believe that he was temperamentally suited for the job of director of the laboratory. Robert Wilson, a nuclear physicist who had been trained at Lawrence's Radiation Laboratory and who knew Oppenheimer from Berkeley, characterized him as an "eccentric, almost a professional eccentric when I knew him before 1940. . . . He just wasn't the kind of person that you would think would be an administrator." At first, Wilson disliked him at Los

Alamos, for Oppenheimer was arrogant: "a smart-aleck [who] didn't suffer fools gladly," but within a few months Wilson found that Oppenheimer had transformed himself into a superb administrator. "He had class and he had style." Wilson also noted that "when I was with him, I was a larger person. . . . I became very much of an Oppenheimer person and just idolized him. . . . I changed around completely" (Palevsky 2000, 134–135).

It was Oppenheimer's capacity to grow, to be able to fuse his mastery of the cognitive, technical, and scientific factors involved in the undertaking with his impressive managerial abilities when he acted at the collective level, and at the same time be able to convey sensitivity and understanding when he dealt at the individual level, that made him the remarkable, charismatic director of Los Alamos.

To further highlight what was responsible for making Oppenheimer such an effective leader we should consult Bethe's eulogy:

> Los Alamos might have succeeded without him, but certainly only with much greater strain, less enthusiasm, and less speed. As it was, it was an unforgettable experience for all the members of the laboratory. There were other wartime laboratories of high achievement. . . . But I never observed in any one of these other groups quite the spirit of belonging together, quite the urge to reminisce about the days of the laboratory, quite the feeling that this was really the great time of their lives.
>
> That this was true of Los Alamos was mainly due to Oppenheimer. He was a leader. It was clear to all of us, whenever he spoke, that he knew everything that was important to know about the technical problems of the laboratory, and he somehow had it well organized in his head. But he was not domineering, he never dictated what should be done. He brought out the best in all of us, like a good host with his guests. And clearly because he did his job very well, in a manner all could see, we all strove to do our job as best we could. (Bethe 1991, 225–226)

One of the factors contributing to the success of the laboratory was its democratic organization. The governing board, where questions of general and technical laboratory policy were discussed, consisted of division leaders. The coordinating council included all the group leaders, about fifty in number, and kept all of the leaders informed on the most important technical progress and problems of the various groups in the laboratory. All scientists with a B.A. degree were admitted to the colloquium in which specialized talks about laboratory problems were given. Each of these three assemblies met once a week. In this manner everybody in the laboratory felt that they were part of the whole and that they should contribute to the success of the program. Very often a problem discussed in one of these meetings would intrigue a scientist in a completely different branch of the laboratory, and he would come up with unexpected solutions. This free interchange of information was entirely contrary to the organization of the Manhattan District as a whole. Oppenheimer had to fight hard for free discussion among all qualified members of the laboratory. But the free flow of information and discussion, together with Oppenheimer's personality, kept morale at its highest throughout the war" (Bethe 1991, 226). Los Alamos has been an example for big accelerator laboratories ever since, and although they are concerned with very different scientific problems, Brookhaven, CERN,[22] and many other places have gained much of their spirit from wartime Los Alamos.

The fact that whenever he spoke Oppenheimer "knew everything that was important to know about the technical problems of the laboratory, and that somehow he had it well organized in his head" was surely a key attribute responsible for his success. It again suggests the analogy with the great orchestra. Oppenheimer was a great conductor. He knew intimately the parts played by all the divisions, and he could coordinate all their actions. He set the tempo of the enterprise, and he maintained the sense of urgency. Similarly, he immediately saw the integrative value of the weekly colloquium that Bethe had proposed in the early summer of 1943 and convinced Groves and Richard Tolman of its potential value for boosting the morale of the staff. The colloquium became open to any staff member with a badge that allowed entrance into the technical area. Since all aspects of the work being carried out in the various divisions were presented and discussed at the colloquium, everyone who attended had a general sense of the progress and status of the project. With everyone in attendance privy to the scope and objectives of the laboratory, the colloquium fostered in each participant a sense of personal moral responsibility for the enterprise. It was one of the reasons that after the war the Los Alamos physicists were at the center of efforts to ensure that nuclear developments would be under civilian control, as a collective action under the aegis of a national organization, the Federation of Atomic Scientists (FAS) (Smith 1965).

Once again Oppenheimer, this Protean personality, had molded himself into a remarkable, charismatic director of a project that was to alter world history. In the process, he became the model for how to direct the research laboratories established after World War II.

On May 1, 1945, Henry Stimson, the secretary of war, had formed an Interim Committee to advise him on atomic policy.[23] At its first meeting on May 8, the committee created a Scientific Panel to advise it on technical matters. On the advice of Bush and Conant, the panel was to consist of Arthur Compton, the director of the Met Lab in Chicago, Enrico Fermi, and Ernest Lawrence, the director of the Radiation Laboratory in Berkeley, with Oppenheimer as chairman.[24] The panel was informed that an assault on the main island of Japan was planned for November 1945 and that the invasion would be very costly in American and Japanese lives unless Japan surrendered unconditionally before then. In its deliberations, the panel came to the conclusion that a demonstration of the bomb's potency on an uninhabited island would prove ineffective, and only its use on a military target in a populated area would end the war in time to avoid an invasion. These conclusions were later formally transmitted to the Interim Committee in the panel's report. Oppenheimer somewhat later regretted that conclusion and in 1962 commented:

> The bombs were used against Japan. That had been foreseen.... It was largely taken for granted. There were questions raised, but I believe there was very little deliberation and even less record of any deliberation there was.... I think ... we do know and at the moment cannot know whether a political effort to end the war in the Far East could have been successful.... The actual military plans at that time for the subjugation of Japan and the end of the war were clearly much more terrible in every way and for

everyone concerned than the use of the bomb . . . they would have involved, it was thought, a half million or a million casualties on the Allied side, and twice that number on the Japanese side. Nevertheless, my own feeling is that if the bombs were to be used there could have been more effective warning and much less wanton killing than took place in the heat of battle and confusion of the campaign. (Oppenheimer 1964a, 59–60)

All the people who knew Oppenheimer report that for the rest of his life, he "remained acutely conscious of the responsibility he bore for his part in developing atomic bombs and in the decision to use them" (Peierls 1970, 216).

Yet another aspect of the wartime project merits some attention here. Nuclear power was the most rapidly developed technology up to that time. Within three years of the discovery of fission a nuclear reactor had been designed and operated. Oppenheimer was deeply aware of the tempo of the transformation. He recognized that in all the wartime projects *science* had been displaced as the source of advances by the *technology* they were developing. He recognized—as undoubtedly did others—and understood why "deeper" levels of understanding were not necessary for designing A- and H-bombs. The description by the then available nuclear theory of the interactions of neutrons and protons with uranium and plutonium nuclei, supplemented and implemented on the basis of experimentally measured cross sections, was sufficient to design the first atomic bombs.

Throughout his life Oppenheimer emphasized that Los Alamos was an engineering project. The physicists who worked on the problems of designing and building the first atomic bombs did so as applied scientists and as high-level engineers. One might, in fact, characterize them as military applied scientists and engineers.

Moreover, Oppenheimer recognized that new means at hand in the area of nuclear weaponry would accelerate the process of giving dominance to technology over science, namely, the computer. At first, the computers, in their primitive form, were the IBM punch card machines that Feynman and others redesigned into efficient computational devices for carrying out the calculations that went into the design of the plutonium bomb. Later, the computers were the machines that von Neumann designed—with the Institute for Advanced Study becoming one of the principal sites for these developments.

The dizzying tempo of change, change brought about by the wartime laboratories, and the consequences of the accelerating pace of change would become one of Oppenheimer's primary concerns after the war.

## The Postwar Years

After Hiroshima and Nagasaki, almost everyone connected with Los Alamos came to the conclusion that a nuclear war must never be waged. Because he bore a heavy responsibility for the bombs being ready by the summer of 1945, and because he had chaired the Scientific Panel that had advised the Interim Committee on how to use the bombs, Oppenheimer was deeply troubled by the obliteration of Hiroshima and Nagasaki.

Oppenheimer and the other members of the panel had been invited to attend the May 31, 1945 meeting of the Interim Committee at which plans for the future developments of atomic weapons were to be discussed. At the meeting there was no discussion of whether atomic bombs should be used against Japan, for there always had been the tacit assumption that, if available, they would be used.[25] The meeting was indeed about future developments. The recommendations of Lawrence, endorsed by Arthur Compton and evidently assented to by a silent Oppenheimer, were a prescription for initiating an arms race with the Soviet Union. As summarized by Stimson, the recommendations were as follows:

1. Keep our industrial plants intact.
2. Build up sizable stockpiles of material for military use and for industrial and technical use.
3. Open the door to industrial development.

Oppenheimer did speak up when the discussion turned to the question of international control. The notes of the meeting record him as stating that the immediate concern had been to shorten the war. The research that had led to this development had only opened the door to future discoveries. . . . He thought it might be wise for the United States to offer the world free interchange of information with particular emphasis on the development of peace-time uses. The basic goal of all endeavors should be the enlargement of human welfare. If we were to offer to exchange information before the bomb was actually used, our moral position would be greatly strengthened. (Rhodes 1988, 643–645)

In early June 1945, the group of scientists working at the Metallurgical Laboratory, who had gotten together at the urging of Leo Szilard to analyze the social and political implications of atomic bombs, issued what came to be known as the Franck Report. Its signatories recommended that the bombs not to be used against Japan in a surprise attack but that a demonstration of its explosive power on a barren island or in a desert site be arranged, to which representatives of the United Nations would be invited. Oppenheimer became aware of the Report at the mid-June meeting of the Advisory Panel, which he had called together to discuss the panel's recommendations to the Interim Committee on the immediate use of atomic weapons. Arthur Compton, the director of the Met Lab in Chicago, had brought a copy of the Report with him and had given it to Oppenheimer.

On June 17, 1945, the memorandum that Oppenheimer had drafted summarizing the panel's recommendations was sent to Stimson. It recommended that the Soviet Union, France, and China be informed of the existence of atomic weapons and that they and Great Britain be asked to make suggestions "as to how we can cooperate in making this development contribute to improved international relations." The panel also informed Stimson of the divided opinion among their scientific colleagues on the use of the bomb against Japan. "Those who advocate a purely technical demonstration would wish to outlaw the use of atomic weapons, and have feared that that if we use the weapons now our position in future negotiations will be prejudiced." But the panel came down on the side of those who saw in its immediate military use the opportunity to prevent the invasion of Japan scheduled for November 1,

1945, and thus save American lives. Oppenheimer had become convinced that use of the bomb during the war might prevent all future wars, and Oppenheimer in turn swayed the panel. They indicated that some of their colleagues believed that its use in the present war might "improve international prospects, in that they are more concerned with the prevention of war than the elimination of this specific weapon." The panel concluded as follows: "We find ourselves closer to the latter views; we can propose no technical demonstration likely to bring an end to the war; we see no acceptable alternative to direct military use" (Rhodes 1988, 697). However, after the successful Trinity test, the panel could not define the conditions of when and where to implement the "direct military use." Moreover, the panel observed that "we as scientific men have no proprietary rights and no claim to special competence in solving the political, social and military problems which are presented by the advent of atomic power."

Shortly after the devastation of Hiroshima and Nagasaki, Oppenheimer wrote Stimson: "The safety of this nation, as opposed to its ability to inflict damage on an enemy power—cannot lie wholly or entirely in its scientific or technical prowess. It can only be based on making future wars impossible."[26]

But how does one make future wars impossible?

President Truman, in announcing the use of the first atomic bomb on August 6, 1945, concluded his address by saying, "It has never been the habit of the scientists of this country or the policy of this Government to withhold from the world scientific knowledge. Normally, therefore, everything about the work with atomic energy would be made public."[27]

The Smyth Report, issued a few days later, was the proof of this assertion (Smyth 1989). Truman continued, "But under the present circumstances it is not intended to divulge the terminal processes of production or all the military applications, pending further examination of possible methods of protecting us and the rest of the world from the danger of sudden destruction." Since the atomic bomb was too dangerous to let loose into a lawless world, Great Britain and the

United States, who had "the secret of its production," would not reveal that "secret" until the means had been found to control the bomb so as to protect the world from the danger of total destruction. The United States would therefore constitute itself as the "trustee" of this new force to prevent its misuse and "to turn it into channels of service to mankind." Thus started the notion of an "atomic secret."

On October 3, 1945, President Truman sent a message to the Senate outlining his position regarding atomic energy. That same day the May-Johnson bill[28] was introduced in both houses of Congress, which created the governmental framework to oversee the control of atomic energy. Upon becoming aware of the content of the bill, the various groups of atomic scientists who had organized themselves in Chicago, Los Alamos, and Oak Ridge to inform and educate the public and government officials about nuclear energy and the threat of nuclear weaponry decided to unite and form the FAS. Its initial goal was to prevent the passage of the May-Johnson bill. That legislation would have given the military the primary responsibility for oversight of atomic energy. It emphasized its military aspects, and it imposed strict controls on the dissemination of information relating to atomic energy. It also created a Commission and delegated to an administrator, appointed by the president and not responsible to the electorate, "the authority and duty of determining and formulating, in addition to enforcing, all national policy in regard to atomic energy." On October 18 the Atomic Scientists of Oak Ridge and Chicago issued a statement indicating that, although they believed that "controls should and must be exercised for the military security and general welfare of our people," these controls should be "subject to international agreements for the preservation of world peace" (Smith 1965).

The previous day, October 17, Oppenheimer had testified in support of the May-Johnson bill, though he remarked that if the power of the Commission was executed unwisely, it could stop the progress of science. By contrast Einstein, joined by fifty-nine other citizens,[29] sent a telegram to President Truman on October 24, which stated that "We are emphatically opposed to the May-Johnson Atomic Energy Bill . . . because the Bill is in direct violation of American principles, would establish totalitarian authority, and would seriously impede scientific research and progress."[30]

Congressional hearings on the May-Johnson bill were held for a single day on May 9, 1946, during which Secretary of War Robert Patterson (Stimson's successor), General Groves, Vannevar Bush, and James Conant, who had all been instrumental in drafting the legislation, testified in favor of the bill. However, the hearings had to be reopened for another day when the FAS vehemently insisted on being heard. Oppenheimer also testified, again in favor of the bill, stressing the need for its quick passage in order to permit nuclear physicists to resume their work and in order to be able to establish guidelines for the United States' position to be presented to the United Nations committee on international control of atomic energy. Oppenheimer had in fact become a technical adviser to the committee that Secretary of State Byrnes had appointed to make recommendations on this issue.

The FAS's opposition to the May-Johnson bill eventually led to its withdrawal and to the introduction of the McMahon bill. The latter legislation satisfactorily addressed many of the FAS's previous objections. It established a civilian commission, appointed by and responsible to the president, which was to be the sole producer and distributor of fissionable materials. It also included weaker security provisions regarding the dissemination of information and required the Commission to issue periodic reports to guarantee openness and accountability. Furthermore, the bill explicitly stated that should international agreements concerning the peaceful uses of atomic energy be ratified later, these agreements would supersede the relevant provisions of the bill. Although the FAS had some reservations regarding the liaison committee between the Commission and the military, the McMahon bill received the full support of the FAS and was enacted.

Oppenheimer's conduct in these matters reflected a basic difference in outlook from that of the FAS. Though frequently in agreement with the FAS's position, he often differed with its approach and he never joined its ranks. He had greater faith and confidence in government officials and in the military than in the members of FAS, perhaps as a reflection of his "deference to authority." He was not sympathetic to the scientists' demands to be involved in policy matters, and apparently he did not think his own role as adviser to the government was at odds with his position regarding other scientists. Nor did he see that it was difficult to draw a sharp boundary "between those issues that scientists might legitimately try to influence at the early policy-making state and those upon which they should express themselves only in public debate" (Smith 1965, 181). He did contribute to *One World or None,* the booklet the FAS issued in 1946 to alert the public to the fact that atomic bombs had made it imperative that humankind summon the wisdom and strength to abolish war as a means of solving international conflicts (Masters and Way, 1946, 1972). As members of the FAS, scientists were collectively stressing the moral, political, and social issues involved. Oppenheimer took on that responsibility as an individual. With his resignation as director of Los Alamos, Oppenheimer became ever more involved in presenting to the public and to the government the problems raised by the mastery of fission.

How to use this nuclear power effectively and safely, prevent the use of atomic energy as a weapon of war, and in the meantime curb the spread of nuclear bombs—weapons of sudden, overwhelming, and horrifying destructiveness as revealed by the leveling of Hiroshima and Nagasaki—became the challenge that faced the newly founded United Nations Organization and its newly created International Atomic Energy Commission (IAEC). The IAEC was directly responsible to the Security Council, and the Council had asked it to report on methods to effect the above goals. In early 1946 Secretary of State Byrnes appointed a committee headed by Undersecretary of State Dean Acheson to formulate U.S. policy regarding atomic energy and to draft a plan that would be presented to the IAEC outlining the American position on these issues. The committee included Groves, Bush, Conant, and John McCloy.[31] To assist them, an advisory panel was set up consisting of Harry Winne, a vice-president of General Electric in charge of engineering, Chester Barnard, the president of the New Jersey Bell Telephone

Company, Charles Thomas, vice-president and director of research at Monsanto Chemical Company, and Oppenheimer, with David Lilienthal, then the head of the Tennessee Valley Authority, as chair. The advisory panel was to draft recommendations to answer the question: "Can a workable, feasible way be found to safeguard the world against the atomic bomb?" The panel was to formulate a position that would safeguard American interests and yet have a good chance of being accepted by the Soviet Union.[32] Oppenheimer became the principal architect of the Acheson-Lilienthal report that would have placed all atomic developments under an international agency—the Atomic Development Authority (ADA)—which would have exclusive control over all "dangerous" aspects of atomic energy. The report stipulated that national activity in these "dangerous" areas would be outlawed. The ADA would separate all U235 and plutonium, have control over all raw materials, and run all reactors. Participating nations would have to submit to a survey of uranium resources. The ADA would also promote the cooperative development of the atom's peaceful potentialities. The Acheson-Lilienthal proposal embodied Bohr's vision for international control of atomic energy; Oppenheimer had been deeply influenced by Bohr during the Bohr's stay at Los Alamos.

But trouble developed after Truman and Byrnes appointed the seventy-five-year-old Bernard Baruch as the head of the U.S. delegation to the IAEC. Lilienthal thought the appointment disastrous by virtue of Baruch's age, "his unwillingness to work, his terrifying vanity." At his first meeting with Lilienthal, Baruch spent most of the time telling him "how smart he is, how he doesn't need to study the facts, and how he will be in this thing to the end, that he isn't senile (he said it just about a half dozen times), and that he would outfox everybody" (Lilienthal 1964, 40).

Oppenheimer became deeply disturbed by Baruch's approach and particularly by his unwillingness to comprehend the technical reasons that had led to the formulation of the Acheson-Lilienthal plan. Until his encounter with Bernard Baruch, Oppenheimer's interactions with people like Bush, Conant, Groves, Stimson, Acheson, McCloy, and

Lilienthal had given him confidence that the destiny of the country was in the hands of people with integrity—disciplined men who had worked hard to understand the technical aspects of atomic energy and to appreciate the global implications of nuclear weapons. He might have differences with the viewpoints expressed by senators May, Johnson, and McMahon over the military or civilian control of atomic energy and atomic weapons in peacetime, and with the position of James Byrnes over sharing atomic secrets with the Soviet Union—but he felt that they were dealing with him with integrity and respect and were open to exploring the consequences of the positions they were taking.[33]

Baruch had recruited four associates, who knew nothing about nuclear matters, and with their help reformulated the Acheson-Lilienthal plan. Baruch's proposal—which eventually won Truman's approval—embodied the outline of the Acheson-Lilienthal report but not its spirit. It placed great emphasis on immediate punishment for violations and insisted that no veto power would exist when levying penalties for violations of the rules of the agreement. The central tenet of his plan, similar to the Acheson-Lilienthal plan, was the assertion that once the international agency was operational, the world uranium deposits surveyed and assessed, the punishment for violations agreed upon, and the Security Council veto waived in nuclear issues, then all manufacture of atomic bombs by the United States would terminate and all existing atomic weapons would be dismantled. But Oppenheimer correctly inferred that the Soviet Union would see the Baruch plan as a way to perpetuate the American monopoly in atomic energy and that it would not give up any of its sovereignty in favor of an international agency if the United States were the only power having atomic weapons. The Soviet Union indeed refused to consider the proposal when Baruch presented it to the IAEC.

When Oppenheimer realized the possible consequences of Baruch's recasting of the Acheson-Lilienthal plan, he went public with a defense of the plan in an article he wrote in the June 9, 1946 Sunday *New York Times Magazine*. He stated:

The heart of our proposal was the recommendation of an international atomic development authority, entrusted with the research, development and exploitation of the peaceful applications of atomic energy, with the elimination from national armaments of atomic weapons, and with the studies, researches and controls that must be directed towards that end. (Oppenheimer 1946, 60)

He stressed that the proposal attempted to deal with two facts that made the problem so difficult:

1. that the development of atomic energy for peace cannot be separated from the development for war
2. that the world has no adequate machinery for international control of atomic weapons

The Acheson-Lilienthal plan

proposes that in the field of atomic energy there be set up a world government. That in this field there be a renunciation of sovereignty. That in this field there be no legal veto power. That in this field there be international law. How is this possible in a world of sovereign nations? There are only two ways in which this can ever be possible: One is conquest. That destroys sovereignty. And the other is the partial renunciation of that sovereignty. What is here proposed is such a partial renunciation of that sovereignty, sufficient, but not more than sufficient, for an atomic development authority to come into being; to exercise its function of development, exploitation and control; to enable it to live and grow, and to protect the world against the use of atomic weapons and provide it with the benefits of atomic energy. (Oppenheimer 1946, 61)

Oppenheimer didn't think that the proposal was radical, "since in any major war, such as we have lived through, [atomic explosives] will most certainly be used, there is nothing modest in this proposal for the future: it is there be no such wars again" (Oppenheimer 1946, 61).

But in contrast to Einstein, Oppenheimer did not believe that the acceptance and establishment of a world government with laws applicable to individuals in all nations would be "directly possible, in their full and ultimately necessary scope" as a one-step process. Nonetheless, it is clear that he had been affected by his exchange of letters with Einstein in late September 1945 and by the Reves letter that Einstein had forwarded to him, which had summarized Reves's views regarding world government.

For a while Oppenheimer refused to join Baruch's delegation as a technical adviser. Oppenheimer had tried in vain to have Baruch present a plan that at least would keep discussions going with the Russians. But Baruch was adamant that the position paper he would present to the IAEC incorporate his views regarding the veto and immediate punishment for violations. Oppenheimer became deeply despondent when he saw that no agreement would be reached with the Russians. He made the following prediction:

> The American disposition will be to take plenty of time and not force the issue in a hurry; that then a 10–2 report will go the Security Council and Russia will exercise her veto and decline to go along. This will be construed by us as a demonstration of Russia's warlike intentions. And this will fit perfectly into the plans of a growing number who want to put the country on a war footing, first psychologically, then actually. The Army directing the country's research; Red baiting; treating all labor organizations, CIO first, as Communist and therefore traitorous, etc. (Lilienthal 1964, 70)

If, on the one hand, Oppenheimer had become aware of "a growing number who want to put the country on a war footing, first psychologically and then actually," he undoubtedly also knew of the pressures from the corporate and business community to expand the available global markets for American products. Already in November 1944 Dean Acheson, addressing the congressional committee on Postwar Economic Policy and Planning, had pointed to the consequences if the war was followed by a slide back into depression:

> We cannot go through another ten years like the ten years at the end of the twenties and the beginning of the thirties, without having the most far-reaching consequences upon our economic and social system.

Acheson noted that

> You don't have a problem of production. . . . The important thing is markets. We have got to see that what the country produces is used and is sold under financial arrangements which make its production possible.

Acheson concluded that short of changing our economic and political system to ensure the domestic consumption of all that is produced, the only way to achieve full output and full employment was through access to foreign markets. (Quoted in Williams 1962, 235–236)

In confirmation of that view, headlines in *Business Week* in early 1946 stated:

> U.S. Drive to Stop Communism Abroad Means Heavy Financial Outlays for Bases, Relief, Reconstruction. But in Return, American Business, Is Bound to Get New Markets Abroad.

Constant references to an international communist conspiracy within the press helped justify efforts to keep the "free world" safe from atheism, open to democracy and to U.S. exports, and justified what would become a permanent arms race.

On July 23, 1946, David Lilienthal met Oppenheimer in the evening and talked with him until 1:30 in the morning. The next day Lilienthal made the following entry in his *Journal:*

> O[ppenheimer] is in deep despair about the way things are going in New York. He sees no hope of an agreement; he doesn't feel that our plan [i.e., the Acheson-Lilienthal plan] is understood by the American delegation, that Baruch's preoccupation with punishment and veto has done a great deal of harm so that there is little or no discussion of the essentials of the plan. The whole business is quite undirected; he [Baruch] makes suggestions and quite uncritically they are accepted. There is no real discussion. The subcommittees are going through motions that induce what he feels is a wholly false sense of encouragement. (Lilienthal 1964, 69)

Lilienthal added: "[Oppenheimer] is really a tragic figure; with all his attractiveness, brilliance of mind." When they had parted early in the morning Oppenheimer, looking so sad, had said: *"I am ready to go any-where and to do anything, but I am bankrupt of further ideas. And I find that physics and the teaching of physics, which is my life, now seems irrelevant"* (Lilienthal 1964, 69; my emphasis).

From that time forward Oppenheimer stopped doing research in physics as a vocation and became primarily concerned with synoptic assessments. But it should be stressed that, although he stopped doing research, to the extent that was possible given his other commitments, he assiduously kept up with developments in quantum field theory and high energy physics.

In a postmortem assessment of the Acheson-Lilienthal (Baruch) plan that Oppenheimer published in the January 1948 issue of *Foreign Affairs*, he summed up his views of what could be learned from its rejection:

> Thus, if we try to examine what part atomic energy may play in interna-tional relations in the near future, we can hardly believe that it alone can reverse the trend to rivalry and conflict [between the Soviet Union and the US] which exists in the present-day world. My own view is that only a pro-found change in the orientation of Soviet policy, and a corresponding re-orientation of our own, even in matters far from atomic energy, would give substance to the initial high hopes. The aim of those who would work for the establishment of peace and who would wish to see atomic energy play whatever useful part it can in bringing this to pass must be to maintain what was sound in the early hopes, and by all means in their power to look to their eventual realization. (Oppenheimer 1948, 252)

For the next few years Oppenheimer devoted much of his considerable energies and talents to becoming the most influential civilian adviser on atomic energy and atomic weapons within governmental circles. His hope was that he might contribute to the formulation of some sort of *modus vivendi* with the Soviet Union and thereby "look to the realiza-tion" of the early hopes.

And so once again the Protean Oppenheimer fitted himself into this new postwar role of scientist-statesman who owed his influence and sta-tus by virtue of his expertise in current scientific and technical matters. Oppenheimer not only adopted this new role in American political life and became a national public figure, but he also became one of the senior statesmen of American physics after accepting the directorship of the Institute for Advanced Study in 1947. He might have believed that if a few wise people at the top had the right *ideas,* they could effect change. In both politics and physics he was a member of that elite, and perhaps he assumed that therefore things would be fine. For a time, he may in fact have believed that only he had the pertinent insights and appropriate answers in matters of nuclear policy, and that only his participation could bring about a safe nuclear world. Perhaps his singular role at Los Alamos made him feel that he had a unique responsibility for finding solutions to the threat nuclear energy and nuclear weapons posed for humankind. But that role came to an abrupt end when his security clearance was revoked in December 1953.[34]

## Hydrogen Bombs

Relations between the United States and the Soviet Union deteriorated precipitously after the war.[35] The failure to reach an agreement to place atomic energy under international control, the takeover of Czechoslovakia, the Berlin blockade, the victory of Mao Tse-Tung over Chiang Kai-Chek on the Chinese mainland, the detection of the detonation of Joe 1, the first Soviet atomic bomb, detonated in late August 1949, and the arrest of Klaus Fuchs in early January 1950 for having given atomic secrets to the Soviet Union, created an atmosphere that made it impossible for Truman not to order the AEC to go full speed ahead with the development of a hydrogen bomb, against the recommendation of the General Advisory Committee (GAC) to the AEC.

After the detonation of Joe 1, the question of whether the United States should intensify its effort to develop a "super" bomb became the focus of intense debates within the U.S. government. In the fall of 1949 Oppenheimer was opposed to a crash program to build a hydrogen bomb, and in this stance he was strongly influenced by James Bryant Conant's position on the morality of using such weapons. In early October Conant had written Oppenheimer that such a bomb would be built "over my dead body."

Oppenheimer believed that the atomic bomb had helped prevent further Soviet expansion into Western Europe, but it did not seem necessary to him for the United States to develop more powerful weapons to deter Russian aggression even if the USSR had atomic bombs. Whereas in April 1947 the United States had seven fission bombs, by the fall of 1949 the U.S. stockpile totaled over 200 such weapons. Oppenheimer believed that by producing more A-bombs, refining their design to include tactical uses, and having a better delivery capability, the United States would be able to keep its military superiority over the USSR for the indefinite future. Moreover, when the GAC was confronted with the issue in the fall of 1949, it was uncertain whether a hydrogen bomb could be made. The method then proposed had been under theoretical development for seven years, and "in the end turned out to be unpromising, if not useless." At their meeting of October 28–29, 1949, the GAC hoped that the development of fusion weapons could be avoided for both practical and moral reasons. On the practical side, as the report the GAC issued indicated, a fusion bomb was not expected to be an improvement over a fission bomb based on cost, that is, the criterion of damage area per dollar. Furthermore, Teller's conception and design of his "classical" H-bomb, if it could be made to work, required a large amount of tritium, which would require using the present facilities that produce plutonium for tritium production. It was estimated that making one H-bomb of the Teller design with its complement of tritium would mean giving up making over 100 fission bombs, and that it would take several years to produce the necessary amount of tritium needed for a single H-bomb.

Although the GAC opposed a crash program, it was not against the exploration of thermonuclear problems. Oppenheimer himself had written to Conant on October 21, 1949, before the fateful GAC meeting: "On the technical side [the H-bomb is] a weapon of unknown design, cost, deliverability and military value. . . . [However,] it would be folly to oppose the exploration of this weapon. We have always known it had to be done; and it does have to be done."[36]

At its October 29 meeting the GAC recommended to the AEC[37] that the commission intensify efforts to develop tactical weapons, and also "to give attention to the problem of integration of bombs and carrier design in this field" (York 1989, 131).

Truman's order to go ahead with a crash program for the development of an H-bomb also marked the adoption of the policy of immediate and massive retaliation as a deterrent against any Soviet aggression in any quarter of the globe. Since long-range bombers were to be the vehicles of delivery, the Strategic Air Command (SAC) commanded by General Curtis LeMay became the essential component of the policy. Oppenheimer, on the other hand, insisted on greater emphasis on defensive strategies and helped write the conclusions of the 1951 Project Vista, which asserted that small tactical atomic weapons could check any Soviet aggression in Western Europe. These differences of opinions between the armed forces, and the SAC in particular, and an important segment of the scientific elite that was seen as being influenced by Oppenheimer led to sharp confrontations.

In the spring of 1951, Edward Teller and Stanislaw Ulam discovered a mechanism to use the pressure generated by a fission bomb to ignite fusion reactions and thus make hydrogen bombs feasible. Following their discovery, the pressure to establish a second laboratory to pursue the development of thermonuclear weapons became irresistible. Livermore was established in 1952 at the insistent urging of Lawrence and Teller, as well as their allies and backers in the Air Force and in Congress. Livermore stands in sharp contrast to Los Alamos—and to Oppenheimer's administrative legacy there. It was Lawrence, his vision and his administrative style, that set the tone at Livermore. And his protégés—Herbert York, John Foster, and Harold Brown, all of whom had obtained their Ph.D.'s working at the Berkeley Radiation Laboratory—became the directors of the Livermore Laboratory and from there went on to important and influential governmental positions in Washington. In contrast was Oppenheimer, who though he had helped pick Norris Bradbury as his successor as director of Los Alamos, had no further voice in the selection of subsequent directors.

From its inception Livermore was a hybrid collective of human beings embedded in conventions, personal relationships, tools, equipment, material and technical devices, algorithms, and computer codes, "in which *action* including its reflexive dimension produce[d] meaning"[38] (Callon 2005). Its workings involved delineated groups of human beings collaborating with each other, with devices, computers, instruments, creating new tools, new atomic and thermonuclear devices, and new groupings. These assembled groups of scientists, engineers, mathematicians, and administrators developed cognitive properties that were radically different from the cognitive properties of any individual. Lawrence was cognizant of this insight and capitalized on it in building the Radiation Laboratory during the 1930s. It was exported to Livermore, which in fact was part of the Berkeley Radiation Laboratory during its initial stage. At Livermore, Lawrence and Teller had extended their particular cognitive powers by creating an environment in which they could exercise those powers (Hutchins 1995).

Livermore Laboratory's working philosophy called for a constant pushing at the technological extremes. Everyone working there was faithful to this philosophy. Its staff's commitment to thermonuclear weapons was summarized by Herbert York, its first director:

> We set out from the start to construct nuclear explosive devices that had the smallest diameter, the lightest weight, the least investment in rare materials, or the highest yield-to-weight ratio or that carried the state of the art beyond the currently explored frontiers. We were completely confident that the military would find a use for our product after we proved it and that did indeed usually turn out to be the case. [Livermore's] leadership was continuously engaged in efforts to sell its ideas, to anticipate military requirements, to suggest to the military ways in which its new design could be used to support the United States' nuclear strategy. It operated under a doctrine of technology first, requirements after the fact. (York 1987, 77–78)

Oppenheimer had striven for a different kind of relationship. For awhile the physicists had the necessary knowledge, skills, and tools, and Oppenheimer therefore believed that there could be a true partnership in securing what seemed to be the common goal: to preserve the values inherent in American democracy and the American way of life. With the failure to adopt and implement the Bohrian vision in the Baruch version of the Acheson-Lillienthal plan, Oppenheimer came to recognize the transformation that the invention of nuclear weapons had wrought in American foreign policy and in the role of the military. Until 1954, Oppenheimer immersed himself in politics, trying to "pacify" somewhat the processes he had helped set in motion: the nuclear arms race, the intensification of the rivalry and distrust between the United States and the USSR, the militarization of many aspects of public life—in particular, the agenda of the physical sciences.

Teller, Lawrence, and the military saw the Soviet threat differently. The arbitrariness of the Soviet totalitarian regime with Stalin as its undisputed dictator—an arbitrariness and irrationality that had the outstanding Soviet theoretical physicist Lev Landau, almost murdered during the 1938 purges—made Teller equate Soviet communism with Nazi national socialism and Stalin with Hitler. For him, the threat Soviet Russia posed was greater than that of Germany during the 1930s, precisely because of the existence of nuclear weapons. Though strongly anticommunist and a cold warrior, Oppenheimer believed that some form of accommodation and restraint could be reached.

It is thus not surprising that Teller should be so strongly opposed to Oppenheimer's chairmanship of the GAC, to his influence in weapon policy matters, and to his more reasoned stand. He believed that Oppenheimer's and the GAC's October 1949 position on the H-bomb had been deeply flawed, and even though following the Ulam-Teller breakthrough Oppenheimer supported the development of an H-bomb and even considered returning to Los Alamos, Teller thought that Oppenheimer did not grasp the urgency in creating the bomb.

A famous anonymous article in the May 1953 *Fortune* gave a blow-by-blow account from the SAC point of view of "The Hidden Struggle for the H-bomb: The Story of Dr. Oppenheimer's Persistent Campaign to Reverse US Military Policy."[39] The article accused Oppenheimer of being the leader of a conspiracy to hold back the development of the H-bomb.

The final sentence in the article concisely stated the differences between Oppenheimer and some of the scientific elite and the political and military elites: "There is a serious question of the propriety of scientists trying to settle grave national issues alone, inasmuch as they bear no responsibility for the successful execution of the war plans."

Oppenheimer answered these charges in his July 1953 *Foreign Affairs* article. He deplored the futility of the "rather rigid commitment to use [atomic bombs] in a very massive unremitting strategic assault on the enemy and of stockpiling a larger number of atomic weapons than the Russians since our twenty-thousandth bomb . . . will not in any deep strategic sense affect their two thousandth" (Oppenheimer 1953, 528). Oppenheimer further emphasized the ineffectiveness of the policy since "relatively little [was being] done to secure our defenses against the atom." Moreover, he foresaw a time when the "art of delivery and the art of defense will have much higher military relevance than supremacy in the atomic munitions field itself." But Oppenheimer's main point was political rather than technical. Although he had witnessed a great deal of exchanges of opinions in and between many diverse and complex agencies of the government that contributed to the making of policy, the fact was that "a public opinion which is based on confidence that it knows the truth did not exist, for secrecy veiled the deliberations." Oppenheimer therefore recommended that the government exhibit candor in order to strengthen the democratic process and the will of its citizens to confront the challenges to come. "We do not operate well when the important facts, the essential conditions, which limit and determine our choices are unknown. We do not operate well when they are known, in secrecy and in fear, only to a few men "(Oppenheimer 1953, 530).

Oppenheimer's lack of enthusiasm during the early stages of the hydrogen bomb project, his support of explorations to ascertain whether the USSR would be willing to conclude an agreement that neither side test an H-bomb, and his emphasis on the development of tactical fission weapons eventually led to the revocation of his security clearance and his trial. A majority of both the Personnel Review Board and of the AEC commissioners who examined his appeal found that his personality was too complex to carry the responsibilities entailed by his scientific advisory position. He had fabricated stories, that is, lied, in 1943 in order to protect his friend, Haakon Chevalier, who had approached him with a proposition that he himself had considered treasonable, yet had dined with him in Paris in December 1953.

The Gray Board found Oppenheimer's susceptibility to influence a threat to security. Although the Board did not doubt Oppenheimer's loyalty, it denied his appeal to have his security clearance reinstated because of his lack of enthusiasm in support of the security program and for a crash program to develop a hydrogen bomb. The Board also expressed concern with "his highly persuasive influence in matters in which his convictions were not necessarily a reflection of technical judgment nor necessarily related to the strongest offensive military interests of this country." In arriving at this position, the Board had accepted the contention that the only viable strategy was one based on massive retaliation with hydrogen bombs, in contrast to Oppenheimer's commitment to a more balanced, defensive strategy that relied on tactical atomic weapons and an air defense system (Oppenheimer 1970).

Oppenheimer's past incriminating behavior—that he had lied, had fabricated stories—was of course held against him. But he had been cleared in the past. Undoubtedly, what counted heavily in arriving at the final verdict was Oppenheimer's inability to withstand the battering that Roger Robb, the counsel for the Gray Board, subjected him to during the trial. As a result of Robb's long hours of brutal cross-examination relating to the different versions of the report Oppenheimer had given to security officers concerning Haakon Chevalier's visit with Oppenheimer in the spring of 1943, during which Chevalier had mentioned the possibility of sharing atomic information with the Soviet Union, Oppenheimer bent over and, wringing his hands, had blurted out that he had been an "idiot" (Oppenheimer 1970, 137). Under the cross-examination he had lost his "sense of self" and confessed that this had been a problem in the past. Clearly a person who had fabricated stories, had been an "idiot," and had lost his sense of self could not be trusted.

The hearings revealed another disturbing aspect of Oppenheimer's character. During the war he had made very damaging statements to security officers of the Radiation Laboratory concerning the political views and character of some of his former students—notably, Bernard Peters, David Bohm, Giovanni Rossi, and Joseph Weinberg. His deposition subsequently became available to the House Un-American Activities Committee. In 1949 Bohm and Rossi were subpoened to testify before the committee, and both pleaded the Fifth Amendment when asked whether they had been members of the Communist Party. Shortly thereafter a Rochester newspaper revealed the deeply disturbing, very damaging, and inaccurate testimony Oppenheimer had given the committee regarding Bernard Peters—which elicited an open reprimand from his colleagues Hans Bethe and Victor Weisskopf. It seems that to secure and later save his own position, Oppenheimer was willing to sacrifice his students. His testimony effectively destroyed some of their careers.[40] The revelation of these shameful actions was undoubtedly a factor in Oppenheimer's reaction to the hearings and the decision of the Board. His actions surely must have raised questions in his own mind about his character—or lack thereof.

The loss of his clearance and the ensuing appeals trial constituted another deep crisis for Oppenheimer. His apathetic defense reflected the reopening of past wounds: the fragmentation of self that he had experienced at Harvard, and the breakdown that he had experienced in Cambridge. Moreover, he loved his country and could not understand how his loyalty could be doubted or how his willingness to contribute to its strength and welfare could be questioned. That he had been permitted to contribute to the unraveling of the destructive powers of the atom, but could not contribute in an official manner to the avoidance of war and to the realization of the potentialities he believed the science of nuclear physics possessed—for peaceful applications, for communication, and for understanding among men—was undoubtedly one of the most disheartening experiences of his life.

The hurt and suffering he endured following the revocation of his clearance was movingly conveyed in talks he gave in the fall of 1954. In a radio address delivered on December 26, 1954 on the occasion of Columbia University's Bicentennial celebration, Oppenheimer painted a bleak and despairing overview of the world of the arts and sciences.[41] Although he saw the arts and sciences as flourishing, he stressed the diversity of language and techniques that separated "science from science and art from art, and all of one from all of the other." If each art and each science were thought of as a village, then a "high altitude picture" would reveal innumerable villages with no paths between them.

> Here and there passing near a village, sometimes through its heart, there [is] a superhighway, along which windy traffic moves at high speed. The superhighways [the mass media] seem to have little connection with the villages, starting anywhere and ending anywhere, and sometimes appearing almost by design to disrupt the quiet of the village. (Oppenheimer 1955b, 143)

In any village, that is, in any science, there is harmony between practitioners. Each practitioner of that science, as a professional, is a member of a community where common understanding combines with common purpose and interest to bind men together in both freedom and cooperation. Their world and work are "objectively" communicable. But in their relations with a wider society, there is neither the sense of community nor the sense of objective understanding. Even though the sciences developed out of the practical arts, the language and the knowledge of science have become so specialized that communication is possible only among initiates. Only the artist retains as the "end of his work" communication with an audience, which "must be man, and not a specialized set of experts among his fellows." Only the artist can speak to his fellow men in "intimacy,... directness and ... depth." But the artist is bound to fail because "the traditions and the culture, the symbols and the history, the myths and the common experience, which it is his function to illuminate, to harmonize, and to portray, have been dissolved in a changing world." And in the new world "the unity of knowledge, the nature of human communities, the order of society, the order of ideas, the very notions of society and culture have changed and will not return to what they have been in the past. . . . The very difficulties it

presents derive from the growth of understanding, in skill, and power."
And if the growth in knowledge is responsible for the present evil, it is
futile to seek to eradicate what has been learned. It is not only futile,
"but in a deep sense, wicked. We need to recognize the change and learn
what resources we have." It is a new, open world, and the character of its
openness stems from the irreversibility of knowledge (Oppenheimer
1955b, 145–146).

Salvation for the individual scientist lies in pursuing his vocation
fruitfully. His place is thus not in the larger society but in the villages—in
the communities of artists and scientists bound in freedom and cooperation by the common bond of creativity. The primary responsibility of
the creative man is not the well-being of the general society but the keeping of the gardens in his village—the true community—and to keep them
flourishing in this great open, windy world.

With the loss of his security clearance, Oppenheimer's life changed.
On the one hand, a great responsibility had been lifted, and he could
now cultivate his wide interests. He kept up with developments in
physics, and though he no longer mastered the technical details, he
nonetheless could contribute by virtue of his remarkable critical facul-
ties. He continued to chair the weekly theoretical physics seminars at
the Institute in his capacity as the senior physics professor, and he
helped maintain the Institute as an outstanding center in theoretical
physics. In 1971 Dirac made the following observations concerning Oppenheimer's qualities as a discussion leader:

> I knew him for more than forty years. There was a time when we were
> young students together at Göttingen. We both stayed in the same pen-
> sion. We both had the same interests, going to the same lectures. . . . Since
> these early days, I have met Oppenheimer on many occasions and I have
> been able to see what admirable qualities he had, particularly as a chair-
> man for a discussion or a colloquium. He had a very quick mind which en-
> abled him to pick on the main point at issue and if there was something
> which the lecturer couldn't explain very well, or if some member of the au-
> dience was asking a question he could not formulate very clearly, Oppen-
> heimer would frequently jump into the breach and explain in lucid

language just what was needed in order to bring the point clearly home to everybody and enable the discussion to proceed on clarified lines. (Dirac 1971, 10–12)

After 1954 Oppenheimer also devoted more time and energy to his duties as director of the Institute. He tried to build bridges between the various disciplines. Perhaps the greatest tragedy of Oppenheimer's life was not the ordeal he went through over the issue of his loyalty but his failure to make the Institute for Advanced Study the intellectual community he had envisaged. Kennan observed that Oppenheimer was often discouraged and in the end deeply disillusioned by the fact that

> the members of the faculty of the Institute were often not able to bring to each other, as a concomitant of the respect they entertained for each other's scholarly attainments, the sort of affection, and almost reverence, which he himself thought these qualities ought naturally to command. His fondest dream had been [Kennan thought] one of a certain rich and harmonious fellowship of the mind. He had hoped to create this at the Institute for Advanced Study; and it did come into being, to a certain extent, within the individual disciplines. But very little could be created from discipline to discipline; and the fact that this was so— the fact that mathematicians and historians continued to seek their own tables in the cafeteria, and that he himself remained so largely alone in his ability to bridge in a single inner world those wholly disparate workings of the human intellect—this was for him [Kennan was sure] a source of profound bewilderment and disappointment. (Kennan 1972, 19)

What emerges from my account of Oppenheimer's public life is that one can narrate his biography in terms of the tensions that he embodied: as a teenager, growing up in a very well-to-do secular Jewish family of German descent but wishing to escape his Jewishness; as an amazingly gifted student at Harvard, dealing with the tensions between his humanistic ambitions and the demands of technical specialization becoming a physicist entailed; at the Cavendish Laboratory in Cambridge confronting the ambiguities of his sexual life and his ineptness in J. J. Thomson's laboratory; in Göttingen, dealing with his arrogance and self-importance; in Berkeley, coping with the demands of being the leading theorist and teacher there, confronting the commitments of personal involvements, and balancing the demands of his professional, personal, and political involvements; after the war, managing the tensions between his roles as scientific statesman, public intellectual, aspiring philosopher, director of the most prestigious postdoctoral educational institution in the world, and trying to counter the declining role of the scientist, the artist, and the humanist in the contemporary culture.

But what stands out is the absence of a lifelong project that could have given coherence to the tasks he undertook. Giving up doing research in physics after the war took away the gratification that comes from the sense of mastery, creativity, and accomplishment in work well done and the accompanying pride. On many occasions after the war, Oppenheimer spoke of the fleeting, fragmented character of the new world he was living in and of his inability to master it all. He could not get the various pieces to mesh coherently, nor could he mold himself to become the kind of person who is extraordinarily good at doing lots of things and is at one with himself.

## Epilogue

In his eulogy for Franklin Roosevelt at the memorial service in Los Alamos in April 1945, Oppenheimer quoted a verse from the *Bhagavad Gita:* "Man is a creature whose substance is faith. What his faith is, he is" (Smith and Weiner 1980, 288). What was Oppenheimer's faith? Surely, the tenets of the Ethical Culture movement left their mark on him. I believe they helped mold his moral outlook and his sensitivity toward moral issues. They shaped his unusual consideration for others, his sense of *noblesse oblige,* and what he believed were his responsibilities towards others. The Ethical Culture School also inculcated in him the belief that he should aspire to become a reformer, a leader, a person who would be supremely competent to change the world—and that this change would entail greater harmony with moral ideals. Change the world he did. No one saw more clearly the global dangers engendered by the mastery of atomic energy, and no one was more profoundly aware of the dilemmas entailed by this new power over nature, a power that seemed to be out of proportion with man's moral strength. And no one was more passionate in his desire to be useful in averting the calamities that atomic weapons threatened to bring to humankind.

Similarly, the tenets of Hinduism deeply influenced him. Already as an undergraduate at Harvard, Oppenheimer was conversant with the classical Sanskrit literature, but at that time he could not read it in the original. In 1932 Oppenheimer began taking lessons from the Berkeley Sanskrit scholar Arthur W. Ryder, who had translated the *Gita*[42] and studied the *Gita* with him.[43] The *Gita,* Oppenheimer excitedly wrote to his brother in October 1933, was "very easy and quite marvelous"(Smith and Weiner 1980,165).[44] This is the earliest direct evidence of the impression the book made on Oppenheimer, and it was a lasting impression.

In his introduction to his translation of the *Gita,* Ryder summarized the song as follows: "The great epic relates the events of a mighty struggle between two families of princely cousins, reared and educated together. In manhood they quarrel over the royal inheritance, and their difference is sternly solved by war."[45] As the epic opens, Prince Arjuna, whose courage and skill at archery had been displayed in previous battles, rides his chariot onto the battlefield and recognizes in the enemy ranks his own relatives, teachers, and friends to whom honor is due. Distressed by the prospect of killing people close to him, he refuses to fight. But he is willing to take counsel from his charioteer, Krishna, an avatar (embodiment) of Vishnu, who has assumed the incarnation of a man. The intent of the *Gita* is clear: "If it can be shown why a warrior must, on fit occasion, kill his own kinsmen, all lesser and easier cases of duties are obviously included."[46]

In their exchanges, which take up eighteen chapters, Krishna enlightens Arjuna as to why he should join the battle. Krishna tells Arjuna that since he is a soldier, it is his duty to fight. Furthermore, it will be Krishna, and not Arjuna, who will determine who lives and who dies, and Arjuna "should neither mourn nor rejoice over what fate has in store but should be unattached to such results. Finally, the most important thing is devotion to Krishna—faith will save Arjuna's soul" (Hijiya 2000, 131).

There are many bas-reliefs in Hindu temples depicting Vishnu as a charioteer delivering his sermon to the vacillating Arjuna. Some of them include all the elements of the story: the huge armies in position for battle, the chariot, the troubled Arjuna, his bow and quiver beside him, kneeling when he realizes his charioteer is Vishnu, with Vishnu's hand held in the typical mudra (gesture) indicating a caring, teaching manner: the outstretched left hand with thumb and forefinger touching.[47] This mudra is a symbol of teaching, of discussion, and is called the *vitarka* gesture in statues of the Buddha. As we shall see in Chapter 5, in several photographs Einstein adopted this mudra.

Oppenheimer later called the *Gita* the most beautiful philosophical song existing in any known tongue and quoted from it at singular, extraordinary moments. He kept a well-worn copy of it conveniently on hand on the bookshelf closest to his desk and often gave the book (in translation) to friends as a present. He continued to browse in it while director of the Los Alamos laboratory. Thus, as is well known, upon witnessing the Trinity fireball in August 1945, Oppenheimer later claimed that the line from the *Gita*, "I am become Death, the destroyer of worlds," burst into his consciousness. Undoubtedly, Oppenheimer's affinity to the *Gita* and its philosophy played a role in shaping his views and actions.

In an article on Oppenheimer and the *Gita*, James Hijiya (2000) suggested that Oppenheimer was deeply affected by three tenets of Hinduism: duty, fate, and faith. Hijiya believes that while at Los Alamos, Oppenheimer, following the dictates of duty in Hindu thought, accepted that he had a job to do, that he should do it because it was *his* job to do as a scientist, and that he should do it without any intent of self-aggrandizement. Moreover, perhaps Oppenheimer believed that following these principles would bring a measure of serenity into his tormented existence and fragmented self. In the Hindu ethical system, when one's obligations as a member of a caste and one's duties stemming from being in a certain stage in life conflict with universal duties obligatory for all human beings, the particular duties prevail. Hijiya adduces this tenet to explain Oppenheimer's role in the decisions to bomb Hiroshima and Nagasaki (Hijiya 2000).

It is ironic that the Vedantic philosophy upon which the *Gita* is based teaches that there is but one reality—and this a spiritual one—the "self," and that the self is the Absolute. As Ryder put it in the Introduction to his versification of the *Gita*, "Such infinite expansion of the self until fear and desire vanish, offers a vision of extraordinary emotional power; it is what we should all believe if we could. And even those not mystically gifted have occasional glimpses, in aesthetic joy and other strong, pure experiences, of what the leveling of all barriers might mean" (Ryder 1929, xi–xii). The possibility of this vision must have been especially attractive for Oppenheimer with his sense of fragmented self. Add to this that Ryder "was an astounding person, a remarkable combination of austereness thru which peeps the gentlest kind of soul"—the depiction of Ryder given by Oppenheimer's father—and one obtains a measure of the influence of the *Gita* and Ryder on Oppenheimer. Oppenheimer himself characterized Ryder as a scholar who "felt and thought and talked as a stoic," and he credited Ryder with giving him a renewed feeling for the role of ethics. He saw Ryder as one of those very rare persons who have a tragic sense of life, in that they attribute to human actions the completely decisive role in the difference between salvation and damnation. Ryder knew that a man could commit irretrievable error and that in the face of this fact, all others were secondary."[48]

But the *Gita* cannot be the *sole* explanation for the role Oppenheimer played in the decisions to bomb Hiroshima and Nagasaki. The views of key associates he respected—Bohr, Conant, Tolman, Groves, Rabi—those of his colleagues and friends in the Los Alamos community, and his ambition, were surely also factors. His later faith, to the extent that it can be characterized, was an amalgam of many things. It had components derived from his Ethical Culture upbringing with its emphasis on human welfare and *noblesse oblige*, from Christian thought with its emphasis on *caritas*, from the *Gita* and Sanskrit literature and its tenets of duty, from his readings of the Stoics and their notion of discipline, from Spinoza, from Bohr's notion of complementarity, from his readings in political theory.

Oppenheimer came to reject some of the platitudinous, universalistic tenets of Ethical Culture and became very much more concerned with the self-shaping, volitional aspect of ethical conduct. He also came to include contextual factors and culture-specific values and motives in making sense of himself as a moral agent. In his lectures after the war, Oppenheimer often spoke of the break with tradition, of the feeling of novelty he was experiencing in the new world he was living in, and of the vertigo brought about by the tempo of change. I would characterize him as almost postmodern during the last decades of his life. He was a relativist, for he did not believe that one could recognize a detached or valid perspective from which to judge the morality of other societies. Already in a letter in 1951 to George Kennan, who found it difficult to understand his sense of morality, Oppenheimer explained:

> It is not in our judgment of ourselves or our own actions that I would reject moralism: it is rather in our attitude toward the behavior of other peoples. What I question is our ability to put ourselves, as a nation, in the place of these other peoples and decide what is right or wrong in the light of their standards and traditions, as they see them, or even in the eyes of the Almighty. I regard the behavior of other societies as something the morality of which I would prefer not to have to determine. I think it is our business to study that behavior attentively, to measure the intensity of the emotional forces behind it, and to take careful account of the potency of its influence on international affairs; but I feel we would do better not to attempt to classify it as right or wrong, praiseworthy or reprehensible. We Americans have enough, it seems to me, with our consciences and with the necessity, now upon us, to reconcile an individualistic tradition with the centralizing pressures of advanced technology.
>
> It is for this that we are accountable as a body politic, not for the decisions and solutions arrived by others.
>
> Let us conduct our policies in such a way that they are in keeping with our own character and tradition. This means, of course, that the moral element, as we feel it, must be present. (Oppenheimer Papers, Library of Congress, Archives Division)[49]

In the closing comments of the Whidden Lectures he delivered at McMaster University in 1962, Oppenheimer stated what he thought were the duties of the members of scientific communities, such as physics. In the first place,

> To give an honest account of what we know together, . . . to give that information whenever that is possible, [to] give it to our governments in secret when the governments ask for it, or, even if the government do not ask for it, that they should be made aware of it, when we think it essential, as Einstein did in 1939.

Second, and more important,

> To distinguish what we know in the vast regions of science, . . . from all those other things of which we would like to speak and should speak in another context and in another way, those things for which we hope, those things which we value.

And finally,

> To work for the growth of an international community of knowledge and understanding . . . with our colleagues in other lands, with our colleagues in competing, antagonistic, possibly hostile lands, . . . and with others with whom we have any community of interest, any community of professional, of human, or of political concern. (Oppenheimer 1964a, 64)

And the latter, of course, is incumbent on all of us.

As he had come to embrace the variety and diversity not only of the arts and the sciences, but of the cultures that populate our changing world, the concluding remarks of the Whidden Lectures evoke Darwin's last paragraph of the *Origin of Species*. Darwin's tangled bank "clothed with plants of many kind, with birds singing on the bushes" was proof of the wondrous workings of biological evolution resulting in the amazing diversity of living things. But it was from "the war of nature, from famine and death, that the most exalted objects we are capable of conceiving," namely, the production of human beings, "directly follows." For Oppenheimer, "the increasingly tangled, increasingly wonderful and unexpected situation" that our social world presents is the result of cultural evolution. But it is *we* who must contribute "to the making of a world which is varied and cherishes variety, which is free and cherishes freedom, and which is freely changing to adapt to the inevitable needs of change of the twentieth century and all centuries to come, but a world which, with all its variety, freedom and change, is without nation states armed for war and above all, a world without war" (Oppenheimer 1964a, 65).

# CHAPTER FOUR

# J. Robert Oppenheimer and American Pragmatism

Loosely speaking, and in general, it may be said that all things cohere and adhere to each other somehow, and that the universe exists practically in reticulated or concatenated forms which make of it a continuous or integrated affair.

—*William James (1907, 52)*

For his contribution to the *Festspeil* celebrating the twenty-five years of activities of the School of Social Science of the Institute for Advanced Study in Princeton, Charles Taylor (2001) offered an essay on "Modernity and Identity." In this work he attempted to better understand the present-day discourse of identity and to answer the question: "Why would our ancestors have found it hard to understand the contemporary preoccupation with identity?" (Taylor 2001, 139). To do so he distinguished three different contexts in which the word is commonly used. The first is the sense given it by Erik Erikson, namely, "identity" as a definition of oneself, which human beings must be able to elaborate in the course of becoming adults and must continue to elaborate during the course of their lives. It is through one's identity that one knows what is truly important to one self and, conversely, what is less so. Failure to achieve a stable identity may result in an inability to function normally. A threat to one's identity under those conditions can result in a state of crisis. What constitutes an "identity crisis" is the loss of the landmarks that delineate what is truly important.[1] Taylor phrased the issue succinctly: One's identity situates one in the moral sphere and defines the horizons of that space.

One of the characteristics of modernity is that identity became the attribute of an individual. One's horizon was no longer fated by virtue of birth, class, or religion. Identity is something that individuals may mold for themselves: Each human being can shape his or her own mode of being human—and identity therefore becomes "an object of inquiry." Furthermore, achievement of a successful identity requires recognition, as "we cannot define ourselves by ourselves; we need the cooperation of others, 'significant' others" (Taylor 2001, 141–142). And to be recognized implies exertion, great effort, and often struggle.

Taylor also explored a third component of the concept of "identity," the element that relates the individual to collectives. The question to be answered in this case is: "What does a group identity consist of?" All three components are interrelated—and the struggle for identity may involve interrelated contestations at all levels—ethnic, religious, sexual, gender, professional, national, and transnational (Taylor 2001, 143).

I shall use Taylor's essay as a suggestive guideline to structure this chapter, which is concerned primarily with Oppenheimer's molding of himself as an intellectual after World War II and with his struggles for recognition in that role. Taylor's formulation allows me to stress once again the importance of the Ethical Culture School in Oppenheimer's upbringing (Schweber 2000, chap. 2). It left an indelible mark on him, making him sensitive to the horizons of his moral world throughout his life.

Felix Adler, the founder and guiding spirit of the Ethical Culture School, had succinctly summarized its aims in the fall of 1931—two years before his death—in an address to its student body: "What the school is for is to make a better world." He added, "In order to do that, to make yourself fit for such an undertaking, it is necessary that you should make something of yourself. You must cultivate your best talents in relation to those of others" (Friess 1981, 136). Therefore the school was to be an "educational temple . . . to train for the growing life of the world, . . . an altar to the . . . unknown, unpredictable, inconceivable divine things that slumber as yet unborn in the bosom of mankind" (ibid., 137). Its curriculum emphasized rationality and progress, with progress subsuming moral progress: progress was to be sensitive to making the world a better place. Science was considered the paradigm for rationality, and scientific activities were nurtured at the school by outstanding teachers who believed scientific knowledge grew in a cumulative, progressive fashion.

The Ethical Culture School's emphasis on making a better world, rather than just a better nation, is what differentiated it from other progressive schools. All these schools tried to indoctrinate their students with political and moral axioms and principles[2] that would guide their actions as individuals and as citizens. But at the Ethical Culture School education for citizenship also emphasized internationalism and never became identified with patriotism. The means by which the democratic spirit was inculcated in the school was by making sure that it would not be a "class" school. The children of the rich and of the poor, and those of different ethnic background, were to meet there and learn to respect one another, both in their work and play. But Adler never shed a certain *noblesse oblige* attitude that became part of the ethos of the school.[3] Oppenheimer became inculcated with doing the "noble thing," a stance that resonated with the values that permeated his parents' home. Oppenheimer was known to bestow lavish gifts, and whatever other reasons motivated him to do so, I believe that this sense of *noblesse oblige* was always a component: he had the means, monetary and intellectual, to do so, and he felt morally obligated to act accordingly.

My primary concern in this chapter is with Oppenheimer after World War II. My point of departure is the aftermath of Bernard Baruch's rejection of the Acheson-Lilienthal report as the American plan to be presented to the United Nations for the control of atomic energy and atomic weaponry. From that time on Oppenheimer stopped doing physics, that is, doing research as a vocation.[4] Serving as technical adviser to the American delegation to the UN headed by Baruch had resulted in a crisis. Thereafter Oppenheimer molded a new personality for himself and fashioned himself into the role of scientist-statesman, addressing the issues confronting humankind in the new world in which mastery over the atomic and nuclear domains had been achieved and spelling out the hopes and the perils this posed. His influence and status as scientist-statesman stemmed from his role as the wartime director of Los Alamos where the first two atomic bombs had been assembled and from his expertise in these scientific and technical matters. Oppenheimer adopted this new role in American political life and became a national public figure.

That role came to a stop with the rulings of the Gray Board and of the Atomic Energy Commission (AEC) that permanently revoked his security clearance. Thereafter he assumed a new role as public intellectual,[5] and his addresses contained far fewer references to atomic energy and weaponry. He then concentrated on the nature of scientific knowledge and on the relation between science and society. He reverted to being what Foucault had called a "universal intellectual" in contrast to the role of "specific intellectual" that he had created before the revocation of his clearance.

> Perhaps, it was the atomic scientist (in a word, or rather a name: Oppenheimer) who acted as the point of transition between the universal and the specific intellectual. It is because he had a direct and localized relation to scientific knowledge and institutions that the atomic scientist could make his interventions; but since the nuclear threat affected the whole human race and the fate of the world, his discourse could at the same time be the discourse of the universal. Under the rubric of this protest, which concerned the entire world, the atomic expert brought into play his specific position in the order of knowledge. And for the first time, I think, the intellectual was hounded by political powers, no longer on account of the general discourse he conducted, but because of the knowledge at his disposal: it was at this level that he constituted a political threat. (Foucault 1980, 127–128)

For Foucault the "Oppenheimer affair" was one of the crucial and decisive events of the twentieth century.

Central to my presentation in the present chapter are Oppenheimer's role as director of the Institute for Advanced Study in Princeton, his hopes and aspirations for that institution, and his involvement with

Harvard University as an Overseer from 1949 to 1955. These activities not only brought him in contact with people who deepened his knowledge of current social scientific, historical, legal, and philosophical thought but perhaps more importantly, made him read their writings and confront his beliefs in the light of them. I shall concentrate on one set of lectures—the William James lectures that he delivered at Harvard in 1957—which seem to me to be one of the clearest expositions of what he had hoped he might accomplish as a public intellectual.[6] His aim in these lectures was to inform the educated public about what had recently been accomplished in the various sciences and the consequences of that new knowledge. In addition, and more particularly, Oppenheimer wanted to indicate how the philosophical difficulties that stemmed from the quantum mechanical representation of the atomic world had been resolved, especially by Niels Bohr with his emphasis on the role of common language and by his formulation of the complementarity principle.[7] Bohr had noted that in quantum mechanical description of the kinematics of a particle two mutually exclusive representations can be distinctly defined: one that ascribes a definite position to the particle (and thus ascribes it a property familiar from the macroscopic, classical realm), the other a specific momentum (or equivalently a definite wavelength $\lambda$ through the de Broglie relation $\lambda = h/p$ where $p$ is the momentum of the particle and thus ascribes wavelike properties to the particle). In a particular experimental situation, we are limited and can only adopt one or the other representation. Yet for a more comprehensive understanding of the microscopic world, the complementary representation to the one chosen is also needed. Bohr's complementarity principle is the attempt to encompass both representations without fusing them into one.[8]

Oppenheimer tried to apply and extend these insights to help solve some of the problems facing the social sciences.[9] In this effort he was profoundly influenced by his commitment to the views of the American pragmatists—Charles Peirce, William James, and John Dewey. Peirce, James, and Dewey all believed that the acceptance of scientific habits of mind would make any culture more creative and inventive.

Oppenheimer concurred. And like them, Oppenheimer believed that philosophy could be a guide to the solution of real, practical problems. This philosophic outlook was deepened after the war by Oppenheimer's interaction with Morton White, primarily through the reading of his books and essays on pragmatism.

In the same way that the Civil War and Darwin's *Origin of Species* had deeply influenced Charles Peirce, William James, and Oliver Wendell Holmes and had shaped their philosophical outlook, and much as their discussions had been responsible for the emergence of pragmaticism and pragmatism (Menand 2001), in an analogous fashion World War II, quantum theory, and Bohr's writings were responsible for Oppenheimer addressing anew his philosophical outlook. World War II—with its staggering loss of life and destruction; its concentration camps; its gas chambers and crematoria; its fire raids on Dresden and Tokyo and other civilian targets, culminating with the devastation of Hiroshima and Nagasaki by new weapons of mass destruction—convinced Oppenheimer—and most of the people who had worked with him at Los Alamos—that a new worldview was necessary. Quantum mechanics had given an understanding and explanation of a vast domain of atomic and nuclear physics, and Los Alamos and its associated technologies—accelerators, reactors, electronic circuitries, computers, chemical and metallurgical laboratories—had indicated how the mastery of these realms could be translated into the design of practical macroscopic devices. Among these were weapons of mass destruction that could be delivered by a single plane, weapons that, if used, would kill staggering numbers of victims, and leave deleterious genetic aftereffects on the offspring of those exposed to its radiation. What was needed was guidance on how to live in this new world and make sensible, humane use of the new knowledge and technologies it was producing.

Clearly, Oppenheimer felt a particular responsibility to do so. As he had told President Truman, "physicists have known sin" in helping to create this new world. What he meant by this statement was that the physicists at Los Alamos "had intervened in a deliberate and decisive way in human history; probably not such a bad way, but a way that is very unfamiliar for a scientist whose job is not to apply knowledge and exercise the options it gives, but to be sure that the knowledge is understood."[10]

As director of the Institute for Advanced Study, Oppenheimer came in extensive contact with Dean Acheson, Jerome Bruner, Francis Fergusson, Paul Freund, George Kennan, Alfred Kroeber, Erwin Panofsky, Ralph Barton Perry, Ruth Tolman, Edward Tolman, Morton White, and many other leading scholars and statesmen, and through them he continued to be attentive to important new developments in philosophy, psychology, history, the law, the social sciences, and the humanities. Concomitantly, because he was searching for new activities to be sponsored by the Institute, he became the nexus of various multidisciplinary enterprises. I believe that his discussions with these people and his study of their writings led him to believe that he could formulate a new philosophical vision appropriate and relevant for the new times. He became a champion of antifoundationalism, multiculturalism, and a form of neo-pragmatism at a time when most of American philosophy was turning analytic, when most philosophers of science were logical positivists, and when most of his colleagues in physics were reductionists and resonated with Einstein's credo that the task of the physicist was to arrive at fundamental laws from which the cosmos can be reconstructed.[11]

The present chapter should be read with context in mind. When Oppenheimer came to the Institute in the fall of 1947, he had ambitious plans to transform it. Initially, Oppenheimer had the enthusiastic support of the Institute's Board of Trustees, and his buoyant mood at the time is reflected in an interview with Harry M. Davis for an article that appeared in the magazine section of the April 18, 1948 Sunday *New York Times*. Davis could state that a few months after Oppenheimer's arrival the atmospherics at the Institute had changed: there was now a much more energetic atmosphere, as evidenced by the fact that "a discussion group in which young scholars consider the relation of their science to political and social events [meets regularly]; there are lectures on the influence of . . . military power on world politics; [and] there is incessant communication and cooperation with Princeton University" (Davis 1948, 54).

Oppenheimer told Davis of his hopes for the institution. He planned to have fewer life members and more people coming to the Institute for a semester or a year of specific study:

> We have been given a fund to use in experimenting in two directions. First, we expect to invite people who have experience outside the academic field—in business or politics, for example—and who have reached the point where they have something to communicate, to take a year and gather their ideas together, and write them down. Second, we are setting up a standing offer to explore areas which have not been regarded as subject to scientific investigations. (Davis 1948, 54)

In addition to their objections to this possible restructuring of the Institute, what probably irritated the permanent faculty, and possibly some members of the Board even more, were the next few paragraphs in Davis's article:

> Suppose you had funds at your disposal based on a $21,000,000 endowment, with the prospect of getting more by convincing benefactors of the need. Suppose you could use this fund to invite, as your salaried house guest the world's greatest scholars, scientists, creative artists—your favorite poet, the author of the book that interested you so much, the European physicist with whom you would like to mull over some speculations about the nature of the universe.
>
> This is precisely the set-up that Oppenheimer enjoys. He can indulge his every interest and curiosity, because his curiosity corresponds with the whole range of science and culture, and that coincides in turn with the scope of the institute. (Davis 1948, 54)

Seemingly, without consulting anyone, Oppenheimer wanted to change the character of the Institute and had arrogated to himself the power to do so. All this did not sit well with "the life members," especially with the mathematicians. Until Oppenheimer's arrival, the Institute had been dominated by the School of Mathematics, its largest school. The power of the mathematicians devolved from the fact that all professors, that is, all the permanent senior members of the faculty, voted on all new appointments, regardless of which school the appointments were in. Oppenheimer intended to give both theoretical physics and the humanities much greater emphasis and representation, and the mathematicians objected. Their hostility was exacerbated by the fact that Oswald Veblen, an eminent mathematician who, with Herman Weyl and Albert Einstein had been one of the first faculty members of the Institute, developed a deep antipathy to Oppenheimer. Veblen sat on the Board of Trustees as the faculty representative and wielded considerable influence in its deliberations. The mantle of opposition was later taken up by André Weil, an outstanding algebraic geometer, but also a brilliant, sharp-tongued, and difficult individual.

In addition, Oppenheimer had to contend with the ambitions and power of the vice-chair of the Board of Trustees, Lewis Strauss—the very same man who became his nemesis in Washington—whose influence on the Board stemmed from the fact that he was its wealthiest trustee. Increasingly, Oppenheimer found his plans resisted and his initial hopes dashed. Nevertheless, under his aegis, the Institute grew, as did its distinction, importance, and renown. Until the 1960s, under Oppenheimer's guidance, the Institute became *the* outstanding center of theoretical physics. Freeman Dyson, Robert Oppenheimer, Abraham Pais, Tulio Regge, and Chen Ning Yang made up the professorial staff of the School of Physics, and many of the world's brightest young theoretical physicists spent a year or more there as visiting members. At the Institute Oppenheimer duplicated at the international, postdoctoral level what he had achieved at the national level at Berkeley.

The present chapter reports on Oppenheimer as the director of the Institute for Advanced Study and on some of the workshops and conferences he organized there in the early 1950s. It also deals with Oppenheimer's encounters with philosophy; with his role as an Overseer of Harvard College; with the 1956 American Academy of Arts and Sciences (AAAS) conference "On Science and the Modern World View"[12] honoring Percy Bridgman and Philipp Frank on the occasion of their retirement from Harvard; with the relation of Oppenheimer's presentation on that occasion (Oppenheimer 1958a) to the problem of the two cultures that C. P. Snow would address in 1959 in his Rede lectures (Snow 1959); and with the William James lectures Oppenheimer delivered at Harvard in 1957.

## The Director's Fund

After Oppenheimer became director of the Institute for Advanced Study in the fall of 1947, he thought about what novel roles the institution could assume to better understand the new world that had developed after the war. Until 1948 the Institute had three schools: a large School of Mathematics, which included applied mathematics and physics; a relatively small School of Humanistic Studies, which encompassed classical archeology, paleography, and the history of art; and a School of Economics and Politics. The death of several faculty members of the School of Economics and Politics permitted the questioning of whether the Institute was the proper place for continued advanced work in economics, and more generally the direction of the work of the Institute in fields other than physics and mathematics. As a result of extensive discussions, Oppenheimer became convinced that "the unifying and invigorating element of work in history and the humanities must be the conscious and scrupulous use of the historical method; and that a school devoted to this purpose would, for all its heterogeneity, be a proper complement and balance to a school devoted to mathematical and logical analysis."[13] In the autumn of 1949, the academic work of the Institute was reorganized under two schools: a School of Historical Studies and a School of Mathematics.

Although a great deal of important and interesting work could and would be carried out in these two schools supported by the funds available to them, Oppenheimer believed that there were many fields in which new directions were being charted and that people working in these fields would not be able to be invited as visiting members by either of the two schools. In a memorandum to the Board of Trustees of the Institute in the fall of 1947, he pointed to two main kinds of potential efforts:

1. The application of scientific methods to fields in which there are evident, important new beginnings.
2. The encouragement of work by men to whom experience in the creative arts had brought deep insight.

Oppenheimer did not put forward specific programs for such efforts, but suggested that there would be opportunities for exploring new fields "outside and beyond the specific areas of the Schools, which in some cases have narrow interests." He asked that there be appointments of members who would not be members of the two schools. To implement his plan, he requested that the trustees establish a General Fund of $120,000 on a five-year basis, to be used for "stipends, memberships and work not at present part of the activities pursued at the Institute." In addition, he recommended that an Advisory Committee be established for the use of this fund. Oppenheimer hoped that in this way "the Institute may carry out its functions in a more experimental way; and thus a coordinate community of scholars may be created."[14]

At a special meeting of the Board of Trustees on December 16, 1947, Oppenheimer's proposal was "strongly commended," and one of the trustees, Dr. Fulton, suggested "that the motion of acceptance of the plan be a vote of confidence in the new Director. At the suggestion of another trustee the fund was designated the Director's Fund. "The motion was unanimously carried that $120,000 be appropriated to the Director's Fund from surplus over the next five years; that $20,000 be made available for the year 1948; that the Fund be used as the Director sees fit."[15] At the end of 1954, an unexpended sum of $8,500 remained in the Fund. The reason for this surplus was that a large part of the support of the invited members was contributed by the Carnegie, Ford, and Rockefeller Foundations and by funds from the universities of which they were faculty members. Given the value and impact of the program, the Board of Trustees extended the appropriation through June 1960 at the rate of $25,000 a year. The Director's Fund also made possible a few preliminary workshops to determine future courses of action: in law, in contemporary history, and in psychology. These workshops are reviewed in the next few subsections.

*The "Interpretation of History" Conference*

On May 12 and 13, 1949, the first conference supported by the Director's Fund took place at the Institute. It was a conference on "The Interpretation of History," and the participants included

> Rushton Coulborn (rapporteur), Atlanta University
> John Marshall, Rockefeller Foundation
> Elmore Harris Harbison, Princeton University
> Erich Kahler, Institute for Advanced Study
> Alfred L. Kroeber (chair), University of California, Berkeley and
>     Columbia University
> Robert Oppenheimer, Institute for Advanced Study
> Erwin Panofsky, Princeton University
> Curt Sachs, New York University
> Paul Schrecker, Swarthmore College and Bryn Mawr College
> Walter Stewart, Institute for Advanced Study
> Prof. and Mrs. Arnold Toynbee, Institute for Advanced Study and
>     Royal Institute of International Affairs
> Ralph Turner, Yale University
> Robert Warren, Institute for Advanced Study

The letter inviting the participants indicated that the broad description "The Interpretation of History" had been adopted to denote the task that historians had hitherto neglected, namely, "that of investigating morphological processes in history. To distinguish the repetitive from the nonrepetitive in history; to explain the rise and fall of societies and cultures; to discern pattern, and possibly rhythm, in succession of events."[16]

One aim of the conference was to see whether it was possible to develop a common scientific discourse broad enough to include the particular interests and conceptions of different scholars, one that might in time prove to be an effective mode of understanding bodies of historical knowledge that were currently presented primarily in narrative form. In order to exploit the insights and methods of the various social sciences, the anthropologist Alfred Kroeber, then at Berkeley, had been invited and made chairman of the conference.[17] He had been asked to formulate the general problems to be discussed at the conference, and his suggestions were followed. He stated the problematiques as follows: To the extent that civilizations or major cultures are to be construed as the units of macro- or supernational history:

1. Does this definition necessarily imply the concept of cycle or of fixed duration?

2. To what extent are the total forms of civilizations most profitably expressed in terms of (a) culture, (b) society, (c) events, and (d) psychology and character?

3. Does this level of description preclude consideration of persons as exemplars or indices of "superindividual currents or formations?"

4. What are the theoretical presuppositions of the criteria for determining a civilization as an empirical entity? Where is "the profitable middle course between the extreme positions that civilizations are self-sufficient, highly coordinated, essentially impermeable systems (Spengler—and partly true for languages), or alternatively, that the segmentation of history is subjective and perhaps fictitious, since events occur as a continuum and since culture tends to be both transmitted and diffused continuously"?

5. "Art styles . . . seem to have an irreversible course. May a civilization be construed as a style-assemblage or superstyle, also with a one-way and limited course?"

The summary of the conference indicates that a good deal of discussion centered on the dynamic character of development in the arts and its relation to the growth and decline of civilizations and that it went on from there to the dynamics of the growth of scientific knowledge and its place in the life-history of civilizations.

After an exposition of the dynamics of the developments of the arts by Sachs in which he indicated that art styles do move in cycles, doing so

irreversibly, and that he believed that a civilization may in fact be construed as a style–assemblage, Stewart asked Oppenheimer about "movement in science and whether similar evolutionary cycles could be found
in the development of scientific thought." Oppenheimer's answer is the
reason for the somewhat detailed exposition of this conference here.
His first answer to this question was as follows:

> A plain negative: he said that it is best to think that science is not a culture,
> and that science develops as a result of a process intrinsic to itself and not
> through promptings or other influence from the culture at large; the de-
> velopment of science is continuous, one problem succeeding another be-
> cause the solution of one problem in itself poses another problem; this is
> not a matter of style or of the predilection of particular scientists.[18]

In the discussion that followed Oppenheimer's remarks, it was "clearly"
brought out that the history of science manifests two contrasting processes, one that is seemingly linear and "possibly of intrinsic motive
power," and the other proceeding from the general movement of the culture. "There was no disposition to accept Mr. Oppenheimer's conception
of a purely linear development as the only factor in the total development of science and Mr. Oppenheimer himself, with reservations, agreed
to the existence of the other factors" but insisted that science in "our
own Western history reached an autonomy which is quite special."[19]

*The Psychology Committee*

When in the late 1940s Oppenheimer was considering whether the Institute might extend its activities to encompass theoretical psychology,
he set up an advisory consisting of:

Jerome S. Bruner, Harvard University
Paul E. Meehl, University of Minnesota
George A. Miller, MIT
Edward C. Tolman, University of California, Berkeley
Ruth Tolman, Pasadena
Edwin G. Boring, Harvard University

The committee met regularly and through its meetings kept Oppenheimer apprised of developments in the discipline, particularly in the
study of perception, of concept formation, of linguistics, and of learning. Edward Tolman was arguably the most distinguished member of
the panel. Oppenheimer had known him since the early 1930s. He was
the brother of Richard Tolman, the distinguished physicist and physical
chemist at Cal Tech who became a close friend and deeply respected colleague of Oppenheimer. Edward Tolman's researches were concerned
with learning theory. While a graduate student in philosophy and psychology at Harvard in the 1910s, he had been influenced by his interactions with "pragmatist" members of its philosophy department, in
particular by Clarence Irving Lewis, and later by his contacts with Wolfgang Koehler and Kurt Lewin, two of the leading exponents of Gestalt
psychology. In the interwar period he became one of the leading "cogni-
tists" in the United States, holding the view that learning was principally a task of sorting out and organizing knowledge from the top
down. This was in contrast to the views of the associationists, who believed that learning is a bottom-up enterprise of accretion. In a famous
lecture delivered in Berkeley in 1947, which Oppenheimer surely knew
about,[20] Tolman expanded on his pragmatist belief that learning is like
mapmaking and that our cognitive maps are "not mirrors of our hap-
penstance of our encounters with the world but a record of our strivings and what has proved relevant to their outcome. . . . [T]o learn is to
organize things in the light of their utility for achieving ends" (Bruner
2004, 18). The two people Oppenheimer was closest to on the committee, Ruth Tolman[21] and Jerome Bruner,[22] were involved in projects
involving "foundational" matters concerning classification and categorizing, and all the committee members were involved with how to constitute the new science of psychology. All were keenly sensitive to
philosophical issues and were deeply knowledgeable about the historical factors that had helped shape the discipline.[23] Everyone on the
committee evidently profited from the mutual interaction with one
another—Oppenheimer perhaps most of all. His indebtedness can be
gauged by the glowing, long, and detailed review he gave Bruner,

Goodnow, and Austin's *A Study of Thinking* when that book was published.[24] The concluding sentence of Oppenheimer's review contained one of the most important points he wanted to get across in his first William James lecture:

> That man must act in order to know, that he must thereby reject other actions of which he is capable, and loose other knowledge of what is knowable in the world, will not solve the old philosophical questions; but it will alter, deepen, and illuminate them. (Oppenheimer 1958b, 490)

Oppenheimer was suggesting that if it is recognized that under certain circumstances, as in the atomic domain, certain actions are complementary and mutually exclusive, then perhaps Bohr's musings concerning complementarity might illuminate certain problems confronting the social sciences. This was at least Oppenheimer's hope.

*Legal Studies*

During the spring of 1949 Oppenheimer, no doubt concerned about the hardships and tribulations that his brother and some of his former students were undergoing by being investigated by the House Un-American Activities Committee—they had worked at the Berkeley Radiation Laboratory during the war and were thought to be socialists or communists—looked into the possibility of underwriting a project that would deal with matters of security and loyalty. As he wrote to Grenville Clark[25] to enlist his help on the project, he had noted "with grave concern, how high a price in freedom this country is preparing to pay in our attempt to achieve security in the present difficult times."[26] Although some studies had been done demonstrating the unwarranted hardships that the Presidential Loyalty Order, the security clearance procedures of the National Military Establishment and of the AEC, and the activities of congressional investigative committees were inflicting on individuals, Oppenheimer did not think they had come adequately to grips with the legal and constitutional essentials of the problem.

They do not, for example, deal with the profound qualitative changes which have accompanied the increasing activity of the Federal Bureau of Investigation, and the other investigative agencies of the government, and the correspondingly altered relations between these agencies, the Department of Justice and the rest of the Federal Government. They also do not come to grips with the question of what the objectives of our current security and loyalty programs are, nor raise, nor resolve the issue of whether these objectives are in themselves desirable, and whether or not they could be achieved by methods less disruptive to our health and freedom.[27]

Oppenheimer also contacted Paul Freund at the Harvard Law School, who in turn approached the dean of the Law School and got his support for the project to be jointly sponsored by the Institute and the Harvard Law School. A meeting at the Institute attended by David Cavers, Robert Cushman, Paul Freund, Herbert Hart, Edwin Huddleson, John O'Brian, Oppenheimer, and Max Radin resulted in a memorandum drafted by Henry Hart and extensively revised by Paul Freund concerning the study of security measures. The consensus of the meeting had been that the problems of secrecy and of loyalty requirements that had been discussed were susceptible of fruitful investigation, and that the investigation ought to be pursued. Freund's memorandum contained a detailed analysis of what the project should undertake. He noted that an inquiry into secrecy is "basically a study of the process of decision in a democracy, and of the bearing on that process of lack of information at the level both of official interchange and decision-making and of the formation of public opinion"; and that an inquiry into loyalty programs is "basically a study of social morale, in the largest sense, and of the bearing upon it of fear and distrust."[28] Freund remarked that both inquiries touched upon a fundamental principle of democracy, namely, "the principle of maximizing the pool of human abilities which can be drawn upon in the solution of social problems." He envisaged various phases to the investigation, from ascertaining the character and extent of secrecy measures and practices throughout the government, to determining the proper and improper objectives of secrecy, to assessing the effectiveness of secrecy measures in attaining its legitimate objectives, and most importantly to trying to determine the price of secrecy.

I do not know the ultimate fate of the project. In any case, after 1954 these problems were only alluded to in general terms in Oppenheimer's presentations, and in the William James lectures in particular, presumably by virtue of his own awkward position in addressing them.

### Literary Studies

Another area to which Oppenheimer offered support from the Director's Fund resulted in the establishment of the Princeton Seminars in Literary Criticisms, which were initiated during the academic year 1949–1950. These seminars on the interpretation and evaluation of literature were intended to provide "a focus for discussion, study and the exchange of ideas."[29] During the first year, the speakers at the four meetings of the Seminar were Felix Auerbach, Francis Fergusson, Delmore Schwartz, and Mark Schorer. Some twenty-five people usually attended, and the audience included Suzanne Langer, Jacques Barzun, Jacques Maritain, Harold Cherniss, John Berryman, Irving Howe, and the mathematicians Kurt Reidemeister and Morris Kline. The report at the end of the academic year 1949–1950 written by Francis Fergusson stated "that the interest that the Seminars awakened . . . is encouraging." But it also noted the difficulty in finding "common ground, and common methods and terms, whereby a number of men of various diverse backgrounds, training language and habits of thought, may work together in the interpretation and evaluation of our common literature."[30] The Seminars under the joint sponsorship of the Institute and the university continued for several years.

## Philosophy

Not much is known regarding Oppenheimer's philosophical outlook during the 1930s; and not much can be gleaned from either his lecture notes of that decade, or from textbooks like Schiff's *Quantum Mechanics*, which are based on his exposition of the subject matter, except to say he was more interested in conveying technical skills in the use of quantum mechanics than a philosophical outlook. After the war, Oppenheimer began to lecture extensively. One of the recurrent topics of his talks was the nature of science and the relation of science to culture and to society. The views he expressed in these lectures reflected his interaction with Niels Bohr at Los Alamos—such as the international character of science and the idea that secrecy was inimical to the well-being and growth of science. His lectures also demonstrated familiarity with the views and writings of William James, Charles Peirce, and John Dewey.[31] The collective aspects of the scientific enterprise were always stressed; he also drew attention to the fallibility of scientific knowledge, and he rejected all assertions and claims of total or absolute knowledge (Oppenheimer 1954, 54). Like Dewey he believed that the scientific community could serve as a model for a democratic society: "The community of science is a limited but worthy prototype for that tolerant, open-minded community of men which alone can maintain the progress of civilization" (Dewey 1916, 23). Indeed for Oppenheimer, as had been the case for many of the physicists who had been there, wartime Los Alamos represented a realization of that vision both as a social community and as an intellectual community, the cruel reality being that it achieved this when creating a weapon of limitless mass destruction.

I do not know when Oppenheimer was first introduced to pragmatism. As a freshman at Harvard he took the History of Philosophy course, Philosophy A, which introduced him to the writings of the major Western philosophers—Plato, Aristotle, Aquinas, Locke, Rousseau, and John Stuart Mill. One of the lecturers for the course was Ralph Barton Perry, a pupil and later a colleague of William James, whose papers he edited and whose biography would earn him a Pulitzer Prize in 1936. Another lecturer for that course was James Haughton Woods, who also taught a course on Indian philosophy. It is likely that he audited Henry M. Sheffer's lectures on logic and on philosophy of science, as prerequisites for attending Whitehead's seminar on Philosophical Presuppositions of Science (Phil 20h), which he did during the academic year 1924–1925. I do not know whether he audited C. I. Lewis's lectures on Kant and on evolution, or James Haughton Woods's lectures on Indian philosophy.

In his interview with Thomas Kuhn for the History of Quantum Mechanics, Oppenheimer spoke with great warmth of his reading the *Principia Mathematica* (Russell and Whitehead 1910) with Whitehead during his third year at Harvard. Whitehead at the time was shifting his interests in philosophical issues and was turning his attention to the writings of Peirce and James. It is thus likely that Oppenheimer had audited courses that had made him read some of the writings of William James, Charles Peirce, and John Dewey.

As Oppenheimer was also studying with Bridgman, it is very probable that he had become acquainted with Bridgman's notion of operational analysis as an analysis of doing, of processes. It is precisely in refining his understanding of electromagnetism, the course that Oppenheimer took with him as an undergraduate, that Bridgman came to stress that the operations in terms of which a physical concept obtains its meaning need not be exclusively the physical operations of the laboratory but also "pencil-and-paper" ones, that is, mental operations. And most important, Bridgman emphasized doing and *use*, the performative aspects of "operations." Operations must be such that they could be unambiguously performed—and for him this was true of the operations of mathematics, of physical theories, whose outcome had to be expressible in terms of operations applicable in concrete physical situations—as well as of the operations in laboratory settings.

There was a commonality in the writings and views of Peirce, James, Dewey, and Bridgman: they all placed special emphasis on the performative aspect of knowledge, on deeds, on doing things and using tools, whether conceptual ones or physical ones. And Bridgman in particular emphasized the importance of the ways that tools and expertise are acquired and how one's skills are honed by practice.

From Whitehead and his reading with him of the *Principia Mathematica* Oppenheimer would have heard of and become acquainted with

David Hilbert's "formalistic" program of reconstructing mathematics along positivist lines, with proofs serving the analogues of the measuring apparati of the empiricist scientist. Also, if not at Harvard then certainly while in Göttingen or in Zurich, Oppenheimer would have heard of the Vienna Circle, and in particular of some of the views of Moritz Schlick (1918) and of Ludwig Wittgenstein (Wittgenstein 1922; Monk 1990). He probably then or earlier became acquainted with the Vienna Circle distinction between statements of mathematics and logic and those of empirical science: Mathematical and logical statements being true by virtue of the meanings of their terms, whereas those of the empirical sciences were deemed true on the basis of observations of the world or of experiments.(White 1956, 10) The Vienna Circle position—that philosophy was primarily concerned with the logical analysis of scientific language and method—resonated with Bridgman's operationalism.[32] Oppenheimer, too, would come to emphasize the linguistic aspect of thought, what he called "verbalizing."

On the question of whether science was public—as Dewey strongly advocated—or private—as Bridgman held—we know that Oppenheimer sided with Dewey, as indicated by his public statement on the occasion of Albert Einstein's sixtieth birthday. Very probably, it was Oppenheimer's political views that had brought him to that position. He himself indicated that political thought became the focus of his readings in the late 1930s. As he became more deeply drawn into left-wing politics as a consequence of his relationship with Jean Tatlock and the outbreak of the Spanish Civil War, he began to read political literature extensively. Thus, during a train ride from the east to the west coast he read all of Marx's *Das Kapital*.

Besides his immersion in Indian philosophy, I do not know how much philosophical reading Oppenheimer did during the 1930s.[33] It would be interesting to know his thoughts at that time on the various "isms" he had come across: Peircian pragmaticism, Deweyan pragmatism, logical positivism, Marxism. How was logical positivism's rejection of metaphysics to be reconciled with the underlying assumptions of the *Bhagavad Gita* he was studying at Berkeley?

## Harvard Overseer

After World War II, Oppenheimer became deeply involved in the academic affairs of his alma mater by virtue of his friendship with James Conant. In 1946 Oppenheimer received an honorary degree from Harvard at the graduation exercises at which George Marshall delivered the commencement address in which he outlined the Marshall Plan for the recovery of Europe. In 1948 Judge Charles Wyzanski, the chairman of the Visiting Committee for Harvard's department of philosophy invited Oppenheimer to serve as a member of that committee for the academic year 1948–1949.[34] Oppenheimer accepted and took part in the committee's deliberations following its meeting with the department on January 8, 1949.[35]

In 1949 Oppenheimer was "duly elected an Overseer of Harvard College for a term of six years."[36] One of the primary duties of an Overseer is membership on the Visiting Committees that review the activities of the major academic and administrative subdivisions of the university. These academic Visiting Committees have a great deal of influence in determining the structure of departments as they evaluate the teaching and research activities of its faculty members and formally report to the president and the corporation every three years. Also, a representative of the Visiting Committee is usually a voting member of the ad hoc committee that determines appointments and tenure to the department.[37] When he became an Overseer, after being apprised of the duties and responsibilities of the Visiting Committees, Oppenheimer accepted becoming the chair of the Visiting Committees for the departments of philosophy and of physics, and a member of the Visiting Committee of the chemistry department.[38] By becoming chairman of the philosophy department's Visiting Committee, Oppenheimer had the specific responsibility to see to it that every year the committee assessed the department's strength and standing in the discipline, judge its contribution to the curriculum and to the "vitality" of the college, and review its plans for the future.[39] Oppenheimer took his chairmanship assignment very seriously and was responsible for important structural changes within the department of philosophy.[40] A recurrent theme during Oppenheimer's chairmanship was the relationship of the department to the General Education program, and more generally its relations with other academic departments in the natural and social sciences and the arts. Under his influence, the committee[41] came to regard these relations "as of decisive importance, not only for the health of the department itself, but for the realization of the intellectual and educational ideals of the College and the University." And in characteristic language, Oppenheimer in his 1950 draft of the Interim Report of the Visiting Committee asserted "that it was clear . . . that the vigor of a synoptic discipline such as philosophy will always be closely related both to general undergraduate instruction, and to a critical and sensitive awareness of work in more specialized fields."[42] The latter was not the opinion of the department, and the Visiting Committees on several occasions during Oppenheimer's chairmanship expressed disappointment over the degree of specialization that it found in the department and with its relations with other departments.[43] But by January 1955, at the end of his chairmanship, Oppenheimer, "writing for the Committee," indicated that

> the Committee is glad to report that Department has resisted the temptation to cross breed promiscuously with other disciplines, and that it has maintained the integrity of the great classical fields of philosophical study: logic, epistemology, ethics, and metaphysics. In an elementary course, or in a course of undergraduate study, or in a specialized study as a graduate student a man at Harvard may learn the essential traits of philosophical analysis and something of the history of man's long effort to understand his place in the world.

Nonetheless, Oppenheimer was concerned "that in resisting the pressure to become corrupt, to substitute for philosophy the history of ideas, or a social scientist's view of philosophical questions," the department may have missed opportunities for productive connections with other branches of learning or with non-European philosophical traditions. Oppenheimer believed, and the committee evidently concurred, that "the department would be stronger, and its standards not necessarily less high," were it to enrich its program by adding to its staff a philosopher of science—given that Philipp Frank and Percy Bridgman were retiring—and also a philosopher of law and a scholar specializing in Eastern philosophy.[44]

When Oppenheimer became chairman of the Visiting Committee in 1949, the permanent members of the Harvard department of philosophy were Clarence Irving, Henry Sheffer, Rafael Demos, Donald Williams, John Wild, Willard Quine, and Henry Aiken. The two junior members of the department were assistant professors Bugbee and White. Morton White, who had joined the department in the fall of 1948 as an assistant professor, was an exceptionally gifted individual who not only was an outstanding logician, but was also an extremely erudite and remarkably prolific historian of ideas, with a specialty in American thought and history (White 1999). There developed between White and Oppenheimer a deeply respectful professional friendship and personal amity. When White's election as professor of philosophy came up at the 1952 meeting of the Board of Overseers, a meeting that Oppenheimer could not attend, he wrote Bailey, the secretary of the Board of Overseers, that he "most heartily" endorsed White's election. "Professor White is a brilliant and creative philosopher, and an ornament to the Department and to the University."[45] Oppenheimer invited White to the Institute for the academic year 1953–1954, and White recommended that Oppenheimer be continued as a lay member of the Visiting Committee for Philosophy when his term as an Overseer expired.

White's impression of Oppenheimer when he first met him in 1949 was that some of his ways were peculiar, particularly "his tendency not to look you in the eye when talking to you" (White 1999, 137). Also, he thought Oppenheimer "was likely to be fuzzy" when he talked about philosophy and was inclined to support the appointment of people "who would speak movingly about man's condition and the cosmos—someone like Whitehead" (White 1999, 138).[46] White got to know Oppenheimer much better during his year at the Institute and grew to like him more. It seemed to him that Oppenheimer postured less, seemed friendlier and less pretentious there than he did as chairman of the Visiting Committee at Harvard. "He didn't try to demonstrate that he was a universal genius. Oppenheimer the man was more in evidence in Princeton and he was far more attractive there than the quirky darting figure" White had met in Cambridge (White 1999, 139). But that was the year that Oppenheimer was undergoing his trial, and he emerged a transformed man from that crisis.

It is my contention that Oppenheimer acquired a critical overview of contemporary philosophical inquiries and practices during his membership on the Visiting Committee of the Harvard philosophy department. As chairman he became acquainted with the textbooks that were being used in the undergraduate and graduate courses and had the opportunity to study the scholarly books and articles being written by the faculty. I do not know whether he read Quine's *Two Dogmas of Empiricism,* but he certainly looked at the articles and books that Morton White had written.[47] A good deal of White's *Toward Reunion in Philosophy* was written while he was a visiting member at the Institute in 1953–1954. The preface of the book concludes with the following statement:

> And finally I should like to thank the Institute for Advanced Study in Princeton, New Jersey, and its brilliant and humane director, Robert Oppenheimer, for having provided me with a year of peace and quiet in 1952–3 when the main draft of this book was concluded. It was one of the most profitable and pleasant years of my scholarly life. (White 1956, xii)

White's book is a remarkably lucid presentation of the philosophical tenets of the American pragmatists—Peirce, James, and Dewey—those of Bertrand Russell and G. E. Moore, and those of the logical positivists—Wittgenstein, Schlick, Carnap, and Reichenbach—all approaches that replaced the idealism and the neo-Kantianism of the nineteenth century. The aim of White's presentation was to deal "with some of the fundamental questions that emerge out of the intermingling of these tendencies in the first half of the twentieth century" (White 1956, 7). White examined in depth three fundamental concepts: existence, a priori knowledge, and value. In so doing, among other things, he expounded at length and with great clarity on why he, Nelson Goodman, and Willard Quine gave up the sharp distinction between analytic and synthetic as a result of their confrontation with the Duhem thesis. Pierre Duhem[48] had asserted that scientific explanation and prediction had put to the test a whole body of scientific beliefs, rather than merely the one supposedly being tested. Quine generalized the thesis so that not only scientific principles became implicated, but also the logic and mathematics used in the explanatory and predictive reasoning became drawn in—hence the difficulty in maintaining a sharp separation between analytic and synthetic statements and the method of establishing logical as opposed to empirical truths. Quine's argument was primarily limited to the case where a system of scientific discourse was related to empirical evidence that ultimately could be related to sensory experience. Quine also emphasized the quality of simplicity in choosing a scientific theory, and Nelson Goodman called attention to a factor that he described as the counterpart of conscience in the philosophical reconstruction of science. Both of these notions introduce what White called the quasi-ethics of belief (White 1956, 278). White broadened the Duhem-Quine thesis further so that it not only applied to scientific statements but also included ethical statements. Under those circumstances, besides scientific, logical, and empirical statements, White had to include "moral feelings of approval, revulsion, loathing, etc, toward actions."

> The resulting picture is that of the scientist bringing a system of logical and scientific theory to experience and that of the moralist bringing a combination of logic, scientific theory, and moral principle to bear on his moral problems. By traditional standards it involves a breakdown of the epistemological differences between the logical, the physical, and the ethical. By positivistic standards it involves a breakdown of the semantic walls between the analytic, the synthetic and the moral. (White 1956, 256–257)

In White's view, it follows that to the extent that philosophy is partly a critique of science and of morality, to that extent it is a normative discipline (White 1956, 279).

I believe that Oppenheimer's interactions with White, Quine, and the other members of the Harvard philosophy department, as well as his contacts with Ralph Barton Perry, who was a member of the Visiting Committee and who, with White, was a visiting member of the Institute for Advanced Study during 1953–1954, led Oppenheimer to carefully restudy Peirce, James, and Dewey. He had read some of their writings before, but they now acquired new significance. Many of his public lectures in the late 1950s and early 1960s echo the views that Peirce, James, and Dewey had expounded but now deepened and amended by what he had learned by reading White's essays and books. These lectures reflect a holism supported by what he understood of the Duhem-Quine-White thesis and by the notions of complementarity that Bohr had advanced in order to interpret the formalism of quantum mechanics.

Bohr and Pauli were frequent visitors to the Institute[49] after World War II, and their presence there occasioned both private and public discussions of the "Copenhagen interpretation" of quantum mechanics. Oppenheimer hosted many of these discussions—in particular, some of the discussions between Bohr and Einstein. Another visitor to the Institute in the fall of 1950 was Marcus Fierz, with whom Pauli had an extensive correspondence regarding philosophical issues—on the interpretation of quantum mechanics, on realism, and on metaphysics (Laurikainen 1988). These issues certainly came up in conversations with Oppenheimer.

The fact that the repetition identically of the *same* experiment involving microscopic particles—for example, the passage of identically prepared electrons through a double slit arrangement—yields in general individual different results, each one seemingly random, but that the results of a large number of such individual experiments yielded collective, stable, repeatable patterns implies that an account of experience in the microscopic realm must invoke probabilistic concepts. This in turn implies a departure from the customary demands of causal explanations encountered in classical physics and gave rise to the question of whether quantum mechanics yielded an exhaustive description of nature. Bohr undertook to answer this question by examining the necessary conditions for the unambiguous use of the concepts of classical physics in the analysis of atomic phenomena. Bohr's essential point was to stipulate that the description of the experimental arrangement and the recording of observations must be given in what he called "common language": "This is a simple logical demand, since by the word 'experiment' we can only mean a procedure regarding which we are able to communicate to others what we have done and what we have learnt" (Bohr 1963, 3). Furthermore, Bohr insisted that the measuring instruments be macroscopic devices that allowed a completely classical description. But whereas in the classical physics description the interaction between the measured object and the measuring apparatus could either be neglected, compensated, or accounted for, in the quantum situation the interaction forms an ineradicable part of the phenomenon. And thus an account of quantum phenomena must, "in principle, include a description of all relevant features of the experimental arrangement" (Bohr 1963, 4). Furthermore, Bohr emphasized the irreversible nature of a measurement involving microscopic entities since "all unambiguous information concerning atomic objects is derived from permanent marks left on the bodies which define the experimental conditions," such as a track on a photographic plate that results from the impact of a cosmic ray particle on it.

In classical physics, all the characteristic properties of a given object can in principle be determined by a single experimental arrangement, though it may be convenient to use different set-ups to study different aspects of the phenomenon. But in quantum physics, the results obtained by different experimental arrangements exhibit a novel kind of relationship that Bohr called complementary. Although the results obtained may appear contradictory if one attempts to combine them into a single picture, these different results exhaust all conceivable knowledge about the measured object.

> Far from restricting our efforts to put questions to nature in the form of experiments, the notion of *complementarity* simply characterizes the

answers we can receive by such inquiry, whenever the interaction between the measuring instruments and the objects form an integral part of the phenomena. . . . These circumstances find quantitative expression in Heisenberg's indeterminacy relations, which specify the reciprocal latitude for the fixation, in quantum mechanics, of kinematical and dynamical variables required for the definition of the state of a system in classical mechanics. (Bohr, 1963, 4–5)

It is important to emphasize that Bohr's complementarity argument deals with the conditions for observation and description and is therefore a statement of epistemology.[50] As is well known, Bohr suggested extending these notions of complementarity, that is, of protocols for observation and description of mutually exclusive properties that complement one another, to other fields of knowledge.[51] Bohr did so himself, and we will see that Oppenheimer also took up that challenge in his William James lectures.

## The William James Lectures

It was the custom that the departments of philosophy and psychology at Harvard would choose the speakers for the prestigious William James annual lectures in alternate years, and that they would normally choose a speaker in their respective fields. It was therefore highly unusual to have the two departments jointly sponsor Oppenheimer as the 1957 William James lecturer.[52]

When Oppenheimer's appointment as William James lecturer was announced in January 1956, a group of conservative Boston and New York alumni formed the "Veritas Committee."[53] In March 1956 the committee sent a lengthy letter to 10,000 selected Harvard graduates expressing objections to the appointment because as "a man of highly questionable moral background" with "a fundamental defect of character" Oppenheimer was disqualified from this honor. The committee therefore asked them to support a petition to reconsider the appointment.[54]

The tactics that Senator McCarthy had used in his investigations of communist influence had been exposed during televised Senate hearings in January 1954, and at the climax of these hearings he had been rebuked by Joseph Welch, the Army's attorney general, for "having no sense of decency." But despite the fact that McCarthy's public support had declined sharply by the time Edward R. Murrow's highly critical "Report on Joseph R. McCarthy" was aired in March 1954, and even though McCarthy had been censured by the Senate in December 1954, the hysteria he had generated had not yet fully died down a year later. Nor had the sentence imposed on Oppenheimer by the AEC commissioners been recognized as a by-product of that hysteria. The Veritas Committee was evidence of this. Although the committee's appeal was rejected, the Harvard authorities still feared that at the opening lecture there would be public demonstrations protesting Oppenheimer's appointment. There were none.[55]

The lectures were given in Sanders Theater, and all were packed. Morton White, who at the time was the chair of the philosophy department and was Oppenheimer's host, recalled that the students and nonphilosophers on the faculty such as I.A. Richards "loved [the lectures] and applauded wildly, whereas the philosophers were on the whole disapproving" (White 1999, 139). The theme of the lecture series was "The Hope for Order," and the titles of the individual lectures were as follows:

1. Monday, April 8: A Pluralistic Universe[56]
2. Friday, April 12: The Unity of Science
3. Monday, April 22: The Sites of Order
4. Friday, April 26: Instruments of Knowledge
5. Monday, April 29: An Unfamiliar Order[57]
6. Friday, May 3: The Proper Study of Mankind
7. Monday, May 6: Power and Learning
8. Friday, May 10: The Hope for Order

All the lectures were scheduled for 4:30 P.M. The initial lecture was delivered to some 1200 people packed into Sanders Theater, with another 800 listening in a nearby lecture room. Senator Joseph McCarthy had died four days before the first lecture, and his body was lying in state at the Capitol that afternoon. Before he started his lecture, Oppenheimer walked over to a blackboard standing on the stage, wrote in large letters R.I.P. (Rest in Peace), and then walked back to the speaker's podium and began his lecture.[58]

Box 259 of the Oppenheimer Papers in the Library of Congress contains the transcripts of the lectures,[59] including the notes Oppenheimer wrote in preparation for each lecture. This was the usual way Oppenheimer delivered his lectures. They were recorded and transcribed, and he would carefully go over them, amend them, make corrections, and if he found them suitable for publication he would give his approval for his hosts to do so. The William James lectures were never published, even though Oppenheimer went over the transcriptions twice and the available record is the twice-corrected version. He clearly had high hopes for the lectures, for he prepared a carefully thought out set of notes for them: Seven full pages written on a legal size note pad contained the main points of all the lectures. The notes for the first lecture—which was to be an overview of what was to be presented in greater detail in the subsequent lectures—were the most detailed. It is readily apparent why the audience, though mesmerized by the rhetoric and the poetic evocations of the lecture, found it "disconcertedly obscure" (White 1999, 140) and somewhat incoherent and overwhelming, for it was an amalgam of Oppenheimer's "reticulated and ramified"[60] beliefs, views, and insights. His aim had been to create by his lecture "not an architecture of global scope, but an immense, intricate network of intimacy, illumination, and understanding."[61] But the network proved to be too intricate. White recalled that after the first lecture, Oppenheimer came up to him and said: "I could see that you did not like that; but I think you will like what I have to say next time" (White 1999, 140).

I will present the content of the lectures not in sequential form, but rather attempt to give an overview of them and try to delineate their coherence. They can best be understood as stemming from Oppenheimer's confrontation with the constantly growing corpus of knowledge, and of scientific knowledge in particular. They are an attempt on his part to try to make sense of, and come to terms with, the constantly accelerated—seemingly exponential—growth of knowledge, a growth characterized by an ever decreasing characteristic time scale; they represent his attempt to find meaning in this new world. To comprehend the lectures one has to be aware that in these lectures Oppenheimer presented in condensed form much of what he had heard, perused, and read, often without giving their sources. Thus readily apparent are echoes from the Institute for Advanced Study conference on history,[62] from Paul Freund's report on the American constitutional system and its inadequacy in addressing some contemporary problems,[63] from Philipp Frank's lectures on language and logical positivism,[64] from Jerome Bruner's AAAS lecture on Freud,[65] and, in greater detail, particulars of what Oppenheimer had learned from Bruner concerning perception and learning, and more generally, what he heard regarding recent advances in psychology and the analogies and the metaphors that it uses. There are also resonances with what Oppenheimer learned from encounters with the Harvard philosophers, and his perusal and reading of their books.[66] But the closest family resemblance of the lectures are to those given by Bohr, some of which bear the same name and use the same the language. Thus Bohr had given a talk entitled "The Unity of Knowledge" at the bicentennial convocation of Columbia University with Oppenheimer in attendance. It was later published (Bohr 1958) and Oppenheimer undoubtedly read it in draft form.[67] In his lecture Bohr had stressed that "all knowledge presents itself within a conceptual framework adapted to account for previous experience and that any such frame may prove too narrow to comprehend new experiences" (Bohr 1987, 67). It may therefore be necessary at certain times in the growth of knowledge in a particular scientific field to reconsider points of view and concepts that, "because of their fruitfulness and [their] apparently unrestricted applicability, were regarded as indispensable for rational explanation" (Bohr 1987, 67–68). The subsequent "widening" of the conceptual framework—such as is exhibited by the complementarity principle—then becomes essential for restoring stability and comprehensibility.

Similarly, Bohr delivered lectures on "The Connection between the Sciences."[68] It is striking that often Oppenheimer uses the same words as Bohr. But we shall also see that there are notable differences between their views, with Bohr placing greater emphasis on language than Oppenheimer does—perhaps because Bohr found it so difficult to verbalize his thoughts, whereas for Oppenheimer this came without effort. Also, Oppenheimer places greater emphasis on methodology, rules, and sociology. For both of them, however, "complementarity" is at the center of the argument.

As noted earlier, for Bohr the description and explanations of scientific data must be *unambiguous*. This meant that the conceptual framework, that is, the web of interdependent concepts, used in the descriptions and explanations had to depend on the conditions for observations and descriptions within the field of experience under consideration. These conditions vary from field to field. The biological, psychological, and anthropological sciences each introduce quite different conditions for observation and description. Thus for Bohr, "there seems no basis for attempting to reduce psychology to biology, or biology to physics"[69] as the logical positivists had demanded. We will see that Oppenheimer arrived at a similar conclusion but for different reasons.

It is clear that Oppenheimer reread a good deal of William James's writings on pragmatism and on science for these lectures. It should be remembered that he had been committed from much earlier on to the Peircian view of science as a self-corrective process of inquiry and as a communal, cooperative activity. As the lectures make evident, like Peirce, he believed that it was the norms and the rules of inquiry that legitimize claims to scientific knowledge, and that the rules are open to being contested and corrected. All knowledge, including scientific knowledge, is therefore fallible.

But even though he did not refer to him explicitly—perhaps the highest from of compliment—it was the views of Dewey that underlay Oppenheimer's presentation. In the course of his lectures, Oppenheimer made clear that, like Dewey,[70] he believed that scientific values and scientific methodology should be models for historical, philosophical, and social scientific inquiries, and he, too, advocated that these practices take on a more extensive social role in society. Like Dewey, he demanded that the truths of these practices be verified beliefs that had been accepted because they had met certain standards of inquiry, of testing, of confirmation, and of corroboration. Oppenheimer was aware that it was precisely this notion of truth that was the warrant for believing the assertions of experts in a particular field. And this therefore was also the warrant for having believed him when he testified in the past on matters regarding atomic energy and atomic weaponry, or when at present he made statements regarding physics. Like Dewey in his *Democracy and Education*, Oppenheimer believed that science is the agency of progress in action; that it is experience becoming rational; and, that a true appreciation of science cannot be obtained by merely studying its history, but can only be acquired by hands-on experience.

For Dewey, philosophy "was an attempt to comprehend—that is, to gather the varied details of the world and of life into a single whole." And for Dewey, "the love of wisdom" was expressed in the "endeavor to attain as unified, consistent, and complete an outlook upon experience as is possible" (Dewey 1916 [see also 1980, vol. 9]). Oppenheimer would only partially agree. He would have reservations concerning the meaning of "unified," and "complete" would be modified to subsume "complementary" outlooks. It could be said that the James lectures were Oppenheimer's attempt to attain "as unified, consistent and complete an outlook"[71] as was possible, given what he had learned from recent archeological and historical studies; from some of the social sciences and economics; from the psychology of perception and learning; from advances in logic, that is, from the failure of Russell and Whitehead's attempt in *Principia Mathematica* to found mathematics on logic and from Gödel's incompleteness theorem; and from what he had learned from physics when it had reconciled, on the basis of Bohr's complementarity principle, the seeming contradictory aspects of the description of the atomic world.

It is Dewey's *Reconstruction in Philosophy*, the book Dewey wrote shortly after World War I, that was perhaps closest in spirit to Oppenheimer's undertaking in his lectures. Just as for Dewey the world had changed after World War I, the world had changed for Oppenheimer after World War II. Any reconstruction of philosophy had to reflect this fact and had to be based on the new understanding that had been brought about by quantum mechanics, by science in general, and by the new context created by the new technologies[72] that had been introduced, many of them fashioned by science. It was no longer a "closed world" or a "fixed world," but one that exhibited a new hierarchical order, defined by certain constants of nature—Planck's constant, the velocity of light, the charge of the electron—a world that seemingly was infinitely diverse and infinitely open in its possibilities. And like Dewey, Oppenheimer attempted in his lectures "to face the great social and moral defects and troubles from which humanity suffers," to clarify "the causes and exact nature of these evils," and to develop "a clear idea of better social possibilities" (Dewey 1920 [see also 1980, vol. 9]).

Thus Oppenheimer opened his William James lectures with an account of "our predicament[s] in the middle of the twentieth century" and a statement of some of the resources that seemed to him available for coping with them.[73] Oppenheimer's list of predicaments included retaining our liberties in the face of an enemy that does not share our notions of liberty and development; living in a society in which power is more "complicated, vaster, and less understood than it has been in any past society," but with a constitutional system not designed to meet this challenge; living in a world that is largely very poor, in skills, in experience and in wealth, "and that looks to us both for help and with envy; living in a world "where the possibility of an apocalypse is omnipresent, a very different kind of world than men have known for a very, very long time."

But in fact it was not about these worldwide problems that Oppenheimer wanted to talk but about our cognitive predicament:

> The fact that there is a great deal to know; that it is not well ordered; that it does not follow from a few comprehensible principles; that it is offered to us in diverse tongues; that it is shared by rather small groups of very specialized people; that it is growing apace; and that we do not know very much of it.

It was to the order of the cognitive world that Oppenheimer wanted to devote his lectures, for he believed that only if our cognitive house was put in order could we do the same for our other houses. However, he warned that the hope that he alluded to in the title of his lectures was not that we can suddenly revert to a simple life as lived "in simple villages," but that "we will recognize the order available and that we will meet it with glad hearts." He did believe that there was an order, a unity to the world, but it was not a "closed, or global, or architectonic unity, but rather that of a network, a reticular unity."[74]

Oppenheimer maintained that our cognitive predicament was exacerbated by the marked imbalance between "the intimate, familiar, and old" knowledge—whose meanings were known "in terms of tradition and experience," our common, collective knowledge, "what we can all talk with one another about"—and that which is new, possessing enormous richness and beauty in information, but known "in intimacy" to but a very few people, or known rather superficially and without intimacy to most. And he added that there was a general point that needed to be emphasized, namely, that James's pragmatic definition of meaning and of the tests of truth, statements that he agreed with, were only meaningful because novelty was circumscribed, that "ideas change a little at a time; that there is an enormous background of belief and consensus and familiarity, and that what changes is a small part of this." And this was still the case.

These, and other references to James, elicited from Oppenheimer the following acknowledgment during the middle of his first lecture: "[I] admire James because he fought against limitation and preconception; against the applicability of ideas from which specifics could be deduced; against certitude, against doctrine, against system." This characterization was equally applicable to Dewey, who in fact was much more explicit in his stand against certainty, system, and preconception.[75] Perhaps Dewey's liberal political views were what led Oppenheimer to refrain from referring to him, for he was well aware that when his appointment as the William James lecturer was announced, "it was attacked by right wingers among the alumni" (White 1999, 139).

Oppenheimer continued: "[I admire James for] his values and personal strength and courage. It was really in these that he found grounds for rejecting the phrase 'pluralistic universe' and could say of it 'It still makes a universe possible, because of the mediated connections.' "

It could not have been lost on the audience that Oppenheimer had called his first lecture "A Pluralistic Multiverse." Like James who had coined the word "multiverse,"[76] Oppenheimer was presenting not a world that seemingly lacked order or a single ruling and guiding power but "a cognitive jungle." Nonetheless, he was going to speak about the unity and order that he discerned in it. And to do so he pointed to two characteristics of the present situation that helped define the problem. One of these distinctive features—at least in the West—was that "there is no hieratic sorting out of different parts of knowledge, saying that some are unimportant and derivative." It was not wrong for anyone to know anything, and knowledge was not the property of a small class of people. But the pace of change and the accelerated growth of knowledge entrained the imbalance that he had mentioned earlier and posed a problem that was aggravated by the fact that the acquisition of most of the new knowledge required much learning and training, and furthermore that the knowledge was highly instrumental in character. This had as one of its consequences "a certain grim and great danger" when the common man tried to obtain a synopsis of modern science by reading about it: the danger that Philipp Frank had spoken of in his 1956 AAAS presentation—namely, that the words used in genetics or in cosmology may have a familiar sound and may suggest analogies to what was commonly known, but more often than not "they have almost the characteristics of a pun in which the same word has two quite differing meanings." In addition, the problems of fragmentation and rapid change were intensified by the fact that it is history and our institutions that make us and our ways of knowing, that determine what we can and what we cannot see, "and [are] remaking us right now as society and individual man." Knowledge and understanding involves the knower in an active, decisive way, and what he can know depends in a crucial way on his education. Education has a dual role: it transmits culture, and it is the means of learning to learn more. And present-day education leaves much to be desired. Given that schools are sites that one must turn to for resources for the hope for order, society at large must address this grave deficiency.

But there was another, more fundamental point that Oppenheimer wanted to make, one that revealed what he desired his lectures to accomplish. It was something he had learned through his interactions with Jerome Bruner, Edward Tolman, Ruth Tolman, and the other psychologists he had conferred with at the Institute, and more recently by reading the manuscript of Bruner, Goodnow, and Austin's *A Study of Thinking*.[77] Bruner had indicated that

> rational life begins with the selective practice of ignoring differences, failing . . . to perceive them; rational life begins with the failure to use discriminatory power in anything like its full potentiality. It lies in the selection, arrangement, and appropriate adequation to the objects of perception and thought, of limited traits, of a small residue of potential wealth.

Oppenheimer made it a pervasive and universal feature of cognition because it also manifested itself in many of the sciences: in psychology, in linguistics, in many parts of biology, "and deep in atomic physics." In perception, the coding that allows recognition, problem solving and cognition is a process of "selecting out."

> No man ever notices more than a minute fraction of those things which are for him to notice; and that is the price for him noticing anything at all. . . .
>
> It means that the potential that may be known is very greater than what is really known. It means that the potential always transcends the real. . . .
>
> The things that are cut out are possibilities that are lost in order that one may perceive at all.

These remarks echoed the viewpoint that Heisenberg had advanced in his various articles and talks on the interpretational problems of quantum mechanics, in which he adduced the Aristotelian concepts of "potential" and "actual."[78] The mathematical description of the microscopic world given by quantum mechanics in terms of abstract operators and vectors in a Hilbert space relate to the potential attributes of the atomic world. The "actual" properties of the world are conveyed in the data obtained in experiments, which is expressed in "common," objective, readily communicable language. And most importantly, the situation in atomic physics is such that what becomes actual and "real" is determined by the specific apparatus used in the preparation of the system, by the specific apparatus used in the performance of the experiment, and by the specific devices used in the final means of observation and detection—all determined by the experimentalist doing the investigation to the exclusion of other possibilities. And so Oppenheimer concluded that "in order to know at all, man must act and choose; in doing that he loses the opportunity of other knowledge." This is a generalized formulation of the notion of complementarity that Bohr had introduced. It was to be the principle that would help unravel the cognitive chaos, and characterize and help understand the order and unity of the new, open world that Oppenheimer was talking about. But this crucial point came at the end of a long lecture—the session had gone on for close to two hours at this point—and it may have been lost to the majority of the audience, which had been held spellbound by Oppenheimer's lyrical delivery. And those who understood the point he was making probably thought that it was too inconsequential and superficial an instrument to resolve the difficult problems of the modern, open world.

In his second lecture, Oppenheimer amplified his remarks concerning the specialized communities whose members are in possession of much of the new knowledge and noted that "they live an objective life in which they can talk with each other without ambiguity; and . . . the life they live reveals a great deal of order."[79] But the knowledge that they possess does not have a global character of the kind envisaged by Newton or LaPlace: it is applicable in a limited, clearly delineated domain. The unity of these sciences is of a complex kind. "It has a ramified, gentle, complex character in which there are no insuperable elements of disunity." In fact, Oppenheimer believed that "no part of science will be in any kind of contradiction to another part." Although in contemporary physics there were parts that were in contradiction with other parts, for example, quantum mechanics and special relativity, this was because of the incompleteness of today's physics.

Even though there is no unity of the sciences of the kind envisaged by the Newtonian picture, what has been learned suggests that no part of our experience or our knowledge is wholly "inaccessible" from another." And even though physics may "imply" chemistry, and chemistry may "imply" biology, and biology may "imply" a good deal of human nature, Oppenheimer thought that it was highly unlikely that physics would predict these things. "Physics will make very few predictions outside its own domain."

The next few lectures were devoted to the history of physics, from Newton to Heisenberg, Schrödinger, and Dirac, with particular emphasis on the advent of quantum mechanics in order to emphasize Bohr's correspondence principle and to expound at length on Bohr's notions of complementarity. Much of the material can be found in Oppenheimer's Reith lectures (Oppenheimer 1954). They culminated with a succinct statement of Oppenheimer's views of the quantum mechanical representation of the atomic world. After describing a nucleus undergoing alpha-particle decay and an electron passing through a two-slit screen, Oppenheimer emphasized that[80]

> every event is in part unpredictable [and] writes its own unexpected history; every new observation closes a chapter of the past. . . .
>     . . . the description of atoms is not a description of things, it is a description of things in interaction with our means of finding out about it. . . . It is a description in terms of the information we have gotten; and "we" here is not people . . . it is the practicing physicist or perhaps chemist, who is trying to study the situation.

But though conceptually revolutionary, quantum mechanics retained all the relevant physical experience of the past. It is a transcendent theory in the sense that it did not force physicists to throw away anything they had known in the past nor discard the ideas used to describe that experience. Quantum mechanics refined that experience, deepened it, and generalized it, and delimited clearly the domain of validity of the

ideas and the representations used. Oppenheimer placed great importance on the fact that there existed limiting situations in which classical
and quantum mechanical description overlapped—for example, in the
case that very large quantum numbers characterize the state of an
atomic system such as a hydrogen atom, the domain to which the Bohr
correspondence principle applies. This was an example of continuity in
the face of novelty and an example of transcendence, in that the new
theory encompasses both the old and the new. Oppenheimer believed
physics, in fact all of science, to be strictly cumulative. As far as Oppenheimer was concerned, there was no incommensurability—in the sense
of Kuhn—in the transition from classical physics to quantum mechanics. A similar case could be made for the special relativistic and the
Galilean description of physical phenomena.

The rest of Lecture 6 was devoted to examining the differences between history and science—in the way either physics or astronomy is a
science—and to apply these insights to say something about the social
sciences. "The [physical] scientist is a great agent in creating new experience, at the same time that he is trying to order it" and in the process
has produced "overwhelming vistas of order, of simplicity and of harmony." The historian is also concerned with ordering experience, but
deals mostly with particulars. But his ordering is not concerned with
finding propositions from which special cases follow or constructing
theories from which one can deduce observations. Rather, the historian
focuses on sorting out, identifying, categorizing, and recognizing
things that it makes sense to talk about.[81]

Oppenheimer concluded Lecture 6 with the statement that he believed that what will be learned about man is not going to constitute a
single science of man. Rather, it is going to be many sciences put together "with bridges, and glue, and sealing wax," the latter allusions to
the experimental physics. Moreover, he avowed:

> I firmly and deeply believe that the years ahead will see an increased un-
> derstanding of man by methods which are no way a parallel to those of
> atomic physicists have used; but by methods which are not, primarily,

historical or descriptive, which are both experimental and logical. I believe that this experience already foreshadows countless issues where we
have the phenomenon of complementarity, but do not have the knowledge to resolve the complementary phenomena in a harmonious whole,
as in physics, we have had.

A good deal of Lecture 7 was devoted to seeing whether extended notions of complementarity could help resolve some of the different
points of view in the various approaches to psychology. Oppenheimer
was pessimistic in his conclusion.

> In many of these cases I believe that some kind of hard application of the
> ideas of complementarity may eventually be possible. The approaches
> that largely appear to exclude each other[82] as having no visible common
> basis have, in no case, a very clear theory on either side or very good or
> clear operational definitions of what they are talking about; and in no
> case is there now a transcendent view including both of them.

In Lecture 8, entitled "The Hope for Order," Oppenheimer recapitu-
lated the high points of the previous lectures and reminded his audience that

> the only kind of order that I have been addressing myself to in these
> talks; the order to hope for; the order that is hard enough, . . . is a partial
> one, and a mutable one, and one that we must expect will always be tran-
> scended and never quite finished.

His closing words were normative and hortatory:

> We have, I think, a dual duty: a duty of faithfulness and firmness and
> steadfastness in the things which by accident in our own time, in our
> own place, are our knowledge, our skill, our arts; [and] we have a duty of
> great openness to others . . . , learning to welcome the strange and being
> glad to learn something that is new, and that we have not suspected,
> and that does not fit, that is different and in some sense alien. And both
> duties are part of any love of order and of truth that can exist today.
> Both are helped when we not only love other truth but each other who
> purvey it.

It is not surprising that after the lectures the audience was somewhat bewildered and the philosophers were disappointed. Complementarity invokes mutually exclusive modes of description that complement one another, and thus seemingly asserts the existence of contradictory elements in reality. Nor was it clear how this approach opened new perspectives on the social sciences that would enable them to solve concrete problems. Thus it was difficult to come away from the lectures with a definite notion of what one was supposed to get out of them. The words Oppenheimer used in Lecture 8 when he described the unity he ascribed to the sciences characterize his lectures. He had commented there that the subsciences; different as they were in terms of practices, they nonetheless gave "an evocative indication . . . of very varied, often quite subtle, sometimes substantial, sometimes formal, sometimes tenuous sets of bonds between the sciences which give to them a kind of over-arching, largely potential unity." His own lectures had a similar quality: they were very varied, often quite subtle, sometimes substantial, sometimes formal, sometimes made tenuous connections and analogies, and spoke of potential applications of the ideas of complementarity in the social sciences. His hopes for the relevance of Bohrian complementarity notions to transcend the antinomies of the social sciences were just that, hopes. He could not indicate in any particular case how observable consequences could be drawn from the application of such ideas, nor could he specify how they were to be used. A critic might have pointed out that Bohr had advanced complementarity *after* the formalism of quantum mechanics had been put into place and after many substantial problems such as calculating the bound states of atoms and the description of scattering experiments had been solved.

But a careful reading of the edited version of Oppenheimer's talks indicates that he presented a weltanschauung that was at variance with the views of his colleagues in physics,[83] one held or defended by but very few philosophers. During the 1930s, philosophy in the United States had been influenced by the arrival of Carnap, Feigl, Hempel, Reichenbach, Frank, and the other Austrian and German logical positivists who had been dismissed from their posts by Hitler coming to power and by the Anschluss. After World War II logical positivism merged with British analytic and linguistic philosophy, and the combination became the dominant American approach to study and resolve philosophical problems. The pragmatism of Dewey and James was considered soft and not rigorous, and so was largely ignored. Furthermore, the Cold War was not conducive to endorsing uncertainty and fallibility.

But this was precisely what Oppenheimer did, perhaps more successfully, in his later talks than in his William James lectures. In his later lectures, he made much more explicit, using simpler language, how history, contextual factors, and culture-specific values and motives play a role in making sense of one's world and one's moral horizon.

## Epilogue

In concluding this chapter I want to come back to Charles Taylor's essay and to the problem of identity. Being an American, identifying himself as one, was the one continuous, constant aspect of Oppenheimer's changing identities. The William James lectures were meant to give a philosophical underpinning to the enterprise of making this a better world based on an American philosophical outlook.

# CHAPTER FIVE

# Einstein, Oppenheimer, and the Extension of Physics

Having understood by the middle 1970's, to a large extent, all the four forces of nature in the remarkable successful standard model, attention has returned to Einstein's dream of unifying all the forces with gravity. The goal of unification has been at the forefront of fundamental physics for the last three decades.

—*David Gross (2008, 294)*

## Unification

Eugene Wigner, one of the leading contributors to quantum theory, made a perceptive comment regarding the genesis of quantum mechanics in a lecture he delivered in the early 1980s at an international school of subnuclear physics:

> While I was studying chemical engineering [during the early 1920s] and visited the Berlin University's physics colloquium each week, I gained the impression that many of the participants had doubts whether the human mind is strong enough to extend physics to the microscopic domain. . . . The first change in this pessimistic attitude came with Heisenberg's paper, modest as the aim of that was. . . . An equally large change came with Schrödinger's paper. . . . [But] we were . . . modest, we accepted and worked with a theory which we knew to be incomplete, the limitations of which we recognized. We were both proud and surprised when we could extend it to new sets of phenomena. (Wigner 1983)[1]

Two sets of phenomena were addressed. One set was the extension of quantum mechanics to the macroscopic realm (i.e., explaining the structure of atoms and molecules, the nature of chemical bonding, the structure of solids, the electrical and thermal properties of metals, the cohesion of solids, etc.). Its successes led to the belief "that we possess the basic information from which all chemistry could be deduced" (Wigner 1983, 769).

And indeed, quantum mechanics resulted in the unification of physics and chemistry in the sense that a *foundational* theory of remarkable accuracy and robustness had been apprehended. The other extension was toward the subatomic realm. The confidence gained from mastery of the atomic and molecular realm, and later during the 1930s of the nuclear domain, cannot be overestimated. The success of the quantum mechanical explanations in these domains—the collective accomplishment by the physics community—together with Einstein's formulation of his general relativity theory—the outstanding exemplar of individual creativity drawing on the mathematics and physics resources available at the time—cast aside doubts "whether the human mind is strong enough to extend physics to the microscopic domain."

The state of affairs in "elementary particle" physics after World War II was summarized by John Archibald Wheeler in an important address delivered at a joint meeting of the National Academy of Sciences and the American Philosophical Society in the fall of 1945. Wheeler observed that the experimental and theoretical research of the 1930s had made it possible to identify four fundamental interactions: (1) gravitation, (2) electromagnetism, (3) nuclear (strong) forces, and (4) the weak interactions responsible for the radioactive decay of the neutron and nuclei. The belief that these were four independent realms governed by distinct forces and dynamics was prevalent until the early 1960s—the period of relevance to our story.

Except for a brief period after the advent of general relativity stretching from 1915 until roughly 1926–1927, the physics community was primarily concerned with the elucidation and representation of these four separate realms, and even then primarily the first three. After the advent of quantum mechanics, Einstein was one of the few physicists who attempted to give a unified description of the forces of nature—in his case a unified description of gravitation and electromagnetism, and this at the classical level. A few other researchers investigated this possibility: Oskar Klein in the late 1930s, and after World War II, Julian Schwinger, who on several occasions attempted to give a unified quantum field theoretic formulation of the strong and the electromagnetic interactions using spin 1 gauge bosons. He reported on these efforts in the 1964 issue of the *Reviews of Modern Physics* dedicated to Oppenheimer on the occasion of his sixtieth birthday.

The program of unification traces its origins to Newton and his realization of the universality of gravitation. Different aspects of the program were anchored during the nineteenth century. Hans Christian Oersted and Michael Faraday gave credibility to the quest by providing the first experimental indications of the connection between electricity and magnetism, thus showing that the program of unification had validity. Thereafter, James Clerk Maxwell constructed a model for a unified theory of electricity and magnetism and provided a mathematical formulation that was able to explain much of the observed phenomena and to make predictions of new ones. Joseph von Fraunhofer, Gustav Kirchhoff, and others demonstrated that the laws of physics discovered here on earth also apply to stellar objects. By the end of the century, attempts were being made to give a unified representation of the physical world based on electromagnetism.[2]

With Einstein the vision became all-encompassing. For Einstein, recognizing "the unity of a complex phenomena ... which appear as distinct things" had already become a passion for him when he was a student at the Eidgenössische Technische Hochschule (ETH), the Swiss Federal Polytechnic (RTH) in Zurich. One can look at all his works as part of a "search for a foundation of the whole of physics," as part of a program striving to achieve a *Weltbild* that encompassed all physical events. In an essay written in 1940, he indicated that he had always attempted to find a unifying theoretical basis for all of physics, a basis from which all the concepts and relationships of the subbranches of physics might be derived by logical processes. "The confident belief that this ultimate goal may be reached is the chief source of the passionate devotion which always animated me as a researcher" (Einstein 1954a, 324). But it should be stressed that when investing enormous efforts into formulating a unified theory of electromagnetism and gravitation, Einstein always looked at what he was doing as only a first step in charting a path. This was true for his unified theories of electromagnetism and gravitation—and this was also true for his theory of general relativity as a theory of gravitation: General relativity might well be replaced by a deeper theory that would encompass all of general relativity's results and predict new observable phenomena.

Einstein came to advocate unification coupled to a radical form of theory reductionism. In 1918, in his address celebrating Max Planck's sixtieth birthday, he stated:

> It is my belief [that] . . . the general laws on which the structure of theoretical physics is based, claim to be valid for any natural phenomenon whatsoever. With them it ought to be possible to arrive at the description, that is to say the theory, of every natural process, *including life*, by means of pure deduction, if that process of deduction were not far beyond the capacity of the human intellect. The physicist's renunciation of completeness for his cosmos is therefore not a matter of fundamental principle. The supreme test of the physicist is to arrive at those universal elementary laws from which the cosmos can be built up by pure deduction. (Einstein 1954, 221; my emphasis)

And in his 1949 "Autobiographical Notes" Einstein was even more explicit:

> I would like to state a principle, which cannot be based upon anything more than a faith in the simplicity, i.e., intelligibility, of nature; . . . that . . . nature is so constituted that it is possible logically to lay down such strongly determined laws that within these laws only rationally, completely determined constants occur (not constants, therefore, whose numerical values could be changed without destroying the theory). (Einstein 1949a, 63)

After the formulation of electroweak theory by Steven Weinberg, Sheldon Glashow, and Abdus Salam in the late 1960s, reduction and unification became the two tenets that dominated a large segment of the theoretical physics concerned with high energy physics and the foundational assumptions of physics. Reduction characterizes the hope of giving a unified description of all physical phenomena; unification, the aspiration to reduce the number of independent concepts necessary to formulate the fundamental laws. The conceptual—idealistic—component in reductionism has taken precedence over the materialistic one of reducing the number of so-called elementary particles.

This had not been the case between 1945 and 1965. Yet the hope of unification, to which Einstein had devoted the last twenty-five years of his life, always loomed in the background. However, the notions of unification that Einstein had nurtured and the forms of unification that the post–World War II generations of theorists sought only converged to a common viewpoint—at the mathematical level—after the 1970s when gauge symmetry and symmetry breaking became the foci of further developments.

Before considering the Massachusetts Institute of Technology (MIT) Centennial Celebration colloquium on "The Future of Physics" to indicate what some of the leading theorists thought about unification in the early 1960s—just as symmetry breaking and gauge theories, and in particular, Yang-Mills theory were becoming of great interest to theorists—let me briefly review what has been said about Einstein's program of unification.

## Einstein and Unification

Peter Bergmann, who was Einstein's assistant from 1938 to 1941 and who worked with him exploring different versions of field theories that unified general relativity, that is, gravitation, and electromagnetism, believed that Einstein's endeavors in formulating unified field theories ought to be seen as a continuation of the program that took him from special relativity to general relativity.[3] Bergmann (1982) emphasized the crucial turn that Einstein took in 1907 when he extended his considerations of the description of physical phenomena to accelerated, noninertial frames of reference.

Einstein had noted that whereas the force on a charged particle in an electromagnetic field depended on its charge $e$, and its acceleration on the ratio of its charge to its inertial mass, $e/m$, no analogous parameter entered into the expression for the acceleration due to a gravitational field.[4] Therefore, if at any given point the gravitational acceleration is the same for all bodies, then it vanishes locally for an observer who undergoes that same acceleration. Einstein was thus led to accelerated frames rather than inertial frames and to the equivalence principle. But he came to realize that whereas in special relativity an inertial frame of reference could be assumed to extend over all space, the free-falling, accelerated frame of reference in which the gravitation vanished was defined only locally. The issue became that of answering the question: "What is the relationship between this free-falling frame of reference to free-falling frames in adjacent regions?" It turned out that Bernhard Riemann and other mathematicians—Elwin Bruno Christoffel, Gregori Ricci, and Tullio Levi-Civita—had already given the answer to that question: these relations were subject to field laws, which, in fact, could be chosen according to principles of formal simplicity. From considerations of the implications of the equivalence principle and the conventionality involved in labeling space-time point in Riemannian geometry—general covariance—Einstein arrived at his theory of general relativity by imposing the requirement that for weak gravitational fields the deductions of his theory should agree in lowest approximation with those of Newtonian theory. In his new theory, the gravitational field became conceptualized as a dynamical field that manifested itself in the geometrical properties of space-time, that is, by geometrization. After his success with gravitation, Einstein believed that electromagnetism— the only other force and field theory known at the time—also demanded geometrization and to be unified with gravitation. But whereas general relativity was the result of the impossibility of devising a theory of gravitation that met the requirements of special relativity and of the equivalence principle, this was not the case for Maxwell's theory of electromagnetism, which was invariant under Lorentz transformation and guaranteed that the velocity of light was the same in all inertial frames.

The conceptual elegance and simplicity of general relativity and its success in accounting for the accurate description of the motion of Mercury created a bandwagon effect that led others to also search for a geometrized theory that would include electromagnetism, and thus unify gravitation and electromagnetism. The hope was that in the unified description only in the weak field limit would the two fields be decomposable into two independent sets of equations governing gravitational and electromagnetic phenomena as the lowest approximation. Einstein hoped that the unified description—which would naturally contain terms that reflected the mixing of the two fields—would have some new, observable consequences.[5]

A second aspect of Einstein's program concerned the dualism in the representation of particles and fields: the existence and properties of the "charged" particles in Maxwell-Lorentz theory (where the "charge" was the electric charge of the particle) and in Newtonian gravitational theory (where the "charge" of a particle was its gravitational mass) were independent of the fields. The equations of motion of the particles as determined by the forces on them due to the fields were an independent set of equations, which together with the field equations of the fields—with the motion of the "charged" particles as their source—determined the evolution of the field–matter system.

In 1912, Gustav Mie proposed a very attractive way to overcome the matter–field dualism by giving up the linearity of the Maxwell equations. Such nonlinear generalizations offered the possibility of having solutions that concentrated, in a stable fashion, energy and momentum in small regions of space, which would be identified as particles. Recall that until the late 1920s the only known "elementary particles" were the proton, the electron, and the (massless) photon. The basic properties of matter to be accounted for were thus the existence of the electron and the proton. Einstein hoped that a unified field theory of gravitation and electromagnetism would do this.[6]

Valia Bargmann, during a discussion at the Princeton Centennial meeting in 1979, reflected on this issue while reminiscing on his work as Einstein's assistant in the late 1930s:

> In his various attempts at creating a unified field theory, Einstein always talked in fact only about gravitation and electromagnetism. What about other forces? What about other elementary particles? During the time Peter [Bergmann] and I were working with Einstein, new particles were being discovered, and yet it seemed that Einstein did not show any great interest in them. And once he said to us, "You know, it would be sufficient to really understand the electron." He explained to us more clearly what he meant. The electron was, of course, entirely alien to any previous theory, starting with the [Lorentz] theory of electrons. Consequently, in order to understand the electron properly, one would have to have a deeply modified new theory. And if you were to succeed in this, then the chances are you would also understand the other elementary particles. (Bargmann 1982, 481)

After 1930 the physics community thought Einstein was wrong. One could not attempt to unify electromagnetism and gravitation without taking the quantum theory into account. Moreover, most of the practitioners accepted the irreducible, ontic, nature of probability at the microscopic level and doubted Einstein's hopes of deducing the probabilistic and nondeterministic aspects of the quantum mechanical description from an underlying nonlinear gravitational theory.

## The MIT Centennial Celebration

In April 1961, MIT celebrated its hundredth anniversary with a week-long conference on scientific and engineering education. The morning session of the last day of the gathering was devoted to a panel discussion on "The Future of the Physical Sciences."[7] Francis Low, a professor at the Institute, chaired the session, and the speakers—in the order of their presentations—were John Cockcroft, Rudolf Peierls, Chen Ning Yang, and Richard Feynman. All five were physicists and except for Cockcroft, theorists. The absence of any other kind of physical scientist on the panel is indicative of the status that physicists then enjoyed, and the preponderance of theorists similarly reflected their standing within the scientific community.

The Nobel laureate Cockcroft, who had been director of the British Atomic Energy establishment and had become Master of the newly founded Churchill College in Cambridge, talked about the likely advances during the next decade or two in nuclear, high energy physics and astrophysics that would result from the new accelerators, storage rings, and radio telescopes that were being built.

Peierls, who at the time was professor of mathematical physics at Birmingham University, had gone to the library to prepare his talk to see what physics was like in 1861 and 1911 in order "to get some feeling for how much we could have predicted [150 years ago] about physics today."[8] His examination of the 1861 and 1911 journals convinced him that, although it would have been foolish a hundred years ago to attempt detailed predictions of the physics of the 1960s, it might have made sense to try to do so in 1911. However, in view of the accelerating pace of developments, he did not think that anything one could say today "can be expected to have validity more than perhaps twenty five years ahead." But he suggested that "if we think about the foundations, i.e., the front line of physics, some of the problems which worry us today will continue to be exciting. . . . We shall be looking for connections which will unify the very broad range of basic concepts and facts which we now have to use to describe what we know in physics." Although Peierls felt that there had to be "some new kind of connection which interprets the many fields and couplings [of the then known elementary particles] as different aspects of some common principle," he thought that "we should not expect, as we expected at one time [during the 1930s], that the unification will come with one go. . . . There will be many stages of partial clarification before everything fits together, and, of course, . . . the last stage may never be reached."

Yang, who had recently won the Nobel Prize with Tsung Dao Lee for their work that had led to the demonstration that parity is not conserved in the weak interactions of the elementary particles, was then a professor at the Institute for Advanced Study in Princeton. He began his presentation by asking what made physics so unique an intellectual endeavor. The answer, he believed, lay in the fact that it offered the possibility of formulation of concepts out of which, in the words of Einstein, "a comprehensive workable system of theoretical physics can be constructed. Such a system embodies elementary laws from which the cosmos can be built by pure deduction." Yang noted that the three great conceptual revolutions of the twentieth century—special relativity, general relativity, and quantum mechanics—which unified much of physics, gave hope for the possibility of further unification, and in particular the hope for "integration of the weak, the electromagnetic and the strong interactions." But he also interjected a discordant note and warned his audience that, although he believed that "new levels of penetration" would be achieved, the task of unraveling and understanding the new levels would become harder. Even though it is very natural in the day-to-day work of a physicist "to implicitly believe that the power of the human intellect is limitless and the depth of natural phenomena finite," for this faith gives courage, nonetheless Yang asserted that "the belief that the depth of natural phenomenon is finite is inconsistent, and the faith that the power of the human intellect is limitless is false."[9]

It was Feynman who made the most provocative statements. Although he concurred with "almost everything" that Cockcroft, Peierls, and Yang had said regarding the near-term prospects, he did not agree with "Professor Yang's idea that the thing is getting too difficult for us. I still have courage, . . . I think it looked difficult at any stage." But he shared some of Yang's pessimism about the long term, and it was the long term that he wanted to address, and in particular, the long-term prospects of discovering the fundamental laws in physics, what Peierls had called "the front line of physics." He cautioned his audience that were he talking "about that which is behind the front line, things like solid state physics, and other applications of physics," he would be saying very different things, and he urged them to keep the limitation of the discussion in mind.

Feynman observed that the previous speakers had wanted to be safe in their predictions and had therefore limited their forecast to the next twenty-five years. He said he "was going to be really safe, by predicting 1,000 years ahead." The other speakers were not safe because in twenty-five years "you will see that they are wrong." Were he to follow Peierls's suggestion, Feynman noted that he should then look at the physics of 961 and compare it to that of 1961 "to see the difference between the physics in the era when Omar Khayyám could come out of the same door as in he went, to the physics of today as we open one door after the other and see rooms with treasures in them, and in the backs of the rooms five and six doors obviously going into rooms with even greater treasures." But Feynman thought this would not do, because "this is a heroic age, this is an exciting age of very vital development in the fundamental physics, in the study of the fundamental laws." It is therefore not fair to compare it to 961. One should find another heroic age in the development of physics, and he suggested the third century B.C., that of Archimedes and Aristarchus. But Feynman warned that one could not extrapolate from these developments those events that would take place a thousand years down the road because one cannot "predict the future of physics alone without the context of the political and social world in which it lies." Therefore to answer the question of what physics might be like a thousand years from now one has to answer the question: "Is it possible that it is all going to be over?" What if there is a "terrific war and a collapse"? Feynman, we should remember, was talking during the height of the Cold War—the Cuban missile crisis would occur a year later—and the threat of a nuclear holocaust was a realistic possibility. If that was the case, if there was a "terrific war and collapse," Feynman believed that physics, fundamental physics, would not recover—even if the rest of humanity were to recover—because

in order to have this heroic age, this exciting one, one must have a series of successes. If you look at the successes of the grand ages of different civilizations, you see that they have an enormous confidence in success, that they have a new thing that is different, that they are developing for themselves. If one were to slide back, you would find for a while no great successes. You are doing experiments which were done before. You are working on theories which the ancients knew very well. What can happen is a lot of mouthing and philosophizing, a great effort to do physics in the sense that one *should* do it to be civilized again, but not really do it. To write instead commentaries, that disease of the intellect, which appears in so many fields.

But what if there is no collapse. What, Feynman asked, would likely happen to fundamental physics if, "though it may seem ridiculous," we supposed a society somewhat like our own continuing for a thousand years. One possibility, Feynman conjectured, was that "a final solution" would be found. In contrast to Yang, who believed that "it was self evident that [this] is impossible," Feynman thought this was potentially possible. What Feynman meant by a final solution is that every new experiment would only result in checking the fundamental laws. Under these conditions he thought that it would get "more and more boring as we find that time after time . . . nothing new is discovered that disagrees with the fundamental principles." He concluded that "of course, the attention would then go to the second line." Furthermore, if this were to happen, Feynman thought that it would mark the end of a vigorous philosophy of science, for it seemed to him "that the reason that we are so successful against the encroachment of philosophers and other fools on the subject of knowledge and the way of obtaining new knowledge is that we are not completely successful in physics."

What if, on the other hand, as Yang wished, there are an infinite number of layers? "Then there will be a continual exciting unfolding." Extrapolating from the three revolutions of the present century that Yang had cited, Feynman noted that

three unfoldings in sixty years is fifty unfoldings in a thousand years. Is the world going to have fifty levels of excitements? Is there that much treasure in it? If there is it will become somewhat boring. It will become boring [to discern] . . . that things change always when you look deeper.

Therefore Feynman did not believe that fundamental physics could last through 1,000 years of investigation. Feynman concluded that under those circumstances the enterprise "will slow up, [and] the questions will become more difficult. "Far fewer experimental results would become available, new discoveries would be made ever more slowly, the problems would become harder and harder, and more and more people [would] find it a relatively uninteresting subject, and so it [would] be left in an incomplete state with a few working [on understanding some lower level]." But Feynman alluded to another possibility: Physics may expand "into the studies of astronomical history and cosmology." Feynman stated the situation in physics at that time as follows:

> There is no historical question being studied in physics at the present time [i.e., in 1961]. We do not have a question, "Here are the laws of physics, how did they get that way?" We do not imagine, at the moment, that the laws of physics are somehow changing with time, that they are different in the past than they are at present. Of course, they *may* be, and the moment we find they *are,* the historical question of physics will be wrapped up with the rest of the history of the universe, and then the physicist will be talking about the same problems of astronomers, geologists, and biologists.[10]

In closing his presentation, Feynman reiterated that "we live in a heroic, a unique and a wonderful age of excitement. It is going to be looked at with great jealousy in ages to come," and he summarized what he had said by stating that he believed "that fundamental physics has a finite lifetime." What Feynman meant by fundamental physics in 1961 was essentially high energy and particle physics. He concluded with the caveat that "in these modern times of high speed change, what [he] had suggested would happen in a thousand years will probably occur in a hundred."

Feynman could speak of living in "a heroic, a unique and a wonderful age of excitement," even though at that particular time—the early 1960s—particle physics was in the midst of a crisis. He could characterize as "unique and wonderful" what was happening in high energy physics because it was being propelled by a constant stream of new empirical discoveries stemming from the plethora of high energy accelerators and detectors that had just been built, and were being built. And indeed the end of the decade saw major theoretical advances culminating in the Weinberg-Salam model unifying electromagnetism and the weak interactions, and shortly thereafter the establishment of the standard model, giving a unified description of the electroweak and the strong interactions.

Feynman was also prescient. Physics has become more concerned with historical questions and has become "enwrapped in the cosmological problem."

Each of the speakers at the MIT celebration had addressed the ultimate goals and the future of physics. In 1964, in an effort to gain support for a "superhigh energy accelerator in the energy region of multi-hundred BeV," the high energy community issued a little volume indicating the "Purposes of High Energy Physics" by presenting the views of the leading theorists—all "active in the field of high energy physics, ... to whom collectively, the theoretical progress of the [post–World War II period] ... is largely to be attributed" (Yuan 1965, 2).[11] Oppenheimer wrote its foreword. That he was asked to write the foreword exemplifies his paradoxical standing as icon and as outsider. What is striking in Oppenheimer's remarks is that he did not claim, as did most of the other contributors, that "High energy physics is the essence of today's science of physics ... and is the domain [in which] the fundamental laws of nature are now being discovered" (to quote Robert Sachs, for example; Sachs in Yuan 1965, 20). The consensus, as expressed by Hans Bethe, was that high energy physics was "indeed the most basic field of knowledge in the physical world ... no other field will give us such deep understanding. . . . [P]article physics deserves the greatest support among all the branches of our science because it gives the most fundamental insights" (Bethe in Yuan 1965, 9). It provided, so Henry Primakoff in turn put it, "in some sense the fundamental building blocks ... out of which the material universe is constructed" (Primakoff in Yuan 1965, 18)—sentiments echoed by Victor Weisskopf (Yuan 1965, 26), Gian-Carlo Wick (Yuan 1965, 29), and Richard Dalitz (Yuan 1965, 41). Although Oppenheimer agreed that seeking to understand the structure of matter had been astoundingly intellectually fruitful in the past, he refrained from giving his assent to Abraham Pais's conviction that the very goal of high energy physics was to find a fundamental, unitary theory of the interactions between elementary particles. To Pais's mind, on the contrary, an "interruption of the pursuit toward high energy machines would be disastrous" for just that reason, as "[w]e may be stranded without their synthesis. . . . We may stop short of . . . integrating such disjointed forces into a new unification."

Though clearly supportive of the aims of the enterprise, for Oppenheimer its justification was not the further elucidation of the nature of matter. Rather, for Oppenheimer it was to ensure the continuance of the Enlightenment project:

> The last centuries of science have been marked by an unabating struggle to describe and comprehend the nature of matter, its regularities, its laws, and the language that makes it intelligible. The successes in this struggle . . . have inspired the whole scientific enterprise, and lighted the world of technology, and the whole of man's life. They have informed the education and devotion of young people. . . . We are now, despite tempting and brilliant successes, deep in the agony of this struggle. . . . [W]ithout further penetration into the realm of the very small, the agony this time may not end in a triumph of human reason. That is what is at stake. (Oppenheimer in Yuan 1965, 5; my emphasis)

To the best of my knowledge, Oppenheimer never publicly stated his views regarding the possibility of a "final theory." Given his Sanskrit and Eastern predilections, it is likely he would have been receptive to the views Yang expressed at the 1961 MIT celebration. He did not believe—and so stated—that one could reconstruct the world from the knowledge of a foundational theory in physics. Chemistry, biology, and psychology are all different realms of nature.

Einstein and Oppenheimer had differing views regarding unification. As noted in the previous chapters, they were committed to different metaphysics. Einstein by virtue of education and intellectual milieu had to address Kant and his legacy,[12] and he came to disagree with many of the Kantian concepts. On various occasions he indicated his affinity with Leibniz's notions of a pre-established harmony, and he expressed his admiration of Spinoza's views of determinism and the deity (see Cassidy 1995).[13] Oppenheimer, by virtue of his education and intellectual milieu, was drawn to pragmatism, operationalism, and Bohr's brand of positivism. In the 1930s, like Bohr, Oppenheimer believed that the resolution of the difficulties encountered in relativistic quantum field theory needed a conceptual revolution of the same magnitude as quantum mechanics. But by the 1950s, with the successes of the renormalization program and the plethora of new particles that cosmic rays and the new accelerators were revealing, he viewed the issues differently. It was clear to him that the insights of quantum mechanics and of quantum field theory into the dynamics of the atomic and subatomic world—uncertainty principles, vacuum fluctuations, relations between energy scales and distance scales, new symmetries and ontologies—were such that the next unification had to be expressed within a quantum mechanical framework. Einstein, on the other hand, believed that the nonlinearity of the general theory of relativity would yield solutions that corresponded to localized energy-momentum distributions in space-time, that is, particle-like solutions, and that the interactions between such "particles" would be determined by the gravitational field equations. He also hoped that the unification of electromagnetism and gravitation would similarly yield particle-like solutions endowed with electric charge. And he even believed that this approach might yield the quantum mechanical description, with its statistical feature emerging from an "averaging" of the space-time description.

For Oppenheimer, their contrasting views were irreconcilable because of the new empirical data that had been obtained from cosmic ray experiments and high energy accelerators. The chasm went even deeper, for Oppenheimer was not convinced that Einstein's formulation of general relativity was necessarily the correct theory of gravitation. In the early 1960s Oppenheimer thought that Carl Brans and Robert Dicke's extension on general relativity might be a better theory. It is to this issue that we now turn.

## A Bird's-Eye View of General Relativity, 1915–1960

The reception and reaction to general relativity can be better understood in terms of the following rough periodization of twentieth-century theoretical physics:[14]

1. The period from 1900 to 1927, during which the subdiscipline of theoretical physics became established. This "glorious period" witnessed three great revolutions: special relativity, general relativity, and quantum mechanics.

2. The period from 1927 to 1940, which saw the development of quantum chemistry, quantum mechanical models to describe phenomena and processes in nuclear physics, and quantum field theory, the quantum mechanical description of systems with an infinite number of degrees of freedom.

3. The period from 1945 to the mid-1970s, which culminated with the establishment of relativistic quantum field theory as the foundational formalism for the representation of the microscopic and submicroscopic world. Its end point is marked by the formulation of the standard model, the solution of the phase transition problem, Kenneth Wilson's renormalization group methods, the reinterpretation of the renormalization procedures developed by Feynman, Schwinger, and Dyson in the late 1940s by taking the introduction of cut-offs as physical, reflecting the lack of knowledge of the physical processes and dynamics at very high energy or very short distances, and Anderson's challenge to reductionism with his landmark article, "More Is Different," in 1972.[15]

4. The period from the mid-1970s to the present, which is characterized by the changed character of theoretical physics. On the one hand, string theory was developed and touted as indicative that a "theory of everything" is possible; yet string theorists are in search of possible observable consequences of the theory. On the other hand, it is a prevalent and widely accepted view that all known physics theories—for example, the standard model,

general relativity—are "effective" field theories valid in certain domain. Thus certain condensed matter theorists have claimed that they are already in possession of a "theory of everything" for condensed matter phenomena, and assert that that domain of physics has been "finalized" in the sense that a robust, stable *foundational* theory of remarkable accuracy has been secured and that the "elementary entities"—electrons, nuclei, and photons— that populate that domain are known and their properties ascertained (Laughlin and Pines 2000).

This change in the character of physics reflects a deep cultural change. Paul Forman, in a lengthy and influential article, has proposed that "the abrupt reversal of culturally ascribed primacy in the science-technology relationship—namely, from the primacy of science relative to technology prior to circa 1980, to the primacy of technology relative to science since about that date"—be taken "as a demarcator of postmodernity from modernity: modernity is when 'science' could, and often did, denote technology too; postmodernity is when science is subsumed under technology" (Forman 2007, 1). In his article Forman also noted that a simultaneous transvaluation had occurred:

> In modernity, the cultural rank of science was elevated by that epoch's most basic cultural presuppositions—not merely the presupposition of the superiority of theory to practice, but more importantly the elevation of the public over the private and the disinterested over the interested, and, more importantly still, the belief that the means sanctify the ends, that adherence to proper means is the best guarantee of a "truly good" outcome. Today, on the contrary, technology is the beneficiary, and science the maleficiary, of our pragmatic-utilitarian subordination of means to ends, and of the concomitants of that predominant cultural presupposition, notably, disbelief in disinterestedness and condescension toward conceptual structures. (Forman 2007, 2)

That something dramatic happened in the science/technology relationship during the 1970s has been widely observed. It has manifested itself in universities becoming entrepreneurial,[16] in the restructuring of the intellectual agenda of their departments, and in a further partitioning of universities into those whose primary function is teaching and those where research is also carried out.[17]

There is little doubt that a deep change occurred. The explanations for the change have thus far principally been concerned with political and economic factors, with less attention paid to cognitive factors internal to the various scientific and engineering disciples. The fact is that in the physical sciences in the 1970s advances in quantum field theory justified a view of the world that segmented it into different levels—the submicroscopic, the nuclear, the atomic and molecular, and the macroscopic. Each having its own set of "elementary" entities (quarks, gluons, in the submicroscopic; electrons, photons, nuclei, in the atomic and molecular), and each having a foundational theory—its "effective" theory—that describes the dynamics of the entities that populate that level. These "effective" (field) theories explicitly take into account the range of energies in which the described systems can be probed, and therefore the accuracy with which they can be represented. The excluded high energy, small-distance effects are taken into account by appropriate local interactions that are calibrated by experimentally determined parameters. What is all important is that this highly accurate and stable description of the microscopic world is *not* destabilized by the incorporation of new small-distance effects provided by higher energy results.[18]

The implication is that a form of finalization has occurred in the description of atoms, molecules, and condensed matter. The people working in these fields are principally concerned with the creation of novelty—entities or effects that did not previously exist in the world—and not with establishing the foundational theory that governs the interactions and determines the evolution of the structures that populate that domain. Thus except for a very small component of the practitioners in these fields, the intellectual agenda is set by external factors, by the demands of specific novelty, by usefulness or efficiency, or by expectations, as is the case with nanotechnology, photonics, and quantum computation. Moreover, it is difficult to differentiate these activities from applied science and in many cases from technology.[19]

At the center of all these developments stands the computer. All the disciplines—physics, chemistry, biology—seek to model the systems they investigate or design using computers. For the most part, their collective aim is to build codes and algorithms that mimic real physical and biological processes. Physics, chemistry, biology, and the entities, objects, and "things" they model, design, and create will have an independent reality *in silico,* that is, in the silicon chips that are an integral parts of the ever-present computers.

The generations trained or coming of age during the periods outlined earlier reflected their differing contexts. The generation of Bohr, Born, Debye, Einstein, Ehrenfest, Ewald, Fermi, Langevin, Schrödinger, Sommerfeld, and von Laue was the generation that matured during the first period. These physicists made theoretical physics a discipline. Einstein is of course the towering theorist of that generation, the off-scale contributor to the advances in the field, first by explaining gravitation as a property of the space-time continuum and second by pointing out the universality of the quantum of action, and giving specific examples of its effects in physical systems in thermal equilibrium, for example, as manifested in the specific heat of solids at low temperature, in the low temperature properties of gases made up of atoms with zero or integral spin quantum numbers, and in particular, the strange properties of helium at very low temperature.

The second period is dominated by the generation that includes Pauli, Dirac, Heisenberg, and Jordan, and the theorists who were trained just slightly before or slightly after the advent of quantum mechanics: Bethe, Bloch, Bronstein, Gamov, Goppert-Mayer, Landau, London, Oppenheimer, Peierls, Rosenfeld, Solomon, Tomonaga, Weisskopf, Wentzel, Wigner, and Yukawa. Oppenheimer can be taken as a paradigmatic figure for this second period. As compared to some of the other leading theorists of that generation, he was more drawn toward problems in nuclear physics, cosmic ray physics, and astrophysics by virtue of his association with Lawrence's Radiation Laboratory in Berkeley and the experimental nuclear physics research group and the astronomers at Cal Tech.

Oppenheimer is emblematic of both the accomplishments of that generation and the limits of its accomplishments. Relativistic quantum field theory was riddled with divergence difficulties, and only a very partial solution to these problems was adumbrated. Nonetheless, during the 1930s, Oppenheimer embodied the power that quantum mechanics and special relativity had given physicists by virtue of their understanding, control, and mastery of the atomic and nuclear realms. World War II demonstrated how that knowledge could be translated into the design and construction of macroscopic gadgets.

One of Oppenheimer's most creative research efforts during the 1930s was the work he did with his students Michael Volkoff and Hartland Snyder on the evolution of massive stars that did not become white dwarfs after they had exhausted their thermonuclear sources of energy. Under those conditions, gravitation would induce a collapse of the star to dimensions and densities where general relativistic effects become pronounced. They found that, as seen by a distant observer, general relativity predicted that the star would asymptotically shrink to its Schwarzschild radius as it closed itself off from the rest of the universe except for its intense gravitational field. (Oppenheimer and Volkoff 1939; Oppenheimer and Snyder 1939; Hufbauer 2005). They had in effect discovered what John Archibald Wheeler later called black holes. But the importance of their work was not recognized at the time, neither by them nor by the astrophysical community at large, because it could not be observationally corroborated and because with the outbreak of World War II other problems took precedence with physicists. It is now known that black holes exist, ranging in size from a few solar masses to a few billion, and play an important role in the economy of the universe. It is not clear whether Oppenheimer ever communicated his results to Einstein at the time.

This episode points to a characteristic of the chronicle of general relativity from 1915 to 1965. In terms of the periodization given above, general relativity thrived from 1915 until 1927 (period 1), was dormant during period 2 except for the work of Oppenheimer and his students on black holes and some important work by Einstein and Infeld on the equations of motion of point masses in general relativity, and acquired new vigor and importance during period 3.

The period from 1915 to about 1925 demarcates the first phase of research activities in (classical) general relativity (GR) on the part of physicists and mathematicians after Einstein's papers had appeared. Its high point is of course the Dyson-Eddington observation of the bending of light by the sun. The publication of Pauli's *Relativitätstheorie* article (Pauli 1922) in the *Encyclopedie der Mathematische Wissenschaften,* Herman Weyl's magisterial *Raum, Zeit, Materie* (Weyl 1922a), and Eddington's *The Mathematical Theory of Relativity* (Eddington 1924) established GR as a subdiscipline of theoretical physics. However, the number of people actively working in the field never exceeded a few dozen or so.

The advent of quantum mechanics and its applications to solid state physics, to nuclear physics, and to cosmic ray phenomena in its special relativistic field theoretic extension shifted the interest away from GR during the 1930s. The absence of further experimental tests, beyond the three classic ones Einstein had initially formulated—the gravitational redshift of spectral lines, the secular motion of the perihelion of planetary orbits, and the deflection of light rays by a gravitational field—together with the paucity of accurate new data were also factors in the lack of further attention being devoted to GR, though everyone agreed that GR was a most beautiful theory and an amazing creation of the human mind. The reawakening of interest in the field can be dated to the meeting in Bern on July 11–16, 1955, celebrating the fiftieth anniversary of the creation of special relativity and concerned exclusively with topics in special and general relativity. Pauli was the chair of the organizing committee,[20] and André Mercier its secretary. Mercier communicated to Einstein the plans for the meeting and hoped he might be able to attend. Replying to Mercier, Einstein expressed his joy and gratitude for what appeared to him to be a very promising conference, adding: "It will demonstrate, that the expectations, that were bound to the general principle of relativity, were extraordinarily varied. This is good, because among us researchers the philosophical saying that 'War is the father of all things' [did not entail] the fatal smack [Beigeschmack], that usually sticks to it" (Mercier and Kervaire 1956, 10).

Einstein died in April 1955 and the meeting thus became a gathering at which to come to terms with the fact "that Einstein is no longer with us" and an occasion to bid "farewell to the man."[21] Over eighty people from twenty-two countries attended the conference. Two years later the first of the General Relativity Conferences was held in Chapel Hill organized by Bryce de Witt.[22] Its theme was "The Role of Gravitation in Physics." By the end of the decade, one could point to several "schools" of general relativity: John Archibald Wheeler at Princeton University,[23] Peter Bergmann at Syracuse University,[24] Leopold Infeld in Warsaw,[25] and Bryce de Witt at North Carolina University, as well as various individual workers such as Herman Bondi, Jürgen Ehlers, Vladimir Fock, Henri Lichnerowitz, Christian Møller, Achile Papapetrou, Felix A. E. Pirani, Roger Penrose, and Marie-Antoinette Tonnelat. There thus came into being a community of researchers in general relativity addressing a common set of problems. The end of the decade was marked by new insights—by Martin Kruskall, George Szekeres, and Tulio Regge—into the Schwarzschild solution of the GR equations, which paved the way for the modern version of black hole theory. The early 1960s also saw new experimental developments, by Pobert Pound and Glen Rebka, by Robert Dicke and by Erwin Shapiro, which promised feasible new tests of general relativity and thus stimulated greatly increased activities and new research directions for GR.[26]

The research activities of Dicke, in particular, should be singled out. Dicke was a leading experimenter and theorist at Princeton University who not only had undertaken new experiments to determine the validity of some of general relativity's claims, but also, with one of his graduate students, Carl H. Brans, had formulated a version of general relativity that introduced a new scalar field that simulated possible variations of the (inverse of the) gravitational constant with position and in time.[27] For a few years after its introduction, it was considered a possible challenge to general relativity as proposed experiments by Dicke and Oldenberg of measurements of the oblateness of the sun could discriminate between its prediction and those of GR for the deflection of light in its passage near the sun.

In addition to these developments, the problems connected with developing a consistent formulation of quantum gravity were also receiving attention by some of the leading physicists of the day: Wigner (1957), Schwinger (1962, 1963), Feynman (1963), Wheeler, and some of their students: de Witt, Richard Arnowitt, Stanley Deser, and Charles Misner. The fact that Feynman offered a lecture course on GR and its quantized version at the California Institute of Technology during 1962–1963 helped make GR and quantum gravity respectable fields of investigations for graduate students elsewhere. In connection with his course, Feynman had studied the quantization of the Yang-Mills field theory and the difficulties encountered in quantized GR. He wrote a paper on the subject that proved to be very influential, for it focused on the particular problems that arose in quantizing gauge theories such as Yang-Mills theory and general relativity. Ryoyu Utiyama (1956), who was a visiting member of the Institute for Advanced Study during the academic year 1955–1956, had written an important paper while there that clarified both the Yang-Mills theory and general relativity as gauge theories.[28]

By the late 1960s, general relativity was abuzz with activity and acquired a new status within the physics community, as can be gauged by the fact that in 1968 the first half of the Brandeis Summer School in Theoretical Physics was devoted exclusively to general relativity and was attended by over 100 graduate and postdoctoral students![29]

Oppenheimer was certainly aware of all these developments, for activities at Princeton University and at the Institute were central to the resurgence of interest in GR.

## Epilogue

As I have repeatedly indicated, Einstein did not believe that quantum mechanics charted the correct path to a deeper understanding of nature. After he intimated, and later was able to give a mathematical proof, that the equations of motions of particle-like concentrations of energy and momentum were determined by the nonlinear field equations of general relativity,[30] Einstein's passion—if not his obsession—became how to formulate an approach to a unified description of physics based on general relativity's metaphysical conception of the physical world. And the simplest case to address was the unification of electrodynamics and gravity, in the hope of explaining the properties of electrons. Pauli once remarked that if he could have presented Einstein with a synthesis of his theory of general relativity and the quantum theory, then the discussion with him regarding quantum mechanics would have been considerably easier and seminal. However, since general relativity and quantum mechanics seemed incompatible, Einstein believed that he could ignore advances made in the quantum mechanical description of the microscopic and submicroscopic domains.

In the 1970s it was established that the strongly interacting particles, the hadrons (protons, neutrons, $\Lambda$s, $\Omega$s, and mesons), are "composed" of more basic building blocks (the quarks) and that the forces between the hadrons can be accounted for by the exchange of gauge bosons (gluons) between their constituent quarks; also, that the weak and electromagnetic interactions can be understood in terms of the exchange of gauge bosons ($W^{\pm}$, $W^0$, $\gamma$) between the leptons (electrons, muons, and $\tau$s and their associated neutrinos) and the quarks. The fact that all the forces between the presently conceived elementary entities (quarks, leptons, and gauge bosons) were mediated by gauge quanta described by gauge fields led to a unified description of all these forces, now known as the standard model. One of the consequences of non-abelian gauge theories—asymptotic freedom, the fact that the interactions at very high energy, very small distances vanish, the "particles" behaving like free particles—made it possible to explain the observed large difference in the strength of the strong and the electroweak force as a low energy phenomenon and to show that all forces could have a common origin at very high energy, near the Planck scale of $10^{-33}$ cm. Furthermore, the energy at which the forces described by the standard model appear to unify is approximately that at which gravity becomes equally strong. This is an indication that the next stage of unification should include the amalgamation of the nongravitational forces and gravity. To accomplish this, it is now assumed that Einstein's general theory of relativity is only an "effective theory," valid at large distances, to be replaced by a more fundamental theory at the Planck scale of $10^{-33}$ cm.[31]

String theorists believe that string theory is such an extension of general relativity, although this remains to be seen. Whether or not it fulfills their dreams, it is clear that it has already yielded remarkable new insights into pure mathematics—a modern illustration of the "preestablished harmony between physics and mathematics" that Einstein believed in.

Whatever the outcome—the standard model that gives a unified description of the strong and electroweak interactions at energies below 1 trillion electron volts (1 TeV), the highest energy achievable on a presently constructed accelerator; string theory; or whatever new theory will be corroborated as a result of experiments on the Large Hadron Collider or observations made with the space probes, such as the Gamma Ray Large Area Space Telescope—all attest to Einstein's (and Kant's) belief that "[t]he eternal mystery of the world is its comprehensibility" (Einstein 1954, 292).

# CHAPTER SIX

# Einstein, Oppenheimer, and the Meaning of Community

> I sometimes dream of situations that can't possibly come
> true. I audaciously imagine, for example, that I get a
> chance to chat with Ecclesiastes, the author of that
> moving lament on the vanity of all things. I bow very
> deeply before him, because he is one of the greatest poets,
> at least for me. Then I grab his hand. "There's nothing
> new under the sun": that's what you wrote, Ecclesiastes.
> But you yourself were born new under the sun. And the
> poem you created is also new under the sun, since no one
> wrote it down before you.
>
> —*Wistawa Szymborska, Nobel Speech, 1996*

## The Einstein-Oppenheimer Interaction

Oppenheimer first met Einstein in January 1932 when Einstein visited Cal Tech during his around the world trip of 1931–1932.[1] In January 1935, Oppenheimer went east to attend a meeting of the American Physical Society. While in New York he took the opportunity to go to Princeton to visit the Institute for Advanced Study to which he had been invited to spend a year as a visiting member. He conveyed his reaction to the occupants of Fine Hall—the building where both the mathematics department of Princeton University and, until 1938, the Institute for Advanced Study were located—in a letter to his brother Frank:

> Princeton is a madhouse: its solipsistic luminaries shining in separate &
> helpless desolation. Einstein is completely cuckoo.... I could be of ab-
> solutely no use at such a place, but it took a lot of conversation & arm
> waving to get Weyl to take a no. (Smith and Weiner 1980, 190)

Undoubtedly, Oppenheimer's strong reaction was due to the contrast between Einstein's and his own working style, their sharply differing attitudes to quantum mechanics, and their conflicting views of what constituted the important contemporary problems in physics. As previously noted, although he readily conceded its extraordinary success, Einstein never accepted quantum mechanics as *the* theory demarcating the path to a more fundamental understanding of the physical world.

Einstein's working style is well known. In both in his personal life and in his scientific activities he became what is described as a loner.[2] In an 1931 essay entitled "The World as I See It," Einstein said of himself:

> I am truly a "lone traveler" and have never belonged to my country, my home, my friends, or even my immediate family, with my whole heart; in the face of all these ties, I have never lost the sense of distance and a need for solitude—feelings which increase with the years. No doubt, such a person loses some of his innocence and unconcern; on the other hand, he is largely independent of the opinions, habits, and judgments of his fellows and avoids the temptation to build his inner equilibrium upon such insecure foundations. (Einstein 1954, 9)

His description of himself in 1931 as a lone traveler is interesting inasmuch he then was at the center of a vast network of correspondents. As attested by his *Collected Works,* by 1907, he was already extensively corresponding with Wilhelm Wien, Johannes Stark, Max Planck, Rudolf Ladenburg, Arnold Sommerfeld, and Max von Laue. By 1909 the list of correspondents grew to include Alfred Bucherer, Hendrik Lorentz, Heike Kamerlingh-Onnes, Philipp Lenard, Ernst Mach, Marcel Grossmann, Jean Perrin, Paul Ehrenfest, Fritz Haber, Walther Nernst, Walter Schottky, Marian von Smolukowski, Erwin Freudlich, David Hilbert, and Ernst Zermelo. After the formulation of general relativity, the list was extended yet again to Tullio Levi-Civita, Karl Schwartzschild, Willem de Sitter, Otto Struve, Otto Stern, Hermann Weyl, Constantin Caratheodory, Gustav Mie, Roland Eötvös, Felix Klein, and Arthur Eddington. This list comprises the most eminent physicists, mathematicians, astronomers, and philosophers of the day! And as the handlist of the Einstein Archives indicates, the circle of correspondents and the number of letters to and from Einstein kept on increasing as time went on.

Einstein's self-description as a "lonely traveler" in the sense indicated in the 1931 essay was true during his days in Berlin if it refers to his political stance and his directorship of the Kaiser Wilhelm Institute for Physics in the early 1920s.[3] But this was not the case in his scientific activities. In his student days at the ETH in Zurich, he had developed close friendships with fellow students Marcel Grossmann and Michele Besso and with Mileva Marić, whom he later married. When Einstein worked at the Patent Office in Bern, he developed close, intense interactions with Maurice Solovine and Conrad Habitch in what they called the Akademie Olympia, at whose meetings philosophical issues were thrashed out (Feuer 1974; Fölsing 1997, 99–100). During the *annus mirabilis* of 1905, Einstein had lengthy discussions with Michele Besso, then his colleague at the Patent Office, about the problems he was struggling with. His collaboration with Marcel Grossmann working on the mathematical representations of his insights into the relation between gravitation, general covariance, and geometry, and later his collaboration with Jakob Grommer, were marked by their closeness and length. Also, after Einstein settled in Berlin, there were trips to Holland and meetings with Paul Ehrenfest and his students in Leyden and with Lorentz in Utrecht. And during the 1920s there were discussions with the bright young men who came to Berlin to be where the action in physics was—Szilard, von Neumann, Wigner. But Einstein lectured only intermittently, created no school, and had but one Ph.D. student.[4] There were also the extremely stimulating sessions of the Prussian Academy, which Einstein attended faithfully, as well as the various meetings of the Solvay Congress in which he participated until the early 1930s.[5] But most such interactions ceased after Einstein came to Princeton. He attended the Institute and university physics seminars and colloquia but rarely, and he became intellectually isolated from the physics community.[6] After he came to the Institute, his research was primarily carried out with his assistants, Banesh Hoffman and Leopold Infeld, on the relation of the equations of motion of point particles to the field equations in general relativity, elaborating earlier results with Grommer;[7] with Walter Mayer, Ernst Stauss, Peter Bergmann, Valia Bargmann, and Bruria Kaufman, on unified field theory; and with Boris Podolski and Nathan Rosen, on the completeness of the quantum mechanical description of microscopic phenomena and related philosophical problems such as realism.[8]

His being a "loner" took on a new meaning in the United States. Thus in a letter in 1937 to his friend Otto Juliusburger, a psychiatrist and a Spinoza and Schopenhauer scholar, Einstein wrote:

> I still struggle with the same problems as ten years ago [the unification of electromagnetism and gravitation]. I succeed in small matters but the real goal remains unattainable, even though it sometimes seems palpably close. It is hard and yet rewarding: hard because the goal is beyond my powers, but rewarding because it makes one immune to the distractions of everyday life.
>
> I can no longer accommodate myself to the people here and their way of life. I was already too old to do so when I came over, and to tell the truth it was no different in Berlin, and before that in Switzerland. One is born a loner, as you will understand being one yourself.[9]

Einstein's "lonely" path is to be contrasted with that of Oppenheimer who, at Berkeley in the 1930s, created the most important *school* of theoretical physics in the United States.

But if Einstein was isolated from the scientific community in Princeton, he did maintain fairly close social and intellectual links with historian Erich Kahler, novelist Hermann Broch, philosopher Paul Oppenheim, art historian Erwin Panofsky, and their families. He made his voice heard in the running of the Institute—to the extent that Abraham Flexner, its authoritarian director, allowed anyone to have a voice in the running of the Institute (Batterson 2006). He evidently also had various affairs.

The differences in Einstein's and Oppenheimer's working style were made explicit in Oppenheimer's radio address on the occasion of

Einstein's sixtieth birthday in 1939, an address briefly referred to in the Introduction. He began with laudatory remarks:

> Few men have contributed so much to our understanding of the Physical World, to our ability to predict and follow and control its behavior. And we see in Einstein, especially those who have come to know him a little, all those personal qualities that are the counterpart of great work: selfless-ness, humor, and deep kindness.

He went on to declare:

> But if few scientific workers would quarrel with the fact that Einstein is in many ways a perfect symbol of their work, there are many who would feel that there is something a little false and fabulous in the way that he is thought of.... [T]here is ... a general impression, supported in part by his eminence, that his work has been qualitatively different from that of his fellow workers; that it is abstruse, and remote, and useless. This seems to me a very strange ground for admiration. And of course it is not true; and the truth is much better than the false.
>
> All discoveries in science grow from the work, patient and brilliant, of many workers. They would not be possible without this collaboration; they would not be possible without the constant technological develop-ments that are necessary to new experiments and new scientific experi-ence. One may even doubt whether in the end they can be possible except in a world which encourages scientific work, and treasures the knowledge and power which are its fruits.

Interestingly, Oppenheimer then remarked that one of the most spec-tacular projects of contemporary atomic physics was to tap the sources of the sun's energy—a project made possible by Einstein's early work on the theory of relativity—and then added

> that in this as in every case one would find that the development of Ein-stein's discovery was so closely interwoven with the work of countless other scientists and technicians ... that we would come away [from an analysis of the development] with a deepened conviction of the coopera-tive and interrelated character of scientific achievement.[10]

As remarked earlier, Einstein and Oppenheimer came closest in questions of physics in 1939. Although Oppenheimer's focus had been principally on quantum mechanics and quantum field theory, his contacts with the astronomers at Cal Tech brought him closer to Einstein's concerns. Through general relativity, each became enwrapped in his own way in what Feynman had presciently called "the historical question of physics" (Feynman 1963, chaps. 3, 9). In the late 1930s, after having unsuccessfully tackled the problem of energy production in stars, whose solution was given by Bethe in 1938, Oppenheimer concerned himself—if not with the cosmological problem, though deeply interested in it—with a historical question: namely, what happens to heavy stars when they exhaust their nuclear fuel and collapse under gravitational attraction? With his students, George Volkoff and Hartland Snyder, he discovered that general relativity predicted the existence of strange, dense objects: "black holes."[11] Einstein, for his part, was hostile to the idea of black holes, as he did not believe that any solution of the general relativistic field equations that contained a singularity was physically acceptable. In one of his last scientific publications he explained why not:

> A field theory is not yet completely determined by the system of field equations. Should one admit the appearance of singularities? ... [I]t is my opinion that singularities must be excluded. It does not seem reasonable to me to introduce into a continuum theory points (or lines etc.) for which the field equations do not hold. Moreover, the introduction of singularities is equivalent to postulating boundary conditions (which are arbitrary from the point of view of the field equations) on surfaces which closely surround the singularities. Without such a postulate the theory is much too vague. (Einstein 1955, 164)

Nor did Einstein ever contemplate the physical possibility of black holes. According to Freeman Dyson, Oppenheimer, too, in later life was uninterested in black holes—although Dyson thought that, in retrospect, they were his most important contribution to science (Dyson 1995).

When in 1945 Einstein retired from his professorial position at the Institute, the question of who should fill the vacancy came up. Both

Pauli and Oppenheimer were considered, and Einstein and Hermann Weyl wrote a letter to the faculty recommending Pauli's appointment. They noted that "Oppenheimer has made no contribution to physics of such fundamental nature as Pauli's exclusion principle and [the] analysis of electron spin" (Regis 1987, 136).

Einstein certainly admired Oppenheimer for his drafting and promulgating the Acheson-Lilienthal plan, and thereafter for his efforts to contain the arms race between the United States and the Soviet Union. However, he couldn't understand why Oppenheimer had such a thirst for the corridors of power in Washington. He couldn't understand why Oppenheimer didn't join the Emergency Committee of Atomic Scientists and why in the spring of 1947 he refused to be the dinner speaker at one of its fund-raising events.[12]

In the fall of 1947, when Oppenheimer became the director of the Institute for Advanced Study and a professor of physics there, he and Einstein became colleagues.[13] Einstein acquired a deep respect for the new director as an administrator and described him as an "unusually capable man of many sided education" (Fölsing 1997, 784). On occasion they saw one another socially. Thus David Lilienthal told of his having dinner at the Oppenheimers one evening in 1948, sitting next to Einstein and watching him "as he listened (gravely and intently, and at times with a chuckle and wrinkles about his eyes) to Robert Oppenheimer describing neutrinos as 'those creatures', and the beauties of physics" (Lilienthal 1964, 298). Bird and Sherwin (2005, 381) tell of Oppenheimer having an FM antenna installed on the roof of Einstein's house at 112 Mercer Street and giving Einstein a new FM radio as a birthday gift so that he would be able to listen to the New York Philharmonic concerts that were broadcast from Manhattan some 50 miles away from Princeton.

However, Einstein's and Oppenheimer's interactions became colored by two separate considerations, one intellectual, the other political. After his own brief excursion into the field, Oppenheimer did not believe that either general relativity itself—because of lack of observational data— or quantized versions thereof—because of conceptual and technical difficulties—were areas of physics that would reward theorists' efforts. He went so far as to discourage, if not forbid, people from taking up these problems while at the Institute. Thus in October 1948, after he had just arrived at the Institute, Freeman Dyson could write to his parents:

> The general theory of relativity is one of the least promising fields that one can think of for research at the present time. It is the general view of physicists that the theory will remain much as it is until there are either some new experiments to upset it or a development of quantum theory to include it.[14]

At the same time, the interactions between Oppenheimer and Einstein became polarized by the political context of the times. In 1947 the House Un-American Activities Committee (HUAC) began investigating his former students, Bernard Peters, Joseph Weinberg, Rossi Lomanitz, and David Bohm.[15] His own brother, Frank, who had been a member of the Communist Party, was likewise very probably going to be called before the committee (which he was). With the rise of Senator Joseph McCarthy's influence, Oppenheimer felt vulnerable in view of his and his wife's left-wing past. Given his desire to continue his involvement in governmental matters despite the anticommunist hysteria and the repressive atmosphere that McCarthy had generated, Oppenheimer felt obligated to distance himself from Einstein, who had made clear and evident his opposition to the HUAC's activities, his revulsion of McCarthyism, and his commitment to the protection of civil liberties.[16] Oppenheimer undoubtedly also felt under great pressure to insulate the Institute from charges that it was harboring fellow travelers. Einstein had made his political stand manifest as a matter of principle. Thus he had been prominently in the news protesting the "inquisitions" of the HUAC, and the activities of Senator McCarthy's and Senator Jenner's Senate subcommittees. As noted in the Introduction, in December 1953, Einstein had advised Al Shadowitz, an electrical engineer working on government contracts, who had been subpoenaed by Senator McCarthy to appear before his subcommittee. Shadowitz, on the advice of Einstein, instead of pleading the Fifth Amendment as his defense for not answering questions posed to him, pleaded the First Amendment. This led the *New York Times* on December 16, 1953, to prominently display on its front page a photo of Shadowitz and to feature an article with the headline: "WIT-NESS, ON EINSTEIN ADVICE, REFUSES TO SAY IF HE WAS RED." Similarly, a few months earlier, on June 12, 1953, Einstein had made headlines in the *New York Times* for the advice he had given to William Frauenglass, a high school English teacher, who had been called before the Senate Internal Security Subcommittee and had refused to be a cooperative witness, even though it meant the loss of his job. Frauenglass had asked Einstein for a letter of support. Einstein complied and agreed to have the letter sent to the *New York Times*. In his letter Einstein advised civil disobedience, and his stand merits repetition:

> What ought the minority of intellectuals do against this evil? Frankly, I can only see the revolutionary way of non-cooperation in the sense of Gandhi's. Every intellectual who is called before one of the committees ought to refuse to testify, i.e., he must be prepared for jail and economic ruin, in short for the sacrifice of his personal welfare in the interests of the cultural welfare of his country.[17]

Einstein was clearly prepared to follow this course of action and be incarcerated. Though he surely must have known that his iconic standing would cushion whatever steps he took, the fact is that he was ready to sacrifice his personal welfare in the interests of the cultural and political welfare of his adopted country.

In May 1954, during his security hearing, Oppenheimer received a request from Sol Stein, the executive director of the American Committee for Cultural Freedom (ACCF), the American branch of the Congress for Cultural Freedom, of which Oppenheimer was a member, to persuade Einstein not to attend the meeting of the Emergency Civil Liberties Committee that was to honor him on his seventy-fifth birthday. The ACCF believed that the Emergency Civil Liberties Committee was a communist front organization. In his letter, Sol Stein had informed Oppenheimer that he had received urgent telephone calls from leaders of the American Jewish community, "who are very concerned lest Dr. Einstein be sucked into another Communist-inspired occasion. Such an occasion will again tie up Judaism with Communism . . . [and] will help to spread the notion one hears so often nowadays about scientists being political-babes-in-the-woods."[18] Oppenheimer took time out from his appearance before the Gray Board to persuade Einstein not to participate in the gathering.

Further insight into the relationship between Oppenheimer and Einstein is obtained from an exchange of letters between Oppenheimer and Abraham Tulin. Early in 1954 Oppenheimer had been elected a member of the Board of Governors of the Technion Institute of Technology in Haifa. In August 1953 the Technion had decided to confer honorary degrees on Albert Einstein and James Franck, and the two of them had accepted. Since neither Einstein nor Franck was able to travel to Israel to receive the degree in person, Tulin, the chairman of the convocation committee, asked Oppenheimer's advice on where to hold the convocation and "begged" his active participation in the ceremony. Oppenheimer gave a "negative reply" to the request. Even though it was decided to hold the convocation in Princeton, Oppenheimer at the last minute refused to be present at the ceremony and informed the convocation committee that he had to be out of town the day the convocation was to take place.[19]

These were difficult times for Oppenheimer. The ceremony took place in October 1954, and Oppenheimer was still reeling from the blows he had received during his encounter with the Gray Board and from the ruling by the Atomic Energy commissioners that denied him his security clearance. But his note to the gathering is revealing. Although the printed invitation indicated that honorary degrees were to be conferred to Dr. Albert Einstein and Professor James Franck—in that order—in his message to the convocation Oppenheimer inverted the order and wrote:

> I send my greetings to this convocation. We honor today the great contributions to science and insight we owe to Dr. Franck and Dr. Einstein. We express as well the sense of community between institutions and scholars throughout the world dedicated to the advancement of knowledge.[20]

## Eulogies and Memorial Speeches

A few days after Einstein's death on April 18, 1955, Oppenheimer wrote a eulogy for the *Packet,* the local Princeton newspaper. In it he called Einstein

> one of the great of all time. We live today with physics that he first saw in the golden twenty years from 1905 to 1925. He helped bring into being the quantum theory. . . . He made the special Theory of Relativity [and in doing] so, he changed our understanding of measurements of space and time. . . . In the general Theory of relativity he created perhaps the single greatest theoretical synthesis in the whole of science, giving us a new understanding of the universality of gravitation and a new view of the cosmos itself. Unlike most discoveries in science, Einstein's general theory could well have lain undiscovered but for his genius.[21]

Oppenheimer went on to say that those who had gotten to know him could confirm that the popular image of Einstein as a simple, kindly man, "with warm humor, . . . wholly without pretense" was accurate. Einstein was indeed always ready to help those in trouble, and persistently and relentlessly called attention to the plight of the oppressed and the unfortunate. All his life he protested the abuse of power and authority. And in concluding Oppenheimer noted that

> Even above his humanity and kindliness, even above his immense analytical power and depth he had a quality that made him unique. This was his faith that there exists in the natural world an order and a harmony and that this may be apprehended by the mind of man. He left us not only the greatest contribution in evidence of that faith, but the heritage of that faith itself.[22]

The eulogy in the *Packet* was meant for the citizenry of Princeton. Oppenheimer also wrote a eulogy for the physics community that appeared on page 1 of the January 1956 *Reviews of Modern Physics.* There he gave a thoughtful, tender account of Einstein's life and accomplishments. He characterized succinctly and respectfully the last few decades of Einstein's life as follows:

With single-mindedness, he turned his attention to the discovery of what for him would have been a basic and satisfying account of the atomic nature of matter. This was the program of the unified field. Here he sought to generalize the matter-free field equations of general relativity so that they might also account for electromagnetic phenomena. He sought equations whose solutions would correspond to local aggregations of mass and charge, and whose behavior would resemble the atomic world so well described by quantum theory. He was hard at work on this program until his death. (Oppenheimer 1956b)

But in private, Oppenheimer made much less complimentary statements: that Einstein had no understanding of or interest in modern physics and that he had been wasting his time in trying to unify gravitation and electromagnetism. Furthermore, Oppenheimer complained that even though the Institute had supported Einstein for twenty-five years, he did not leave any of his papers to the Institute in his will—everything was to go to Israel.[23]

Given Oppenheimer's commitment to quantum mechanics and to the interpretation of its formalism based on Heisenberg's uncertainty principle and Bohr's complementarity principle, it is not surprising that in his letter to his brother Frank in 1935 and in his 1939 address, Oppenheimer was critical—and in the letter scoffing—of Einstein. If in his public radio address 1939 and in his public eulogies his criticism was subdued and respectful, it was explicit in the talk Oppenheimer gave in 1965 to commemorate the tenth anniversary of the death of Einstein. Already ill with the throat cancer that was to kill him a year and a half later, he accepted an invitation from UNESCO to speak in Paris. He went there just for a day to deliver the lecture. There were two other speakers that day, Ferdinand Gonseth and Julian Huxley, with Giorgo de Santillana, Gerald Holton, Werner Heisenberg, Bonifatiĭ Mikhaĭlovich Kedrov, and Jean Piveteau[24] making presentations the next day.[25]

Oppenheimer delivered his lecture in the evening of the first day of the colloquium, December 13, 1965. He justified giving it by noting that he had known Einstein for over thirty years and that after he became the director of the Institute for Advanced Study in 1947 they "were close colleagues and something of friends."

In his presentation he once again addressed the issue of the relationship between individual creativity and community and made explicit some of his critical attitude toward Einstein. Thus in his opening remarks[26] he indicated that he thought "it might be useful, because I am sure it is not too soon, and for our generation it is a little too late, to start to dispel the clouds of myth [that surround Einstein's genius] and to see the great mountain peak that these clouds hide. As always, the myth has its charm; but the truth is far more beautiful."[27] What was true of the myth, Oppenheimer was quick to call attention to, was Einstein's extraordinary originality. "Einstein was a physicist, a natural philosopher, the greatest truly of our time." He invented quanta, and he drew the consequences of the fact that no signal could travel faster than the velocity of light. Although others would surely have come to formulate the meaning of the Lorentz invariance of Maxwell's equations in the way Einstein had done and would have understood the necessity of this viewpoint also for mechanics, "this simple, brilliant understanding of the physics could well have been slow in coming, and blurred had he [Einstein] not done it for us."

Undoubtedly having the Brans-Dicke theory in mind, Oppenheimer continued:

> The general theory of relativity, which even today, *may very well prove wrong*, (my emphasis) no one but he would have done for a long time.[28] It is in fact only in the last decade, the last ten years, that one has seen how a pedestrian and hard-working physicist, or many of them, might reach that theory and understand this singular union of geometry and gravitation, and we can do that today only because some of the "a priori" open possibilities are limited by the confirmation of Einstein's discovery that light would be deflected by gravity.

Oppenheimer then pointed to another facet of Einstein's works—the deep elements of tradition that were embedded in them—and he noted that by studying Einstein's readings, his friendships, his correspondence, and his research notes, it was possible to discover how he came to them and how they were incorporated into his work. Oppenheimer enumerated three such elements, not necessarily the only ones. One was the statistical tradition going back to Maxwell and Boltzmann, who had indicated the connection between the laws of thermodynamics and the mechanics of large numbers of particles. It was this statistical tradition that had led Einstein to the laws governing the emission and absorption of light by atomic systems. It was this that enabled him to see the connections between photons, Wein's law, and Planck's law and later on to see the relevance of the statistics of light quanta proposed by Bose.

The second element was Einstein's "total love of the idea of a field and the following of physical phenomena in minute and infinitely sub dividable detail in space and time. . . . It was this which made him see[29] that there had to be a field theory of gravitation, long before the clues to that theory were securely in his hand."

The third element was a philosophic one, which Oppenheimer characterized as a "form of the principle of sufficient reason." Einstein thought that the principle of sufficient reason required that determinism and causality be necessary ingredients of any thought that could be called rational—hence his stance regarding quantum mechanics. Rationality with its demands for consistency also directed Einstein to ask for care when stating what we mean, to ask for exactitude about what we can measure, and for clarity about what elements in physics are conventions. It was Einstein who recognized and insisted that conventional elements could have no part in the real predictions of physics. Oppenheimer believed that this element followed from the long tradition of European philosophy, "you may say starting with Descartes . . . leading through the British empiricists, and very clearly formulated, though probably without any influence in Europe, by Charles Peirce, a rather erratic American philosopher, that one had to ask how do we do it, what do we mean, is this just something that we can use to help ourselves in calculating, or is it something that we can actually study in nature by physical means?"[30] Moreover, the fact is that the laws of nature not only describe the results of observations, but the laws of nature determine the scope of observations.[31]

Oppenheimer went on to state, however, that in the last twenty-five years of Einstein's life, the years he spent at Princeton, in a certain sense his tradition failed him. "And this though a source of sorrow, should not be concealed. He had a right to that failure." Einstein spent those years first in trying to prove that the quantum theory had inconsistencies in it—and Oppenheimer remarked that "no one could have been more ingenious in thinking up devilish examples,"[32] but the inconsistencies were not there—and eventually Einstein could only say that he didn't like the theory, that he didn't like the abandonment of continuity and of causality. "[T]o see them lost, even though he had put the dagger in the hand of the assassin by his own work, was very hard on him." He also struggled with an ambitious program to amalgamate the understanding of electromagnetism and gravitation in such a way as to explain "what he regarded as the semblance—the illusion—of discreteness of particles in nature."[33] But, Oppenheimer averred, "I think that it was clear then, and believe it to be obviously clear today, that the things that this theory worked with were too meager, left out too much that was known to physicists but had not been known in Einstein's student days. Thus it looked like a hopelessly limited and historically rather accidentally conditioned approach. Although Einstein commanded the affection, and that's not strong enough a word, the love of everyone for his determination to see through his programme, he lost contact with the profession of physics, because there were things that had been learned which came too late in life for him to learn them" (Oppenheimer 1966a, 5). Oppenheimer then commented that Einstein "was in an important way "alone" and "lonely"[34]—though a deep and loyal friend, "the stronger human affections played not a very deep or central role in his life taken as a whole. . . . [H]e did not have in the technical jargon, a school. He did not have very many students who were his concern as apprentices and disciples. In later years, he had people working with him. They were typically called assistants . . . [and] did one thing which he lacked thoroughly in his young days. His early papers are paralyzingly beautiful but they are thoroughly corrupt with errors, and this has delayed the publication of his collected works for almost ten years. A man whose errors can take that long to correct is quite a man."

Oppenheimer then elicited laughter in the hall when he commented that Einstein's fame gave him some very great pleasure, meeting people, but in particular of playing music not only with Queen Elizabeth of Belgium, "but what was more with Adolphe Busch, for he wasn't that good a violinist."

In the rest of his lecture, the tone changed. Oppenheimer extolled Einstein for his goodwill and humanity and for his ever present concern for not doing harm. He described Einstein's attitude toward human problems by the Sanskrit word *Ahinsa*, the Hindu doctrine of noninjury, of not hurting living things. And he commended Einstein for his consequential stand against violence and cruelty wherever he saw it, and that in particular that "after the war he spoke with deep emotion and I believe with great weight about the supreme violence of . . . atomic weapons" and the need to "now make a world government."

With his concluding sentence, "In his last years, as I knew him, Einstein was a 20th century Ecclesiastes saying with unrelenting and indomitable cheerfulness 'Vanity of Vanities, all is Vanity,' " Oppenheimer indicated his empathy with Einstein, who early on had come to appreciate "the nothingness of the hopes and strivings which chase most men restlessly through life." It also reflected the fact that both of them had accomplished tasks at the limit of human capabilities and had received a commensurate degree of fame. And both had to confront their inability to match these heights thereafter, yet seeking what turned out to be futile ways to attempt to do so. As Einstein had once told Oppenheimer, "You know, when it's once has been given to a man to do something sensible, afterwards life is a little strange." And it is probably the case that Oppenheimer, when referring to "Vanity of Vanities," also had in mind the *vanitas* tradition in art, a genre of painting that flourished in the Netherlands in the early seventeenth century. In these paintings objects symbolic of the inevitability of death reminded the viewer of the ephemeral, fleeting quality and vanity of earthly achievements and pleasures. Einstein had evidently heard the bell toll in his last years, and the bell had tolled for Oppenheimer.

Oppenheimer's talk had immediate repercussions. On the day following Oppenheimer's lecture, the *New York Times* covered the story headlined by the statement "Oppenheimer View of Einstein Warm But Not Uncritical." Although the article began with the statement that a warm and admiring but not uncritical portrait of Albert Einstein was drawn by Oppenheimer for an audience of about 1,000, the rest of it highlighted Oppenheimer's most critical statements: that "Einstein's early work was paralyzingly beautiful but full of errors" and that their correction had delayed the publication of Einstein's collected work by ten years; "A man whose errors take ten years to correct is quite a man"; that Einstein founded no school and that he did not have many students; that he did not play a vital role in the development of the atomic bomb; and that his letter to President Roosevelt in which he sought to make him aware of the possibility of a fission bomb "was not important."

Upon reading the *Times*'s article, Abraham Pais, who had been a close associate of Oppenheimer at the Institute, wrote him saying euphemistically that it made him "slightly uncomfortable."[35] In the note Oppenheimer attached to the copy of the full text of his speech that he sent to various friends, Oppenheimer commented that a number of his colleagues "suggested that I had been out of my mind" after reading the *Times*'s coverage of his address.[36] Word that Helen Dukas, Einstein's longtime secretary and one of the trustees of his estate, was deeply upset by what she had heard of and had read of Oppenheimer's talk in the *New York Times* must have reached him.[37] Oppenheimer thereafter wrote her saying that "when I saw the story in the *New York Times* last month I shuddered for you." With his note he enclosed his edited version of the talk and added disingenuously: "I hope that this [the edited version], which is what I did say, will seem closer to the truth, and more welcome."[38]

The views that Oppenheimer had expressed in Paris were widely disseminated. The *Gazette de Lausanne* gave the story extensive coverage. The *Nouvel Observateur* translated Oppenheimer's speech verbatim and printed it in the December 22 issue. Gerard Bonnot, a reporter for *L'Express*, had interviewed him after his lecture—while Oppenheimer was drinking some whiskey—and *L'Express* devoted a page of its December 20–26, 1965 issue to their exchange. The views Oppenheimer expressed during the interview were even sharper than in the lecture:

> During all the end of his life, Einstein did no good [Einstein n'a rien fait de bon]. He worked all alone with an assistant who was there to correct his calculations. . . . He turned his back on experiments, he even tried to rid himself of the facts that he himself had contributed to establish. . . . He wanted to realize the unity of knowledge. At all cost. In our days, this is impossible.

Interestingly, however, Oppenheimer stated that "he was convinced that still today, as in Einstein's time, a solitary researcher can effect a startling [foudroyante] discovery. He will only need more strength of character [force d'âme]." Bonnot concluded his interview by asking Oppenheimer whether he had any regrets? Any nostalgia? A smiling Oppenheimer answered, "Of course, I would have liked to be the young Einstein. This goes without saying."[39]

## Roots and Tradition

> The art of renunciation is an act of courage—it requires the
> sacrifice of things universally desired (not without hesitation
> and regret) for matters that are great and incomprehensible.
>
> —*Zbigniew Herbert (1991, 145)*

Simone Weil perceptively noted that:

> To be rooted is perhaps the most important and least recognized need for the human soul. It is one of the hardest to define. A human being has roots by virtue of his real, active and natural participation in the life of a community which preserves in living shape certain particular treasures of the past and certain particular expectations for the future. This participation is a natural one, in the sense that it is automatically brought about by place, condition of birth, profession and social surroundings. Every human being needs to have multiple roots. It is necessary for him to draw wellnigh the whole of his moral, intellectual and spiritual life by way of the environment of which he forms a natural part. (Weil 1952, 41)

In contrast to Einstein, for whom roots without the communal aspects became sufficient, the rootless Oppenheimer was always in need of a community.

Both Einstein and Oppenheimer were born into emancipated, nonobservant Jewish families. The nature of Einstein's ties to Judaism and to the Jewish culture has often been told (Lanczos 1974; Jammer 1999) and need not be repeated here. However, I want to point to the interaction of the young Albert with Gustav Maier, which I believe was deeply influential and which points to a commonality with experiences of Oppenheimer. Gustav Maier was a friend of the Einstein family in Ulm and an avuncular figure to the young Albert. He was the manager of the Ulm branch of the Deutsche Reichbank until 1881, at which time he moved to Frankfurt am Main to become the manager of the Reichbank branch there. Before leaving Ulm, he published a little pamphlet entitled *Mehr Licht! Ein Wort zur "Judenfrage" an unsere christliche Mitbürger* (More light! A word on the "Jewish question" to our Christian co-citizens) in which he defended both reform Judaism and socialism against the charges of hostility to Christianity. In 1886 he was one of the founders of the Frankfurt Peace Union. In 1891 he withdrew from the business world, moved to Switzerland, and devoted himself to writing on social and economic issues. By virtue of common political leanings and cultural interests, he and Jost Winteler met and became good friends. Both of them helped found the Swiss Ethical Culture Society in 1896, and Maier became the editor of its publication until 1919. It was he who helped the young Albert, whom he considered a wunderkind, to obtain permission to take the ETH entrance examination required of applicants without a secondary-school certificate.[40] It was also he who arranged Albert's stay with the Winteler family in Aarhau so that he could complete his high school education. Albert often visited Maier while he was a student at the ETH during 1896–1900. In 1898 Maier published a very influential book entitled *Soziale Bewegungen und Theorien bis zu modernen Arbeiterbewegung* (Social movements and theories until modern labor movements), which went through nine editions. This remarkable little book of nine chapters and a little less than 150 pages gave a succinct account of communism, socialism, and anarchism, narrated the genesis and evolution of the Egyptian, Babylonian, Chinese, Japanese civilizations, described the platonic state, the Roman Empire, and the utopian vision of Thomas More, gave a historical overview of feudalism, the Reformation, the Peasant Wars, and Luther, and from there went on to Colbert, mercantilism, and Turgot. The final three chapters covered the Industrial Revolution, the economic theories of Adam Smith, David Ricardo, Thomas Malthus, and the contemporary colonial, political, and social policy of England, and presented the views and accomplishments of Ferdinand Lasalle, Friedrich List, and other socialists of the first half of the nineteenth century, paying special attention to Saint-Simon and the Saint-Simonians, Joseph Fourier, Etiènne Cabet, and Robert Owen. The last chapter was devoted to Wilhelm von Humboldt's views of the state, socialism, Pierre-Joseph Proudhon, and Karl Marx and concluded with a "Ruckblick und Ausblick" (Retro- and prospective view) in which he commended the communitarian form of socialism to which he was committed.

It was probably Maier and Winteler who planted the seeds of socialism into Einstein—and Einstein repeatedly made his own commitment to socialism well known.[41] Nor were his ties to Maier and Winteler, and theirs to the Ethical Culture Society, forgotten as evidenced by the fact that Einstein accepted the invitation to write a message to be read at the celebration of the seventy-fifth anniversary of the founding of the Ethical Cultural Society in January 1951.[42]

Oppenheimer was educated at the Ethical Cultural School, which in many ways was a parallel development to the Social Gospel movement of American Protestantism—a response to the dismal conditions of the lower classes brought about by the industrialization and urbanization of the last third of the nineteenth century. Walter Rauschenbush and the other theologians of that movement believed that the primary message of the Gospels was a call for social justice and that individual salvation was to be achieved through works for the good of society. Like these theologians, Felix Adler, the founder of the Ethical Culture Society, supported and worked for the rights of workers to a decent wage, safe

working conditions, and good schools for their children. Part of Oppenheimer's experience at the Ethical Culture School was attending school
with the children of workers, and while in high school visiting and working in the settlement houses in lower Manhattan. However, in the United
States the Ethical Culture Society, its support coming from emancipated
middle- and upper-class Jews, never formed ties with socialism. As has
been noted earlier, the Wilhelminian Germany in which Einstein grew
up, the Switzerland[43] he matured in, the Austro-Hungarian Empire in
which he got his first full professorial appointment, as well as Weimar
Germany, all harbored covert—and often explicit—anti-Semitism; all of
these countries maintained barriers to the integration of Jews into the
social and cultural life of the country. By contrast, Oppenheimer grew
up in the United States at a time when, though anti-Semitism and discrimination certainly existed, the notion of the melting pot had wide
currency—at least for the Caucasian component of the population. Identifying as an American meant identifying with the aspirations of the
most successful experiment in democracy until that time. It demanded
solidarity with other Americans and a commitment to make the United
States a better nation for everyone.

The priority in how one identified oneself, whether first as an American and then as a Protestant of a given denomination, posed no problems for the citizens of that faith, for they could think of the United
States as a Christian nation. The situation was somewhat more difficult
for Catholics, who were seen as owing some of their allegiance to the Vat-
ican, and even more so for Jews. Oppenheimer in his professional life did
overcome the discrimination against Jews that existed in the elite American universities—Harvard, Princeton, Yale, and Cal Tech. However, for
Oppenheimer to identify himself as an American Jew presented a conflict that he found difficult to resolve as a teenager and that lingered on
into his adulthood. Recall that when the young Oppenheimer was recu-
perating from a severe case of dysentery his parents had asked Herbert
Smith, his remarkable English teacher at the Ethical Culture School who
became a sort of surrogate parent to Robert, to accompany him on a trip
to the Southwest of the United States. Smith later told Alice Kimball

Smith and Charles Weiner that he became aware of "how deeply Robert felt the fact of being Jewish when he asked to travel west as Smith's brother." And his close friend from the Ethical Culture School, Francis Fergusson, believed that for Robert going to New Mexico was partly to escape from "his Jewishness and his wealth, and his eastern connections" (Smith and Weiner 1980, 9). Oppenheimer did not escape from these tensions at Harvard, as indicated by the fact that Percy Bridgman, his thesis adviser, when writing a letter of recommendation for him to go work with Rutherford at the Cavendish, felt compelled to add a final paragraph referring to the candidate's Jewish background:

> As appears from his name, Oppenheimer is a Jew, but entirely without the usual qualifications of the race. He is a tall, well set-up young man, with a rather engaging diffidence of manner, and I think you need have no hesitation whatever for any reason of this sort in considering his application. (Smith and Weiner 1980, 77)

Nor was the atmosphere at Berkeley and at Cal Tech free from anti-Jewish prejudice. Even after World War II, when Oppenheimer was intent on leaving Berkeley and was considering going to Cal Tech full-time he could write to Charles Lauritsen, a respected nuclear physicist at Cal Tech:

> I proposed twice getting Rabi to the institute [Cal Tech] thinking a good thing generally, and for us in particular a great source of strength. Has this fallen through? If so, is it lack of money, is it reluctance to add another Jew to the faculty, is it a general feeling that he would not fit in? (Smith and Weiner 1980, 299)

Einstein, in an undelivered speech he prepared for a Jewish audience in March 1947 to help launch Brandeis University, perceptively noted:

> The majority is less disturbed by antiquated prejudices than the minority which is suffering more from them. That is why such great ideals for mankind take more easily root in a minority—at least superficially. This often causes the individual to forget or even passionately to deny that he belongs to a minority group and to attach himself to the majority which,

however, does not appreciate his attitude and rejects it. He consequently finds himself in a state of internal insecurity and loneliness, which is caused by his own conduct and not, as he assumes, by the mistake of the group from which he tries to separate himself.[44]

Had the Oppenheimers lived in Germany or Austria, they might have converted and thus opened the channels of upward mobility to their sons. This is what the Peierls, the Weisskopfs, and other liberal, emancipated Jews, had done. Alternatively, had Oppenheimer grown up Jewish in Germany, he might have fallen under the widespread influence of Martin Buber's writing and found some form of inner peace by experiencing what Buber had called *Erlebnis*. By *Erlebnis* Buber meant a form of experience beyond mere knowledge, one more stable and robust than social norms and conventions; it could be translated as one's authentic "identity." Buber meant it to be a historical identity, and in the case of Jews a Jewish identity, that would allow alienated young persons to "feel, in the presence of his Self, the continuity of his Self's unending past."[45] But Oppenheimer was born in the United States where conversion was a less frequently traveled road, one that might have been painful to his parents had he taken it as a young man. In any case, Ethical Culture was to provide an outlet for these inner tensions, but it clearly did not do so for the young Robert.

Identification as a member of the Ethical Culture Society did not prove to be an escape from Oppenheimer being identified as a Jew. Moreover, he had found its ethical messages somewhat shallow. When he was seventeen, he wrote a jingle for his father's birthday, teasing him for having "swallowed Dr. Adler like morality compressed." After Oppenheimer left the Ethical Culture School and went to Harvard (1922–1925), he began seeking a more satisfying approach to religious thought in the Hindu classics, albeit in English translation. When Isidor Rabi met him in Germany in 1929, it seemed to Rabi that Oppenheimer was more interested in those literary works than in physics—but that might very well have been a subterfuge on Oppenheimer's part to try to indicate how easily doing physics came to him.

Einstein (standing, second from right) at the Solvay Congress, 1911. (Courtesy of Solvay Institutes)

Einstein, too, had an affinity to Hindu thought with which he first became acquainted through his readings of Schopenhauer.[46] One of Hinduism's manifestations in Einstein was the striking way that he holds his thumb and index finger together in his poses at the 1911, 1913, and 1927 Solvay Congresses (see the illustrations on the following pages). It is the same way that both Vishnu and, later, the Buddha are represented in many of the sculptures of them: their thumb and forefinger are joined, in what is called the *vitarka* gesture, the sign for compassionate teaching. In a later Bhuddist tradition, the joining of the thumb and forefinger also symbolizes the uniting of method and wisdom.

As was noted in Chapter 3, Oppenheimer had acquired a deep knowledge of the *Bhagavad Gita* in 1933 when, as a young professor of physics at Berkeley with interests ranging far beyond his academic specialty, he studied Sanskrit with Arthur W. Ryder, who taught Sankrit at Berkeley and had translated into English many of the Vedic texts. In later years he would look back on the *Gita* as one of the most important influences in his life. In 1963, the *Christian Century* magazine asked him to list the

Einstein (back row, eighth from right) at the Solvay Congress, 1913. (Courtesy of Solvay Institutes)

ten books that "did most to shape your vocational attitude and your philosophy of life." Along with Shakespeare's *Hamlet* and Eliot's *Waste Land,* Oppenheimer listed the *Gita.*[47] But it is not clear to what extent Sanskrit, the *Gita,* and other Hindu texts were also a mechanism to distance himself from his Jewish roots.

But in contrast to Einstein's relationship to Spinoza and Schopenhauer, whose writings he had carefully studied and whose comportments in life influenced Einstein's, I believe that Oppenheimer only found a partial anchor in the *Gita* and the actions of Arjuna. It was partial in that the lessons of the *Gita* may only have been deeply consequential while Oppenheimer was involved with atomic weaponry, both at Los Alamos and later as governmental adviser, because he then accepted that he had a certain role to play and a given task to perform, and that the play had been written by others.

Einstein had first read Spinoza's *Ethics* with his friends of the Olympia Academy, Besso and Maurice Solovine, while in Bern employed at the Patent Office. To his biographer Carl Seelig, Einstein wrote in 1952 that he had not read much literature while a young man.

Einstein (front row, fifth from right) at the Solvay Congress, 1927. (Courtesy of Solvay Institutes)

"I preferred books whose contents were 'weltanschaulich'[48] and in particular philosophical ones: Schopenhauer, David Hume, Mach, to some extent Kant, Plato and Aristotle."[49] Einstein had studied Hume's *Treatise of Human Nature* "with fervor and admiration shortly before the discovery of the theory of relativity" and added that "[I]t is very well possible that without these philosophical studies I would not have arrived at the special theory of relativity" (Seelig 1956, 67). We do not know whether it was Hume's, perhaps ironic, characterization of Spinoza's "doctrine of the simplicity of the universe, and the unity of its substance, in which he supposes both thought and matter inhere"—and that this was a "hideous hypothesis"—that made Einstein turn to Spinoza. In any case his first reading of Spinoza's *Ethics* dates from that time.[50] He evidently resumed these studies several years later.[51] By 1920, the philosopher whom Einstein admired most was Spinoza,[52] and Leo Szilard related that he studied Spinoza's *Ethics* with Einstein in the early 1920s.[53] In 1936, Einstein wrote to his close friend Maurice Solovine:

Einstein in Philadelphia, 1942.

> I can understand your aversion to the use of the term "religion" to de-
> scribe an emotional and psychological attitude which shows itself most
> clearly in Spinoza. [However,] I have not found a better expression than
> "religious" for the trust in the rational nature of reality that is, at least to
> a certain extent, accessible to human reason. [Einstein 1956b, 102]

Evidently, for Einstein no one expressed more forcefully than Spinoza a
belief in the ultimate comprehensibility of nature.

Jonathan Israel in his masterly *Radical Enlightenment* stresses "the
unity, cohesion, and compelling power" of Spinoza's system, surely a
feature that would resonate with Einstein. Some have suggested that
Einstein's radical reductionism had its origin in Spinoza's view that
"there can be but one substance and therefore but one set of rules governing the whole of reality which surround us and of which we are
part,"[54] and similarly, that his commitment to strict determinism—that
"the future is every whit as necessary and determined as the past" (Einstein 1954, 40)—derives from Spinoza's Proposition XXIX of part I of

the *Ethics,* that "in nature there is nothing contingent, but all things have been determined from the necessity of the divine nature to exist and produce an effect in a certain way." Moreover, "things could have been produced by God in no other way, and in no other order, than they have been proposed" (I, Prop. XXXIII).

But some caution is appropriate when assessing the influence of Spinoza on Einstein. Besides Spinoza, Einstein delighted in reading Schopenhauer. And Schopenhauer, too, was a strict determinist. We know that Einstein had already read Schopenhauer's *Parerga and Paralipomena* in 1901, for he then wrote his friend Marcel Grossmann that he "liked it very much" (Einstein 1987, Doc. 122, 316). A well-worn copy of the two volumes of these essays was in Einstein's library, and a portrait of Schopenhauer adorned the wall of his study. And Schopenhauer was at times critical of Spinoza's methodology. For example, in his discussion of ethics in his *The World as Will and Representation,* Schopenhauer asserted that "[t]he ethics in Spinoza's philosophy does not in the least follow from the inner nature of his teaching, but is attached to it merely by means of weak and palpable sophism, though in itself it is praiseworthy and fine" (Schopenhauer 1966, vol. 1, 285).

Einstein agreed. Ethics was an "exclusively human concern with no superhuman authority behind it." And in *Parerga* Schopenhauer made disapproving statements about pantheism. Einstein attributed his own grappling with what is meant by a cosmic religious feeling to his readings in the Bible while a student in Munich, to his reading of Schopenhauer, and to what Buddhism has to say on the matter. From the Bible, and the Psalms in particular, he obtained the "mysterious" element of his "cosmic religion."

> The most beautiful experience we can have is the mysterious. It is the fundamental emotion which stands at the cradle of true art and true science. . . . It was the experience of mystery—even if mixed with fear—that engendered religion. A knowledge of the existence of something we cannot penetrate, our perceptions of the profoundest reason and the most radiant beauty, which only in their most primitive forms are accessible to our minds—it is this knowledge and this emotion which constitute true

religiosity; in this sense, and in this sense alone, I am a deeply religious man. (Einstein 1954, 9)

Einstein wanted "to experience the universe as a single significant whole"; this was the other element in Einstein's cosmic religion. In this he was deeply influenced by Schopenhauer and Buddhism. In his essay on *Religion and Science* Einstein stated:

> The individual feels the futility of human desires and aims and the sub-limity and marvelous order which reveal themselves both in nature and in the world of thought. Individual existence impresses him as a sort of prison and he wants to experience the universe as a single significant whole. The beginning of cosmic religious feeling already appears at an early stage of development [of religious experience],[55] e.g. in many of the Psalms of David and in some of the Prophets. Buddhism, as we have learned especially from the wonderful writings of Schopenhauer, con-tains a much stronger element of this. (Einstein 1954, 38)

Many Schopenhauer scholars have asserted that Schopenhauer's phi-losophy was deeply influenced by his knowledge of Vedic and Buddhist texts and that many of its characteristic features can be traced back to Buddhist thought.[56] All his writings—*The World as Will and Representation, Ethics, Parerga and Paralipomena*—contain extensive references to Vedic and Buddhist thought. The influence of the four truths that the Bud-dha introduced in his first sermon at Benares can readily be recognized in Schopenhauer.

1. Existence is suffering; all being is impermanent.
2. All suffering originates in desires.
3. Salvation is achieved by the eradication of desires, by nirvana.
4. Extinction of suffering can be achieved by following the Eight-fold Way: right view, right aspiration or intention, right speech, right conduct or action, right livelihood, right effort, right mind-fulness, rightful self-concentration.[57]

Schopenhauer's philosophy is similarly pessimistic, and his pessimism is intimately related to his belief that the world is in a state of perpetual change. At the core of Schopenhauer's philosophy is the tenet that human individuality is an illusion. That illusion is overcome by the way genius apprehends the cosmos:

> The *gift of genius* is nothing but the most complete *objectivity*, i.e. the objective tendency of the mind, as opposed to the subjective directed to our own person, i.e. to the will. . . . [G]enius is the ability to leave entirely out of sight our own interest, our willing, and our aims, and consequently to discard entirely our own personality for a time, in order to remain *pure knowing subject*, the clear eye of the world; and this not merely for moments, but with the necessary continuity and conscious thought to enable and repeat by deliberate art what has been apprehended. (Schopenhauer 1966, vol. 1, 185–186)

In his essays in *Pererga and Paralipomena,* Schopenhauser characterized genius as follows:

> A genius is a man in whose head the *world as representation* has attained a degree of more clearness and stands out with the stamp of greater distinctness; and as the most important and profound insight is furnished not by a careful observation of details, but only by an intensity of apprehension of the whole, so mankind can look forward to the greatest instruction from him. If he develops and perfects himself, he will give this now in one form and now in another. Accordingly we can also define genius as an exceedingly clear consciousness of things and thus also of the opposite, namely of our own self. Mankind, therefore looks to one so gifted for information about things and about its own nature. (Schopenhauer 1974, vol. 2, 76)[58]

Furthermore,

> Always to see the universal in the particular is precisely the fundamental characteristic of genius, whereas the normal man recognizes in the particular only the particular as such. (Schopenhauer 1966, vol. 2, 379)

And:

> It is *perception*[59] above all to which the real and true nature of things discloses and reveals itself, although still in a limited way. . . . But if our per-

ception were always tied to the real presence of things, its material would be entirely under the domination of chance, which rarely produces things at the right time, seldom arranges them appropriately, and often presents them to us in very defective copies. For this reason *imagination* is needed, in order to complete, arrange, amplify, fix, retain, and repeat at pleasure all the significant pictures of life, according to the aims of a profoundly penetrating knowledge and of the significant work by which it is to be communicated may require. On this rests the high value of imagination as an indispensable instrument of genius. (Schopenhauer 1966, vol. 2, 378–379)

Surely these insights must have struck Einstein, and he may well have aspired to apprehend the world in the fashion of a Schopenhauerian genius.

Schopenhauer's insights also allow us to identify a characteristic difference in the way Einstein and Oppenheimer were exceptional. Schopenhauer thought that "The person endowed with [great] talent thinks more rapidly and accurately than do the rest; on the other hand, the genius perceives a world different from them all, though only by looking more deeply into the world that lies before them also, since it presents itself into his mind more objectively, consequently more purely and distinctly" (Schopenhauer 1966, vol. 2, 376). Moreover,

The able, indeed the eminent man, fitted for great achievements in the practical sphere, is as he is precisely through objects that keenly rouse his will, and spur it on to the restless investigations of their connections and relations. Thus his intellect has grown firmly connected to his will. On the other hand, there floats before the mind of the genius, in its objective apprehension, the phenomenon of the world as something foreign to him, as an object of contemplation, expelling his willing from consciousness. On this hinges the difference between the capacity for *deeds* and that for *works*. The latter demands an objectivity and depth of knowledge that presuppose the complete separation of the intellect from the will. The former, on the other hand, demands the application of knowledge, presence of mind, and resoluteness, and these require that the intellect shall constantly carry out the service of the will. (Schopenhauer 1966, vol. 2, 387)

Indeed, Oppenheimer, reflecting his American setting—his association with Lawrence's Radiation Laboratory, his directorship of Los Alamos and of the Institute, his commitment to pragmatism—developed exceptional capacities for *deeds*. Recall the declaration he made as a young man in 1926: "The kind of person I admire most would be one who becomes extraordinarily good at *doing* lots of things but still maintains a tear-stained countenance" (my emphasis). Reflecting his Germanic and secular Jewish upbringing, Einstein's unique capacities were channeled into *works*—the formulation of the special and general theory of relativity, his contributions to statistical mechanics and to the quantum theory, his critical analysis of issues in the philosophy of science and of mathematics.

The parallel between Einstein's comportment and way of life, particularly after 1930, with that of Spinoza has been commented upon in the past.[60] While giving up his religious affiliation with Judaism, Einstein, like Spinoza, nurtured his Jewish roots and identified as a Jew as a member of the cultural Jewish tradition: "The pursuit of knowledge for its own sake, an almost fanatical love of justice and the desire for personal independence—these are the features of the Jewish tradition which make me thank my stars that I belong to it" (Einstein 1954, 184).

Oppenheimer, too, gave up his religious affiliation with Judaism, but in addition he uprooted himself from the Jewish tradition. Perhaps he hoped to find in the *Gita* and the other Sanskrit holy texts he studied the possibility of planting new roots, and he sought to obtain emotional support from adopting aspects of the Hindu tradition.[61] I believe that Einstein's willingness to identify himself as a Jew, to be active in the support of Zionism, and to readily and openly discuss his religious beliefs annoyed, if not exasperated, Oppenheimer.

If he rejected being a member of the Jewish community, Oppenheimer had a need for community, which manifested itself particularly strongly in his political activities during the 1930s. By wholeheartedly committing himself to the activities of the left, Oppenheimer became a member of a community based on a faith in which everyone was equal. It allowed him to shed the limitations of his social world and to "join a fraternity that transcended the division of the world. This was the attraction of Communism for many Jews who no longer thought of themselves in any way as Jewish. And for many, faith remained stronger than interest" (Glazer 1961, 168). Perhaps Haakon Chevalier was right when he observed that

> the fervor that Opje[62] displayed in all his political activity, the importance he attached to it, were, I think we all felt—those of us who were "on the inside"—a projection of an exceptional, almost anguished concern with the fate of man, both individually and in the large. . . . This was the "Hebrew prophet" side of his nature (which co-existed with, and never quite obliterated) the . . . sophisticated, worldly side. (Chevalier 1965)

The need to avoid being identified as Jewish intensified Oppenheimer's nationalism. Identifying himself as an American was the one continuous, constant aspect of Oppenheimer's changing identities. Making American physics on a par and better than anywhere else was part of his mission when he was building his school of theoretical physics at Berkeley during the 1930s.

Oppenheimer's nationalism sheds light on his comportment during World War II. In October 1941, after it had become clear through the work of Otto Frisch and Rudolf Peierls and the MAUD Report that atomic bombs could be assembled, Roosevelt gave the go ahead to develop nuclear weapons to Vannevar Bush and James Conant. On that occasion Bush and Conant signed an agreement that delegated all authority on atomic policy and the use of atomic bombs to the president or his delegated officers. Roosevelt named Secretary of Agriculture Henry Wallace, Secretary of War Henry Stimson, General George Marshall, and Bush and Conant as his delegates, and together they constituted the General Policy Group that was to make all the policy decisions regarding atomic matters. As a result, the atomic bomb project was transformed from a civilian project supported and controlled by the NRDC to a military weapons project under military command with a Military Policy Committee (composed of Bush and Conant and a representative from the Army and one from the Navy), empowered to issue orders to military commanders for the implementation of atomic weaponry policy. Oppenheimer was always aware of this agreement and accepted its scope. De facto the Met Lab at Chicago, Los Alamos, Oak Ridge, and Hanford were military establishments—and Oppenheimer was ready to accommodate to this reality and don with relish the uniform of an Army officer.

The scientists working at these installations were technicians. As General Leslie Groves would assert at Oppenheimer's trial, "Dr. Oppenheimer was used by me as my adviser . . . , not to tell me what to do. . . . We [General Nichols and I] were responsible for the scientific decisions" (Oppenheimer 1970). Oppenheimer was always consistent in his assessment of his role as director of Los Alamos during the war. Thus the following exchange took place in 1964 following a lecture Oppenheimer gave in Geneva:

> *Van Camp:* If you had foreseen the present situation in the world, would you have dared start the researches that led to the atomic bomb?
> *Oppenheimer:* My role was very more modest. . . . My role was to preside over an effort, to make, as soon as possible, something practical. But I would do it again.
> *Weisskopf:* I would like to address Mr. Oppenheimer in a different fashion. Given what has happened these past twenty years, would you in the position you were in 1942, would you again accept to develop the bomb?
> *Oppenheimer:* To this I have answered yes . . .
> *An assistant:* Even after Hiroshima?
> *Oppenheimer:* Yes. (Oppenheimer et al. 1965)

Whatever he did as an adviser to the government and as a public intellectual in the post–World War II period was likewise imbued with a nationalistic feeling: he always meant to transform the potentialities the United States possessed by virtue of its wealth, its institutions, and its know-how into realities and through them make this a safer and better world.

Oppenheimer's identity as an "American" was manifested in many ways. His love of the grandeur of the New Mexico mesas, of horseback riding exploring the wilderness near his ranch, Pero Caliente, his reckless driving, and his dangerous sailing habits were all part of this fashioning of himself. But he was also an intellectual, and like many earlier American intellectuals he distrusted the American city. Like Jefferson, one of his heroes, William James, and John Dewey, to name but a few others,[63] he reacted critically to the American city. When in 1949 Ralph Lapp considered the planning of future decentralized population centers since atomic warfare was a real possibility, he exemplified them with maps of a "satellite city," a "doughnut city," and a "rodlike city."[64] By contrast, Oppenheimer's metaphors always alluded to villages, never to cities.

As noted in an earlier chapter, Oppenheimer delivered an address in the fall of 1954 on the occasion of Columbia University's Bicentennial celebration, shortly after his clearance had been revoked. Oppenheimer gave a bleak and despairing overview of the world of the arts and sciences. Although he saw the arts and the sciences as flourishing, he stressed the chasm that separated "science from science and art from art, and all of one from all of the other." If each art and each science were thought of as a village, then a "high altitude picture" would reveal innumerable villages with no paths between them.

Oppenheimer conveyed a slightly more optimistic vision in the William James lectures. But in the final analysis he agreed with Dewey: "The Great Society created by steam and electricity may be a society but it is no community. The invasion of the community by the new and relatively impersonal and mechanical means of communication is the outstanding fact of modern life" (Dewey 1927, 98).

## Philosophy

> A philosophy in between the pages of which one does not hear
> the tears, the weeping and gnashing of teeth and the terrible din
> of mutual universal murder is no philosophy.
>
> —*Arthur Schopenhauer*[65]

I have indicated some of the factors that might have motivated Oppenheimer in making his critical public statements regarding Einstein, particularly at the UNESCO conference. But I believe there is an additional reason for the sharpness of some of his remarks. Einstein was not only a great physicist, but also an extremely influential philosopher of science. Oppenheimer had aspired to make a mark as a philosopher—but failed—and his philosophical stance was deeply at variance with Einstein's.

The drive toward "unity" was a constant in all that Einstein did, as though he had adopted the Heraclitean saying, "What is wise is one," as a guiding principle.[66] A much-cited letter to his friend Marcel Grossmann in 1900 when they both were still students at the ETH stated: "It is a glorious feeling to recognize the unity of a complex of phenomena, which appear to direct sense perceptions as quite distinct things."[67] In 1932, when questioned about the goals of his work that the Kaiser Leopold German Academy of Scientists posed, Einstein answered: "The real goal of my research has always been the simplification and unification of the system of theoretical physics. I attained this goal satisfactorily for macroscopic phenomena, but not for the phenomena of quanta and atomic structure." He then added, "I believe that despite considerable success, the modern quantum theory is still far from a satisfactory solution of the latter group of problems."[68]

Einstein's philosophical commitment to unity extended beyond science. It was an integral part of his "cosmic religion" and a manifestation of the influence that Buddhism and Schopenhauer had had on him (Howard 1997). Don Howard sees in Schopenhauer the roots of Einstein's position when Einstein spoke of "the momentary overcoming of one's individuality, the recognition of 'the futility of human desires and aims,' the 'sublimity and marvelous order which reveal themselves in nature and in the world of thought, 'the experience of 'the universe as a single significant whole' " (Howard 1997, 97). Einstein put these tenets succinctly in a letter he wrote in 1950 to a grieving father who had lost a beloved son to illness:

A human being is a part of the whole, called by us "Universe", a part limited in time and space. He experiences himself, his thoughts and feelings as something separated from the rest—a kind of optical delusion of his consciousness. The striving to free oneself from this delusion is the one issue of true religion. Not to nourish the delusion but to try to overcome it is the way to reach the attainable measure of peace of mind.[69]

John Archibald Wheeler, who was a colleague and a friend of Einstein in Princeton, movingly summarized an aspect of Einstein's philosophy that he first perceived when he heard Einstein lecture in 1933. Wheeler came to see this aspect ever more clearly as their interactions became more frequent and their friendship deepened.

> Over and above his warmth and considerateness, over and above his deep thoughtfulness, I came to see he had a unique sense of the world of man and nature as one harmonious and someday understandable whole, with all of us feeling our way forward through the darkness together. (Wheeler 1980b, 100)

If unity, the search for a final theory that would fix all the constants of nature, a deep awareness of what constituted his "self," were an integral aspect of Einstein's temperament and personality, Oppenheimer could never fashion a sense of self for himself. After 1947, three somewhat disparate personae coexisted in Oppenheimer: the scientist—as a professor of physics at the Institute; the administrator—as the director of the Institute trying to mold the intellectual agenda of the institution; and the intellectual—as cultural and political activist—until 1954 as the most influential civilian adviser on military and political matters to the U.S. government and after 1954 as a leading member of the Congress for Cultural Freedom.[70]

Starting in the early 1950s in his role as a public intellectual, Oppenheimer decried the specialization that was occurring in the physical and the social sciences as well as in the arts and the humanities, and the consequent fragmentation of culture at large. Communities for him were the bastions that would safeguard the sense of belonging and of interdependence.

The physics community was special for Oppenheimer because mastery of physics was his anchor in integrity, and the esteem of his fellow physicists was deeply important to him psychologically. His standing as a physicist was also essential for the respect he would command as director of the Institute, and therefore for his effectiveness in that capacity. Before 1954, he had had high hopes for the Institute for Advanced

Study as a community. He attempted to overcome the divide and the barriers between the different disciplines and the different intellectual communities that had a home there by trying to engage them in dialogue. He had hoped that dialogue would establish and cement bonds that would enable the institution to flourish as a collective.

After 1954, with the revocation of his clearance and the deep wounds of his "trial," for Oppenheimer, communities became defensive enclaves against the onslaught of the growth of knowledge and against the pervasive intrusion of mass media in all aspects of culture. Thus in the Columbia Centenary convocation in the fall of 1954, the picture of the world he adumbrated was austere:

> This is a world in which each of us, knowing his limitations, knowing the evils of superficiality, will have to cling to what is close to him, to what he knows, to what he can do, to his friends and his tradition and his love, lest he be dissolved in a universal confusion and know nothing and love nothing. (Oppenheimer 1955b, 144)

And he would counsel his audience in this way:

> If a man tells us that he sees differently than we, or that he finds beautiful what we find ugly, we may have to leave the room, from fatigue or trouble; but that is our weakness and our default. If we must live with a perpetual sense that the world and the men in it are greater than we and too much for us, let it be the measure of our virtue that we know this and seek no comfort. (Oppenheimer 1955b, 145)

In time Oppenheimer recovered some of his composure and attempted to take on more fully his role as a public intellectual. As he had done previously in his position regarding the interpretation of quantum mechanics and in his views regarding the control of atomic weapons, he drew on Bohr's viewpoints. In some aspects, Oppenheimer's philosophy could be characterized as more Bohrian than that of Niels Bohr himself—and thus very different from Einstein's.

As noted in Chapter 4, in 1957 Oppenheimer was invited to deliver the William James lectures at Harvard, and he invested a great deal of time in their preparation. Through this forum he had hoped to make sense of, and come to terms with, the constantly accelerated, exponential growth of knowledge, and find meaning in this new world. The lectures were Oppenheimer's attempt to attain, he said, "as unified, consistent and complete an outlook" as was possible under the circumstances.

Recall that his seventh and last lecture closed with these words:

> We have, I think, a dual duty: a duty of faithfulness and firmness and steadfastness in the things which by accident in our own time, in our own place, are our knowledge, our skill, our arts; [and] we have a duty of great openness to others . . . , learning to welcome the strange and being glad to learn something that is new, and that we have not suspected, and that does not fit, that is different and in some sense alien. And both duties are part of any love of order and of truth that can exist today. Both are helped when we not only love other truth but each other who purvey it.[71]

Surely, Oppenheimer must have felt that Einstein during the last twenty-five years of his life had shirked the second duty of the James lectures—"learning to welcome the strange and being glad to learn something that is new, and that we have not suspected, and that does not fit." Oppenheimer's pragmatic stance was at odds with Einstein's search for a unitary theory. It also put his expectations at variance with those of many of his own colleagues, who followed Einstein's vision of unity, if not Einstein's version of its realization.[72]

Oppenheimer's later talks were more successful in making his view clear. In those later settings, using simpler language, he was more successful in making much more explicit how history, contextual factors, and culture-specific values and motives play a role in making sense of one's world and one's moral horizon. Thus in a lecture delivered at the University of North Carolina in 1960, Oppenheimer asserted: "It was one thing to say, along the banks of the Sea of Galilee, Love thy neighbor. It is a different thing to say it in today's world. Not that it is less true; but it has a different meaning in terms of practice and in terms of what men can manage" (Oppenheimer 1960, 13). He also became much more forceful in expressing his distrust of order, which is hierarchical in the sense that "it says that some things are more important than others—that some things are so important that you can derive everything else from them" (Oppenheimer 1959, 39). And he would compellingly avow that as far as science was concerned,

> no part of science follows, really from any other in any usable form. I suppose nothing in chemistry or in biology is in any kind of contradiction with the laws of physics, but they are not branches of physics. One is dealing with a wholly different order of nature.[73]

In one of his last articles, "Perspectives in Modern Physics," written in 1966 on the occasion of the sixtieth birthday of his friend and colleague Hans Bethe, he emphasized the role of contingency: "For every science, much is accident; for every science sees its ideas and order with a sharpness and depth that comes from choice, from exclusion, from its special eyes" (Oppenheimer 1966b, 10).

## Epilogue

It was at Los Alamos, building weapons of mass destruction, that Oppenheimer experienced to the fullest a sense of community and of communal living. Later, as the director of the Institute for Advanced Study, Oppenheimer expended enormous energies and effort to try to make it an intellectual community. In his 1954 report to the Board of Trustees Oppenheimer commented, perhaps too optimistically and not very objectively, that the members of the Institute formed a community because

> many members have . . . a ranging and wide understanding and interest, and some substantial knowledge outside their own field of specialization. They are a community because close friendships contribute to mutual understanding and common interest. The fact that most of the members of the Institute live in the same apartments, eat in the same restaurant, share the same common room and the same library helps to bring them together.[74]

But as time passed, as noted in Chapter 3, Oppenheimer became discouraged and in the end deeply disillusioned by the fact that he had not been able to create "a rich and harmonious fellowship of the mind" at the Institute.

> The members of the faculty of the Institute were often not able to bring to each other, as a concomitant of the respect they entertained for each other's scholarly attainments, the sort of affection, and almost reverence, which he himself thought these qualities ought naturally to command. His fondest dream had been [Kennan thought] one of a certain rich and harmonious fellowship of the mind. He had hoped to create this at the Institute for Advanced Study; and it did come into being, to a certain extent, within the individual disciplines. But very little could be created from discipline to discipline; and the fact that this was so—the fact that mathematicians and historians continued to seek their own tables in the cafeteria, and that he himself remained so largely alone in his ability to bridge in a single inner world those wholly disparate workings of the human intellect—this was for him [Kennan was sure] a source of profound bewilderment and disappointment. (Kennan, 1972, 19)

The contrast between Oppenheimer's communitarian views and those of Einstein as communicated in his message to the Italian Society for the Advancement of Science in 1950 is stark:

> Apart from the knowledge which is offered by accumulated experience and from the rules of logical thinking, there exists in principle for the man in science no authority whose decision and statements could have in themselves a claim to "Truth." This leads to the paradoxical situation that a person who devotes all his strength to objective matters will develop, from a social point of view, into an extreme individualist who, at least in principle, has faith in nothing but his own judgment. It is quite possible to assert that intellectual individualism and scientific eras emerged simultaneously in history and have remained inseparable ever since. (Einstein 1954)

Einstein's message reaffirmed the chasm that existed between Oppenheimer's views regarding collective and individual creativity and Einstein's. When expounding his position, Einstein had in mind not only Galileo, Kepler, Newton, Maxwell, and the like, but also himself. That intellectual individualism is still a potent factor in theoretical physics, for even though all advances in the field grow—as Oppenheimer had said—"from the work, patient and brilliant, of many workers [and that] they would not be possible without this collaboration," very often the initial impetus comes from a single individual who thereafter has a dominant voice in setting the subsequent intellectual agenda of the community. This phenomenon has continued to the present—think of Schwinger, Feynman, Gell-Mann, Chew, Weinberg, Wilson, and Witten, as far as theoretical high energy physics is concerned. Their intellectual contributions and the force of their personalities make understandable why these individuals were able to achieve that position. But only Einstein and Oppenheimer, and later Feynman, attained an iconic standing not only within the physics community as a whole, but with the public at large, and this standing on a global basis. Bohr, who was as influential as Einstein in physics, in philosophy, and in politics, never realized this status.

One of the enabling conditions for achieving this iconic status is communicative skill. Science is also communication, communication with the scientific community and with the public at large. Einstein, Oppenheimer, and Feynman were all great communicators. It is interesting to note the differences between Einstein's, Oppenheimer's, and Feynman's attempts to communicate their science to the public at large. Einstein's popular expositions of his special and general theory of relativity and of the evolution of physics are remarkable in their simplicity, insightfulness, and conciseness and in the resulting profundity.[75] The same is true for Feynman, whether in presenting overviews of physics as a whole, as in his Messenger lectures (Feynman 1965), or in giving an exposition of his approach to quantum mechanics and quantum electrodynamics, as in his QED (Feynman 1985). By contrast, Oppenheimer could only speak of the contributions of other individuals and of the accomplishments of the community as a whole; in doing so he always strove for "depth" and displays of erudition. Furthermore, his mellifluous, mesmerizing style of poetic oral delivery did not transcribe well into written presentations—and most of his essays have their origin in such transcriptions. Part of Oppenheimer's appeal was his absolute control of language, his ability to deliver lengthy lectures without notes, his "weightiness," and the ambiguous, and at times obscure, quality of his messages. To Oppenheimer's "seriousness" contrast both Einstein's and Feynman's delight in being "on stage." In his eulogy for Feynman in 1989 in *Physics Today,* Dyson called Feynman "half genius, half buffoon." Einstein similarly had an element of buffoonery in him. This aspect of their character clearly resonated with the public at large: it humanized otherwise off-scale figures.

Of course, Einstein and Feynman became iconic figures within the scientific community, and thereafter with the public at large, because of their extraordinary scientific contributions. Einstein did so with his theory of general relativity, and Feynman with his novel approach to quantum mechanics and his mathematical reformulation of its formalism as integrals over alternative paths. Oppenheimer made no scientific contribution of this magnitude, nor did he have the self-confidence that derives from such an accomplishment. His exceptional contributions to the war effort were managerial and technical in character and were made possible by his impressive leadership, his charismatic qualities, his remarkable ability to assimilate and understand all the features of the project—and also by the exceptional capabilities of the participants in the project. Los Alamos's technological success was the result of the collective efforts of everyone working there.

Einstein did not need, nor did he seek, the approval of either the scientific community or that of the public at large. Oppenheimer did. Their interactions were affected by this contrast and by the fact that their intellectual, political, and cultural outlooks had been shaped in different times and under different conditions.

But perhaps the disparity in Einstein's and Oppenheimer's views regarding ultimate theories and the goals of physics was not as great as depicted above. At the 1979 Princeton Einstein Centennial celebration Valia Bargmann, who together with Peter Bergmann was Einstein's assistant from 1938 to 1941, commenting on Einstein's predilection of formulating his ideas about various things—scientific and nonscientific—in epigrams and aphorisms recalled the following incident. One of Einstein's best known quotations reads in German, "Raffiniert ist der her

Gott aber boshaft is Er nicht," which Bargmann translated as: "God is subtle, but He isn't malicious." Bargmann then stated its usual interpretation, namely "It might be difficult to find the laws of nature, but it is not impossible." But Bargmann remembered that once Einstein said to Bergmann and him: "I have had second thoughts. Maybe God is malicious," by which Bargmann believed that Einstein meant something very specific: "that God makes us believe that we understand something, when in reality we are very far from it." And Bargmann added, "And Einstein was very much concerned that one should not be uncritical enough to be misled this way" (Bargmann 1980, 480–481).

Perhaps Oppenheimer was never uncritical and never misled as far as the laws of nature were concerned.

# Some Concluding Remarks

> The struggle itself toward the heights is enough to fill a
> man's heart. One must imagine Sisyphus happy.
>
> —*Albert Camus (1955)*

The Preface posed the question: "How did Einstein and Oppenheimer try to remain relevant after they had made their singular contribution?" As we have seen for both Einstein and Oppenheimer, physics was an integral and vital part of their lives—it had been so previously and it remained so subsequently.

Einstein was constant and unshakable in his belief that the nonlinear field equations of his theory of general relativity and those of his "unified" theories of gravity and electromagnetism might yield particle-like solutions that would correspond to known "elementary" particles. He thought that such nonlinear field equations might even provide a mathematical representation of a deterministic substratum that might yield the probabilistic features of quantum mechanics. The self-confidence he had gained from his remarkable success in creating the theory of special relativity, formulating the quantitative explanation of Brownian motion, setting forth the photon hypothesis for the interpretation of the spectrum of black body radiation, providing an explanation of the photoelectric effect, developing a quantum theory of the specific heat of simple solids, and crafting the theory of general relativity set him on a path from which he never strayed. His lifelong goal was to unify gravitation and electromagnetism, and he never vacillated from his stand rejecting quantum mechanics as the basis for further advances in the microscopic domain. He became a "loner." He was perfectly willing to be an "outsider" as far as the physics community was concerned, and he was at peace with this accommodation. This individualism manifested itself not only in his intellectual endeavors but also in his political engagements. In these activities he might at times join hands with a select group of peers, but often he would act singly. And when he did not approve of developments in an enterprise he initially was part of—as in the case of the Hebrew University in Jerusalem or Brandeis University—he would withdraw his support and sever his ties.

Einstein was also an "exile," growing old in a country whose culture, language, manners, and even social organization were foreign to him. Moreover, his elevated status worked to isolate rather than to incorporate him into his relevant communities. Perhaps the ending of T. S. Eliot's *Journey of the Magi* captures something of the feelings associated with the arduous roads Einstein had traveled to and from general relativity:

> . . . I had seen birth and death,
> But had thought they were different; this Birth was
> Hard and bitter agony for us, like Death, our death.
> We returned to our places, these Kingdoms,
> But no longer at ease here, in the old dispensation,
> With an alien people clutching their gods.
> I should be glad of another death. (Eliot 1963, 110)

Oppenheimer was very different. Being part of the physics community mattered deeply to him, as did his status in it. A physicist's "rank" within the theoretical physics community is an important factor in molding his or her "self."

After the war, although he gave up doing research, Oppenheimer kept abreast of developments in high energy physics. Being "in charge" of the theoretical seminars at the Institute for Advanced Study was

Oppenheimer (front row, sixth from left) at the Solvay Congress, 1961. (Courtesy of Solvay Institutes)

essential for him. Similarly, chairing a session at the bi-annual international "Rochester" meetings of the high energy physicists was very important to him.

The animosity that developed between Oppenheimer and Einstein stemmed from their differences in their assessments of quantum mechanics, their ties to the Jewish tradition, their postwar political judgments and opinions, and their general approaches to doing things. The ill feeling was primarily on Oppenheimer's part, though resentment of Einstein's successes did not apparently play any part in Oppenheimer's attitude. Actually, he greatly admired Einstein's accomplishments, as was clear from his interview following his somewhat critical evaluation of Einstein at the 1965 UNESCO conference in Paris. To the question of whether Oppenheimer, looking back on his life and work, had any regrets, any nostalgia, he tellingly answered: "Of course, I would have liked to be the young Einstein. This goes without saying." That is, when he came on the scene as a physicist in 1926, he would have liked the situation to have been the same as it had been for Einstein. In that event, the foundational problems that the community faced would *not* have been solved, and fewer constraints would have been in place than when he was ready to undertake research in theoretical physics. The fact is that Heisenberg, Schrödinger, Born, Jordan, Dirac, and Pauli had formulated the basic concepts and formalism of quantum mechanics by the time Oppenheimer started writing papers in the spring of 1926. I believe that Oppenheimer thought that he could have made contributions comparable to those of Born and Jordan. Moreover, his assertion during this same interview—that "he was convinced that still today, as in Einstein's time, a solitary researcher can effect a startling [foudroyante] discovery. He will only need more strength of character [force d'âme]"—was an admission that he had not possessed this "strength of character" in the 1930s when facing the difficulties encountered in extending the quantum theory to relativistic field systems. He perhaps thought that had he made a fundamental contribution to the formulation of quantum mechanics in 1925–1926—something more fundamental than the Born-Oppenheimer approximation, as important and foundational as that was—it would have given him the *force d'âme*, the self-confidence and backbone, that he evidently lacked.

An ambiguous, uncertain, rapidly changing world had given an enduring place to Einstein, who was unambiguous and certain about himself. Oppenheimer was less certain and confident about himself and the world than Einstein, and he sought to find a meaningful place for himself in this rapidly changing world in the face of his ambiguity and uncertainty. His inquiries into physics, history, the social sciences, law, and psychology were all an effort to obtain insights and resources that would minimize the fragmentation of his self. His personal struggles were similar to those of William James: both of them had broad, profound, and multilayered interests "in human existence as a whole." Like James, Oppenheimer became concerned with establishing guidelines that men generally, and scientists in particular, could use to make right choices to structure their lives in responsible, moral ways. And as was true of James, Oppenheimer produced reflections that were illuminating but often unsystematic.[1]

James's discussion of saintliness in his *Varieties of Religious Experience* gives further insight into Einstein.[2] In the popular imagination, Einstein became seen as a saintly figure. I would go further and suggest that Einstein molded himself into a *secular* saint, willing to make sacrifices and to suffer for others. Thus, when he gave advice to various people on how to deal with the questions being asked by the House Un-American Activities Committee and by Senator McCarthy's committee, it was obvious that he himself was ready to go to jail by taking the stand he was advising others to take. Oppenheimer, in stark contrast, never achieved this degree of selflessness.

If Einstein was lucky to have been born at the right time and in the right place to make his momentous contributions to physics, he was also fortunate in being able to shape himself and to become a secular saint in a world that witnessed so many catastrophic events: World War I, the Great Depression, World War II, and the atrocities committed by Hitler and Stalin.

Einstein's path was no longer open to Oppenheimer, even had he wanted to become a secular saint. Kennan's statement at Oppenheimer's security clearance hearing that "It is only the great sinners who become the great saints" may have been true in the distant past, but it was no longer valid at the time he made it. What Gordon Dean and David Lilienthal said of Oppenheimer to the Gray Board was more accurate. Oppenheimer *was* "a very human man, a sensitive man, . . . a man of loyalty to his country." It was in fact loyalty to and love of his country, and love of his brother, that led him to invent his "cock-and-bull" stories concerning his friend, Haakon Chevalier, and to betray his former students. And it was loyalty to and love of his country, together with his ambition, that explain his statement at the hearing that "I would have done anything that I was asked to . . . if I thought it was technically feasible." If fear of losing what he had gained was partly what was behind this statement, I believe that he also thought that losing his security clearance would damage his country, that he was a voice of reason and clear-headedness on the high-level committees that were recommending nuclear and military policies. And damage his country it did.

Rabi was surely right in his chastisement of the government's actions. In an oft-quoted statement before the Gray Board he asserted

> that the suspension of the clearance of Dr. Oppenheimer was a very unfortunate thing and should not have been done . . . he is a consultant, and if you don't want to consult the guy, you don't consult him, period. . . . So it didn't seem to me the sort of thing that called for this kind of proceeding at all against a man who had accomplished what Dr. Oppenheimer had accomplished. . . . We have an A-bomb and a whole series of it, *** and what more do you want, mermaids? (Oppenheimer 1970, 468)

Although Rabi knew it, the hearing was not the appropriate place for him to say that it was Oppenheimer who had wanted mermaids.

Unlike the J. Alfred Prufrock of T. S. Eliot, the poet J. Robert Oppenheimer had emulated in his youth, Oppenheimer did dare "Disturb the Universe." Oppenheimer, unlike Prufrock, could say that "it was worth it, after all":

> To have bitten off the matter with a smile,
> To have squeezed the universe into a ball
> To roll it towards some overwhelming question,
> To say "I am Lazarus, come from the dead,
> Come back to tell you all, I shall tell you all."

But like Prufrock, Oppenheimer would have had to admit:

> No! I am not Prince Hamlet, nor was meant to be;
> Am an attendant lord, one that will do
> To swell a progress, start a scene or two,
> Advise the prince; no doubt, an easy tool,
> Deferential, glad to be of use,
> Politic, cautious, and meticulous;
> Full of high sentence, but a bit obtuse;
> At times, indeed, almost ridiculous—
> Almost, at times, the Fool.

And after the ruling of the Gray Board and that of the AEC commissioners, like Prufrock, Oppenheimer would have had to say:

> I have heard the mermaids singing, each to each.
>    I do not think they will sing to me.
>    I have seen them riding seaward on the waves
> Combing the white hair of the waves blown back
> When the wind blows the water white and black.
>    We have lingered in the chambers of the sea
> By sea-girls wreathed with seaweed red and brown
> Till human voices wake us, and we drown. (Eliot 1963, 13–17)

# The Russell-Einstein Manifesto

Issued in London, July 9, 1955

In the tragic situation which confronts humanity, we feel that scientists should assemble in conference to appraise the perils that have arisen as a result of the development of weapons of mass destruction, and to discuss a resolution in the spirit of the appended draft.

We are speaking on this occasion, not as members of this or that nation, continent, or creed, but as human beings, members of the species Man, whose continued existence is in doubt. The world is full of conflicts; and, overshadowing all minor conflicts, the titanic struggle between Communism and anti-Communism.

Almost everybody who is politically conscious has strong feelings about one or more of these issues; but we want you, if you can, to set aside such feelings and consider yourselves only as members of a biological species which has had a remarkable history, and whose disappearance none of us can desire.

We shall try to say no single word which should appeal to one group rather than to another. All, equally, are in peril, and, if the peril is understood, there is hope that they may collectively avert it.

We have to learn to think in a new way. We have to learn to ask our-selves, not what steps can be taken to give military victory to whatever group we prefer, for there no longer are such steps; the question we have to ask ourselves is: what steps can be taken to prevent a military contest of which the issue must be disastrous to all parties?

The general public, and even many men in positions of authority, have not realized what would be involved in a war with nuclear bombs. The general public still thinks in terms of the obliteration of cities. It is understood that the new bombs are more powerful than the old, and that, while one A-bomb could obliterate Hiroshima, one H-bomb could obliterate the largest cities, such as London, New York, and Moscow.

No doubt in an H-bomb war great cities would be obliterated. But this is one of the minor disasters that would have to be faced. If everybody in London, New York, and Moscow were exterminated, the world might, in the course of a few centuries, recover from the blow. But we now know, especially since the Bikini test, that nuclear bombs can gradually spread destruction over a very much wider area than had been supposed.

It is stated on very good authority that a bomb can now be manufac-tured which will be 2,500 times as powerful as that which destroyed Hi-roshima. Such a bomb, if exploded near the ground or under water, sends radioactive particles into the upper air. They sink gradually and reach the surface of the earth in the form of a deadly dust or rain. It was this dust which infected the Japanese fishermen and their catch of fish.

No one knows how widely such lethal radioactive particles might be diffused, but the best authorities are unanimous in saying that a war with H-bombs might possibly put an end to the human race. It is feared that if many H-bombs are used there will be universal death, sudden only for a minority, but for the majority a slow torture of disease and disintegration.

Many warnings have been uttered by eminent men of science and by authorities in military strategy. None of them will say that the worst re-sults are certain. What they do say is that these results are possible, and no one can be sure that they will not be realized. We have not yet found that the views of experts on this question depend in any degree upon their politics or prejudices. They depend only, so far as our researches have revealed, upon the extent of the particular expert's knowledge. We have found that the men who know most are the most gloomy.

Here, then, is the problem which we present to you, stark and dreadful and inescapable: Shall we put an end to the human race; or shall mankind renounce war? People will not face this alternative because it is so difficult to abolish war.

The abolition of war will demand distasteful limitations of national sovereignty. But what perhaps impedes understanding of the situation more than anything else is that the term "mankind" feels vague and abstract. People scarcely realize in imagination that the danger is to themselves and their children and their grandchildren, and not only to a dimly apprehended humanity. They can scarcely bring themselves to grasp that they, individually, and those whom they love are in imminent danger of perishing agonizingly. And so they hope that perhaps war may be allowed to continue provided modern weapons are prohibited.

This hope is illusory. Whatever agreements not to use H-bombs had been reached in time of peace, they would no longer be considered binding in time of war, and both sides would set to work to manufacture H-bombs as soon as war broke out, for, if one side manufactured the bombs and the other did not, the side that manufactured them would inevitably be victorious.

Although an agreement to renounce nuclear weapons as part of a general reduction of armaments would not afford an ultimate solution, it would serve certain important purposes. First: any agreement between East and West is to the good in so far as it tends to diminish tension. Second: the abolition of thermo-nuclear weapons, if each side believed that the other had carried it out sincerely, would lessen the fear of a sudden attack in the style of Pearl Harbor, which at present keeps both sides in a state of nervous apprehension. We should, therefore, welcome such an agreement though only as a first step.

Most of us are not neutral in feeling, but, as human beings, we have to remember that, if the issues between East and West are to be decided in any manner that can give any possible satisfaction to anybody, whether Communist or anti-Communist, whether Asian or European or American, whether White or Black, then these issues must not be decided by war. We should wish this to be understood, both in the East and in the West.

There lies before us, if we choose, continual progress in happiness, knowledge, and wisdom. Shall we, instead, choose death, because we cannot forget our quarrels? We appeal, as human beings, to human beings: Remember your humanity, and forget the rest. If you can do so, the way lies open to a new Paradise; if you cannot, there lies before you the risk of universal death.

## Resolution

We invite this Congress, and through it the scientists of the world and the general public, to subscribe to the following resolution:

"In view of the fact that in any future world war nuclear weapons will certainly be employed, and that such weapons threaten the continued existence of mankind, we urge the Governments of the world to realize, and to acknowledge publicly, that their purpose cannot be furthered by a world war, and we urge them, consequently, to find peaceful means for the settlement of all matters of dispute between them."

Max Born, Perry W. Bridgman, Albert Einstein, Leopold Infeld, Frédéric Joliot-Curie, Herman J. Muller, Linus Pauling, Cecil F. Powell, Joseph Rotblat, Bertrand Russell, Hideki Yukawa

# Notes

## Abbreviations

*EA*   Albert Einstein Papers, Albert Einstein Archives, Department of
       Manuscripts & Archives, Jewish National & University Library,
       Hebrew University of Jerusalem. http://www.albert-einstein.org/.
       The correspondence and other materials of Albert Einstein
       available in the Einstein Archives in Jerusalem are identified as
       AE, with the number thereafter referring to the listing in the
       archive.

*EoP*  Nathan and Norden's *Einstein on Peace* (1968).

*I&O*  Einstein's *Ideas and Opinions* (1954).

JROLC  J. Robert Oppenheimer Papers, Manuscript Division, Library of
       Congress, Washington, D.C.

JROWJ  J. Robert Oppenheimer, The William James Lectures, 1957, Box 259
       of JROLC.

RDFA   Robert D. Farber University Archives, Brandeis University.

STWP   Stephen S. Wise Papers, American Jewish Historical Society,
       Waltham, Mass., and New York, N.Y.

## Introduction

1. For an account of the myths surrounding Einstein, see Stachel's essay "Albert Einstein: The Man beyond the Myth," in Stachel 2002, 3–12. To the best of my knowledge there is as yet no account of Einstein's role in creating the myths surrounding him, nor any analysis to explain the tendency to present him in hyperboles.

2. See also JROLC Box 256, Folder "Einstein's 60th Birthday—3/16/39."

3. Needless to say, their perception as icons changed over time. It is not my intent to trace the evolution of the public perception of Einstein and Oppenheimer.

4. After World War II, mimeographed and dittoed preprints of all the theoretical papers that were being submitted to scientific journals were sent out to all the leading physicists in the field and to the major physics departments where research was being carried out. Oppenheimer and the Institute received all these papers, and Oppenheimer read many of them. At the weekly Wednesday afternoon theoretical physics seminar at the Institute over which Oppenheimer presided, the latest developments in all fields of physics were presented and critically discussed. Pais (2006) describes these seminars, as well as the weekly luncheon meetings that Oppenheimer had with theorists from Princeton University, which informed him of the directions theoretical physics was taking.

Proof of Oppenheimer's mastery of the theoretical developments in high energy physics as late as 1965 is given by his contribution to the volume issued when Victor Weisskopf relinquished the directorship of the CERN laboratory in Geneva. Oppenheimer's (1965) article, "The Symmetries of Forces and States," indicates that he had read the most recent articles in the field and had acquainted himself with many of the details. He could indicate the nature of the general problems that workers in the field were attempting to solve but could not suggest further detailed paths to be taken.

5. The numerous conferences and exhibits that were organized in 2005 to celebrate the centennial of Einstein's *annus mirabilis* are certainly proof of this (e.g., the conferences organized by the Einstein Forum in Berlin and the Israel Academy of Sciences and Humanities in Jerusalem, and the Einstein exhibit mounted by Jürgen Renn and the Deutsches Museum in Berlin).

6. It is clear that in the public's mind Einstein's status as a global iconic figure is secure. Thus *Time* magazine in 2000 selected him as "Person of the Century." Frederic Golden in his article on Albert Einstein in that issue noted: "He was the pre-eminent scientist in a century dominated by science. The touchstones of the

era—the Bomb, the Big Bang, quantum physics and electronics—all bear his imprint. He was the embodiment of pure intellect, the bumbling professor with the German accent, a comic cliché in a thousand films. Instantly recognizable, like Charlie Chaplin's Little Tramp, Albert Einstein's shaggy-haired visage was as familiar to ordinary people as to the matrons who fluttered about him in salons from Berlin to Hollywood. Yet he was unfathomably profound—the genius among geniuses who discovered, merely by thinking about it, that the universe was not as it seemed" (Golden 1999, 62). On the other hand, Oppenheimer's status is less clear cut. Several excellent recent biographies have been written about him attesting to the important role he played in the history of the United States during the twentieth century. He is still a widely respected figure among older physicists. However, at the present time he is not a well-known figure among young scientists outside of the United States.

7. Hoffmann 1972, 254n1. The Schopenhauer influence can be seen in Einstein's physics. His insistence on an objective description of the world, one that does not depend on any particular individual's wants, resonates with Schopenhauer's views. Thus in the special theory of relativity Einstein postulates the equivalence of *all* inertial observers for all physical phenomena, and in quantum mechanics he demands that the measuring process not depend on the particular will of the observer. But a word of caution: the explanations are overdetermined. Thus Roman Jakobson (1982) attributes the roots of Einstein's formulation of the special theory of relativity to his exposure while at Aarau to Jost Winteler's theory about sound systems of languages, which stressed the fundamental difference between accidental features and invariant, essential properties.

8. See the foreword by Renn and Schulmann in Einstein and Marić 1992, xxvii–xxviii.

9. Ryder's more poetic, but less straightforward, rendition of the stanza reads as follows:

> Save for the works of sacrifice,
> All labors cling and bind;
> With sacrificial purpose work,
> Attachment left behind. (Ryder 1929, 25)

Both Johnston and Ryder agree that disciplined activity, nonattachment was Krishna's message.

10. This material is based on Dennefink 2005.

11. It was Einstein, shortly after submitting his papers on general relativity to the Prussian Academy, who in papers in which he formulated approximation

methods for handling the general relativistic field equations had pointed out the likely existence of gravitational waves (Einstein 1916).

12. Quoted in Dennefink 2005, 43.

13. Ibid., 44.

14. Einstein and Rosen 1937, 52.

15. Dennefink 2005, 45. It is interesting to note that the freedom to choose particular coordinate systems in the description of gravitational phenomena (coordinate conditions) and the possible singularities introduced thereby were stumbling blocks that prevented Einstein from arriving at general covariant field equations in 1912.

16. The abstract of the Franklin Institute paper reflected Robertson's input and read: "The rigorous solution for cylindrical gravitational waves is given. For the convenience of the reader the theory of gravitational waves and their production, already known in principle, is given in the first part of this paper. After encountering relationships which cast doubt on the existence of *rigorous* solutions for undulatory gravitational fields, we investigate rigorously the case of cylindrical gravitational waves. It turns out that rigorous solutions exist and that the problem reduces to the usual cylindrical waves in euclidean space."

17. The incident is reminiscent of Einstein's interaction with Alexander Friedmann. When Einstein first addressed cosmological questions with his general relativistic field equations, he believed in a static universe. To obtain solutions that corresponded to a static universe, he added another term to his field equations that prevented expansion or contraction for a given energy density. The constant thus introduced has come to be known as the *cosmological constant*. In June 1922 Friedmann, who at the time was teaching physics at Petrograd, sent an article entitled "On the Curvature of Space" to the *Zeitschrift für Physik* in which he demonstrated that the general relativistic field equations admit solutions for which the radius of curvature of the universe could be either an increasing or a periodic function of time. Friedmann wrote about the results of this paper in a book he published a little later (Friedman 2000), and described the results as follows: "The stationary type of Universe comprises only two cases which were considered by Einstein and de Sitter. The variable type of Universe represents a great variety of cases; there can be cases of this type when the world's radius of curvature . . . is constantly increasing in time; cases are also possible when the radius of curvature changes periodically." Einstein quickly responded to Friedmann's article. His reply was received by the *Zeitschrift für Physik* on September 18, 1922: "The results concerning the non-stationary world, contained in [Friedmann's] work, appear to me suspicious. In reality it turns out that the solution given in it does not satisfy the field

equations." On December 6 Friedmann wrote to Einstein: "Considering that the possible existence of a non-stationary world has a certain interest, I will allow myself to present to you here the calculations I have made . . . for verification and critical assessment. [The calculations are given.] . . . Should you find the calculations presented in my letter correct, please be so kind as to inform the editors of the *Zeitschrift für Physik* about it; perhaps in this case you will publish a correction to your statement or provide an opportunity for a portion of this letter to be published."

However, when the letter arrived in Berlin, Einstein had left on his trip to Japan. He did not return to Berlin until March, but he still did not seem to have read Friedmann's letter. Only when Yurii Krutkov, a colleague of Friedmann's from Petrograd, met Einstein at Ehrenfest's house in Leiden in May 1923 and told him of the details contained in Friedmann's letter did Einstein admit his error. He wrote immediately to the *Zeitschrift für Physik:* "In my previous note I criticized [Friedmann's work "On the Curvature of Space"]. However, my criticism, as I became convinced by Friedmann's letter communicated to me by Mr Krutkov, was based on an error in my calculations. I consider that Mr Friedmann's results are correct and shed new light." See the entry by J. J. O'Connor and E. F. Robertson at http://www.gap-system.org/history/Biographies /Friedmann.html (accessed July 2006).

18. See in this connection Howard 1993 and the chapter on Einstein in Howard Gardner's *Creating Minds* (1993).

19. I have in mind the rejection of his suggestions for the structure of Brandeis University and of his nomination of Harold Laski as the man he thought should become its first president. He thereafter refused to have any dealings with that institution. The situation in the case of the Hebrew University of Jerusalem is different. Although he had sharp differences with its administration, he nonetheless deeded all his papers to that institution. See Chapter 2.

20. Diana Trilling spoke of him as "a culture hero for American intellectuals, especially for literary intellectuals," and commended him for his political stand during the 1930s.

21. Robert Crease has penetratingly analyzed the tragic element in Oppenheimer in Carson and Hollinger 2005. But the context and culture dependence of the attribution of "tragic" should be kept in mind. Glenn Most (2000) has insightfully pointed to the difference between the notion of the "tragic" in Greek tragedies and literature in general and the post-Enlightenment, Romantic idea of the "tragic."

22. Philosopher Morton White got to know Oppenheimer fairly well in the late 1940s when Oppenheimer was on the Board of Overseers of Harvard and

the chairman of its Committee to Visit the Department of Philosophy. In his autobiography, White states that he was "quite aware of . . . [Oppenheimer's] arrogance and his pretensions to omniscience [as an Overseer]." But later, when White was a visitor and then a permanent faculty member of the Institute for Advanced Study, it seemed to him that Oppenheimer "postured less than he did when he was running the Harvard Visiting Committee, and he seemed more friendly and less pretentious on his home ground, where he didn't try to demonstrate that he was a universal genius" (White 1999, 137–139). See also Pais's assessment of Oppenheimer's character in Pais 1997.

23. In his reply to Max von Laue, who had invited him to attend the conference to celebrate the fiftieth anniversary of the theory of special relativity as the guest of honor, Einstein wrote back: "Old age and ill health make it impossible for me to take part in such occasions, and also I must confess that this divine dispensation is somewhat liberating for me. For everything that has anything to do with the cult of personality has always been painful to me." Einstein to Max von Laue, February 3, 1955, EA 5-96.00. Quoted in Dukas and Hoffman 1979, 101.

24. See, for example, Hadamard 1945, 96–97; Wertheimer 1959, 213–228; Jakobson 1982; and Miller 1996.

25. See in particular the portrait that Pais paints of Oppenheimer in his *Tale of Two Continents* (Pais 1997).

26. Paul Ehrenfest to Wolfgang Pauli, November 24, 1928, in Pauli 1979–2005, vol. 1, 477.

27. Quoted in Heilbron 2005, 278–279.

28. Oppenheimer to his brother Frank, March 12, 1932 (Smith and Weiner 1980, 155–156).

29. It is interesting to note that Einstein evidently could not appreciate any music written after Mozart, except for Schubert, some of the "smaller" works of Schumann, a few lieder, and chamber works of Brahms. See Dukas and Hoffman 1979, 76–77. He evidently liked Vivaldi, Bach, Mozart, and Schubert best (Bargmann 1982, 488). See also the remarks by Robert Mann, who in the 1950s was the first violinist of the then Julliard Quartet, about playing music for and with Einstein in 1952 (Woolf 1982, 526–527).

30. Quoted in Klein 1986, 329.

31. See in particular Jerome and Taylor 2005.

32. Einstein to Lorentz, July 15, 1923, in Kox 1993.

33. See Schweber 1988.

34. From Goethe's *Faust* as quoted by Einstein.

35. Where Oppenheimer used the word "experience" Bohr advocated using the word "phenomenon" and having it refer exclusively to observations obtained

under specified conditions, which conditions included an account of the whole experiment.

36. See Einstein 1949a, 79 and 81; Infeld 1956; Infeld and Plebański 1960; and references therein. See also Infeld 1941, 277–284; Bergmann 1942, 223–244.

37. See also the articles by Bohr and Einstein in the special issue of *Dialectica* edited by Pauli—*Dialectica* 2(3/4) (1948): 307–422—in which their positions are clearly stated. Bohr stressed the fact that the total experimental situation had to be taken into account; in fact, the total experimental arrangement had to be assumed as given in advance. Einstein on the other hand believed that a reality existed independently of the experimental intervention.

38. Oppenheimer 1950. In many of his later published lectures Oppenheimer would expound on complementarity. See, for example, Oppenheimer 1954.

39. On this issue see in particular Feuer 1974.

40. The freedom is, of course, not total. As Einstein explained: "The liberty of choice, however, is of a special kind: it is not in any way similar to the liberty of a writer of fiction. Rather, it is similar to that of a man engaged in solving a well designed crossword puzzle. He may, it is true, propose any word as a solution; but there is only one word which really solves the puzzle in all its parts" (Einstein 1954, 287).

41. Wilhelm von Humboldt, when creating the blueprint for the structure of the German universities in the early 1800s, characterized science as the "fundamental thing." He believed that it will be correctly and sincerely pursued when pure and that "solitude and freedom are the principles prevailing in its realm" (Paulsen 1906, 53).

42. Einstein 1941. In acknowledging "the co-operative effort as regards the final effect," Einstein was perhaps taking into account Oppenheimer's remarks on the occasion of Einstein's sixtieth birthday in 1939, in which Oppenheimer stated that every scientific discovery was "so closely interwoven with the work of countless other scientists and technicians . . . that we . . . come away [from an analysis of the development] with a deepened conviction of the cooperative and interrelated character of scientific achievement" (Oppenheimer 1939).

43. From Bohr's June 1950 letter to the United Nations (Pais 1991).

44. Quoted in Bird and Sherwin 2005, 355.

45. See Schweber 1994.

46. In 1953 Whittaker put out a new edition of *History of the Theory of the Ether* in which he insisted that everything of importance regarding special relativity had already been said by Poincaré and Lorentz and minimized the role played by Einstein. Whittaker was a colleague and close friend of Born in Edinburgh.

Born wrote how pained he was by the new edition, and Einstein answered him: "Don't lose any sleep over your friend's book. Everybody does what he considers right or, in deterministic terms, what he has to do." Einstein to Born, October 12, 1953 (Einstein 2005, 194).

47. Einstein considered this aphorism his "unfailing well-spring of tolerance" throughout his life (Einstein 1931, 8–9). See also Jammer 1999.

## 1. Albert Einstein and Nuclear Weapons

1. Einstein, January 22, 1947, EA 40–10. The letter appealed for support for the Emergency Committee of Atomic Scientists. For the Emergency Committee see Smith 1965.

2. See Einstein's post-1926 correspondence with Michele Besso (Einstein 1972) and Pierre Speziali's discussion of Einstein's views of quantum mechanics as gleaned from Einstein's letters to Besso (Einstein and Besso 1972, liii–lvi).

3. In private Einstein expressed stronger views. Thus in a letter to Harry Lipkin on July 5, 1952, he stated that the present quantum theory reminded him "a little of the system of delusions of an exceedingly intelligent paranoid, concocted of incoherent elements of thoughts" (EA 15–154.00). Quoted in Fine 1986, 1.

4. Einstein also found objectionable the fact that quantum mechanics entailed the continued "entanglement" of two quantum systems once they had interacted, regardless of how far they may have moved thereafter. See Fine 1996.

5. See, for example, his fairly extensive correspondence with Henri Barbusse during the 1920s (EA 34–36). Barbusse tried to enlist Einstein in many left-wing causes. Einstein agreed to be associated with some of them; he forthrightly refused participation in others.

6. The chairman of the Lincoln Birthday Committee for Democracy and Intellectual Freedom was the anthropologist Franz Boas.

7. Note that Einstein is a "modernist" in Latour's sense. He believed that "even though we construct Nature, Nature is as if we did not construct it"; that "even though we do not construct Society, Society is as if we did construct it"; and most importantly, that "Nature and Society must remain absolutely distinct" (Latour 1993, 32).

8. Thus in Appendix II to the fifth edition of his *Meaning of Relativity* he noted, "At the present time the opinion prevails that a field theory must first be transformed into a statistical theory of field probabilities according to more or established rules. I see this method as only an attempt to describe relationships of an essentially nonlinear character by linear methods" (Einstein 1955, 165).

9. Thus in 1953, upon hearing that Max Born was going back to Germany because the pension that he was receiving from Edinburgh University was in-

sufficient to meet his financial needs, Einstein wrote him: "If anyone can be held responsible for the fact that you are migrating back to the land of the mass-murderers of our kinsmen, it is certainly your adopted fatherland [Scotland and the United Kingdom]—universally notorious for its parsimony." Einstein to Born, October 12, 1953 (Einstein 2005, 195).

10. The entire speech is reproduced in Smith and Weiner 1980, 315–325.

11. It had been Freud, who, in his correspondence of 1932 with Einstein had succinctly pointed out the necessity of an international court: "There is but one sure way of ending war and that is the establishment, by common consent, of a central control which shall have the last word in every conflict of interests. For this two things are needed: first the creation of such a supreme court of jurisdiction; secondly, its investment with adequate executive force. Unless this second requirement is fulfilled, the first is unavailing. Obviously the League of Nations, acting as a Supreme Court, fulfills the first condition; it does not fulfill the second. It has no force at its disposal and can only get it if the members of the new body, its constituent members, furnish it. And, as things are, this is a forlorn hope" (*EoP* 195–196).

12. A word of caution. Nathan and Norden aimed at translating Einstein's German texts into smooth and literate English. To do so they at times took certain liberties with the texts, so that the translation does not correspond to the literal meaning of the original. They themselves in some of their footnotes indicate that they have made certain "revisions." The translations of Sonja Bargmann, which are found in Einstein's *Ideas and Opinions* of 1954, are more faithful to the original texts.

13. See, for example, http://www.dannen.com/szilard.html (accessed June 2006).

14. See Klein 1986 for a particularly sensitive account of their friendship.

15. See Lanouette 1992 for an informative biography of Szilard. See also Telegdi 2000.

16. See http://www.online-literature.com/wellshg/worldsetfree/.

17. Whether the chain reaction could be controlled or would produce an explosion depended on the energy of the produced neutrons, on the probability of the two uranium isotopes that make up natural uranium to capture a neutron and fission, and on the concentration and geometrical arrangement of the uranium isotopes. In either case the amount of energy released per fission was orders of magnitude greater than in a chemical reaction.

18. It was Szilard's knowledge of engineering and of the properties of materials that was responsible for the use of pure (boron-free) graphite as a moderator in the Fermi-Szilard pile. This turned out to be a decisive factor in the

successful construction of a critical assembly. The Germans—who after 1941 concentrated all their efforts on building a reactor—were never able to build one because the graphite they used was impure and absorbed too many neutrons. They thus had to rely on heavy water as a moderator. But the Allies could prevent their acquisition of heavy water by sabotaging and bombing the Norwegian hydroelectric plant that produced it. See Rhodes 1988, part two, for the details of the story.

19. A uranium (U) nucleus contains 92 protons and can contain between 141 and 146 neutrons, with U235 (92 protons and 143 neutrons) and U238 (92 protons and 146 neutrons) the most common isotopes. Naturally occuring samples of uranium contain 99.28 percent of U238, 0.72 percent of U235, and trace amounts of U234. U238 has a small probability to fission spontaneously or to fission when bombarded with fast neutrons. The fact that U235 does fission when bombarded with fast neutrons and does so with the enission of neutrons is what makes it the key isotope in atomic weapons.

20. Einstein and Szilard became friends, and they collaborated in the design of a refrigerator pump with no moving parts for which they were granted a patent in 1930. Their idea was later adopted and used in reactors with liquid metals as coolants. See the articles by Dannen and by Telegdi available on the Web: http://www.dannen.com/budatalk.html, and http://www.aip.org/pt/vol-53/iss-10/p25.html.

21. See, for example, the report in the April 29, 1939 *New York Times* of a session at the Washington meeting of the APS. It carried the headline: "VISION EARTH ROCKED BY ISOTOPE BLAST. Scientists Say Bit of Uranium Could Wreck New York." The article itself stated that "Dr. Nils Bohr [*sic*] of Copenhagen, a colleague of Dr. Einstein at the Institute for Advanced Study, Princeton, N.J., declared that bombardment of a small amount of the pure Isotope 235 of uranium with slow neutron particles of atoms would start a "chain reaction" or atomic explosion sufficiently great to blow up a laboratory and the surrounding country for many miles." Wrecking "an area as large as New York City would be comparatively easy." While the article reported that many physicists thought that it would be "difficult, if not impossible" to separate the U235 Isotope from the more abundant U238, it also reported that Dr. L. Onsager had devised a new apparatus for separating the isotopes when present in gaseous form.

22. *New York Times*, March 14, 1939, 1.

23. The handwritten drafts (e.g., of Einstein's proposed letters to the Belgian ambassador and to Roosevelt), are in the Einstein Archives. It should be noted that at the time Einstein was at the center of the attempt to remove Abraham

Flexner as the director of the Institute for Advanced Study, and Einstein had intimated that he would leave the Institute should Flexner remain as director. See Batterson 2006, chap. 9.

24. Szilard to Einstein, July 19, 1939, EA 39-461, 1.

25. Szilard to Einstein, July 24, 1939, EA 39-474.

26. Szilard to Einstein, August 2, 1939, EA 39-464. In the letter Szilard indicates that he was also contacting Gano Dunn, the president of Cooper Union, for other suggestions.

27. Incidentally, it was only in late September, after having spoken to him, that Szilard came to the conclusion that Lindbergh was "in der Tat nicht unser Mann." Szilard to Einstein, September 27, 1939, EA 39-471.

28. Sachs saw Roosevelt on October 11, at which time he made an oral presentation and submitted Einstein's letter dated August 2, 1939; the memorandum by Szilard dated August 15, 1939; copies of articles in scientific journals on uranium research; and a memorandum by Sachs dated March 10, 1939 entitled "NOTES ON IMMINENCE WORLD WAR IN PERSECTIVE ACCRUED ERRORS AND CULTURAL CRISIS OF THE INTER-WAR DECADES" (EA 39-488). The folder EA 39-488 contains Alexander Sachs's "Documentary Historical Report. From early history atomic project in relation to President Roosevelt, 1939–1940. From inception and presentation of idea to the President by Albert Einstein and Alexander Sachs to the transfer of the project for execution to the National Defense Committee established by President June 15, 1940."

29. On October 19, Roosevelt acknowledged receiving Einstein's communication. He thanked him for his "recent letter and the most interesting and important enclosure" and informed him that "I found this data of such import that I have convened a board consisting of the head of the Bureau of Standards and a chosen representative of the Army and Navy to thoroughly investigate the possibilities of your suggestion regarding the element of uranium" (*EoP*, 297).

30. Sachs in EA 39-488, 7.

31. EA 39-488, 7.

32. The report of the committee noted that "if the chain reaction could be controlled, . . . it might . . . be used as a continuous source of power in submarines" and "if the reaction turned out to be explosive in character, it would provide a [possible source of bombs with destructiveness vastly greater than any now known" (EA 39-488, 8).

33. Thus Sachs visited Einstein in February 1940, at which time Einstein evaluated for Sachs the work being carried out by Joliot-Curie and his group in Paris. During the visit, Einstein stated that the "work being carried out at Columbia was more important" and that conditions should be created for its extension.

Thereafter Sachs wrote to Watson recommending this and indicated that "Prof. Einstein would give a favorable evaluation of the work being completed at Columbia." Sachs to Watson, February 15, 1940, EA 29–488, 11. Similarly, Sachs sent him the *Aide-mémoire* he wrote for conferences with governmental authorities on "Current and prospective needs of universities- national defense projects". See EA 39–488, 28.

34. Einstein to Sachs, March 7, 1940, EA 39–488, 12.

35. Roosevelt to Sachs, April 5, 1940, EA 39–488, 16, Exhibit 8a.

36. Watson to Sachs, April 8, 1939, EA 29–488, 17.

37. Sachs to Watson, April 19, 1939, EA 29–488, 18.

38. Sachs, EA 39–488, 15.

39. Einstein to Briggs, April 25, 1939, EA 29–488, 24.

40. See Feld 1979.

41. Bohr Political Papers. I thank Finn Asserud for pointing me to this entry.

42. Wallace Akers was the director of the Tube Alloys, the British Atomic Bomb project. M. W. Perrin was an important manager at Imperial Chemical Industries and early on became associated with Tube Alloys and became Akers's deputy. See Gowing 1964.

43. See, in particular, Oppenheimer's (1964b) account of "Neils Bohr and Nuclear Weapons" and Gowing 1964.

44. Felix Frankfurter, memorandum to Secretary of War Stimson, April 26, 1945, Box 34, JROLC.

45. Finn Aaserud (1999) has given the most complete account of Bohr's activities during World War II regarding atomic weapons.

46. Although he continued to live in Pittsburgh—he was a member of the faculty at Carnegie Tech—he traveled every six weeks to Chicago for the information meetings held there. It seems that he knew a fair amount about the state of developments of atomic bombs.

47. See Smith 1965 for details.

48. In 1921–1922 Otto Stern and Walther Gerlach carried out a crucial experiment in which a beam of silver atoms was deflected by an inhomogeneous magnetic field. It indicated that electrons and atoms have intrinsic quantum properties.

49. Clark in his biography of Einstein gives the date of the meeting and of the letter as December 11 and 12, respectively. The letter to Bohr in the Einstein archive is not dated. Clark gives as his reference for both the letter and the dates a "diplomatic source" (Clark 1973, 645).

50. Einstein to Bohr, mid-December 1945, EA 08–094, 1.

51. Ibid. The relevant paragraph in the original German reads as follows: "Gestern war nun Stern wieder da, und es schien uns, dass es doch einen Weg gibt, der—wenn auch geringe—Aussichten auf Erfolg gibt. Es gibt in den hauptsächlichen Ländern Wissenschaftler, die wirklich einflussreich sind und bei den politischen Leitern Gehör finden können. Da sind *Sie* mit Ihren internationale Beziehungen, A. Compton hier in U.S.A., Lindeman in England, Kapitza und Joffe in Russland etc. Die Idee ist, diese zu gemeinsamer Aktion auf die Leiter der Politik in ihren Ländern zu bringen, um eine Internationalisierung der Militärmacht zu erreichen—ein Weg, der als zu abenteuerlich schon geraume Zeit fallen gelassen worden ist. Aber dieser radikale Schritt mit all seinen weitgehenden politischen Voraussetzungen betreffend übernationale Regierung scheint die einzige Alternative gegen das technische Geheimwettrüsten zu sein."

52. James Franck was awarded the Nobel Prize in Physics in 1925 for experimental work that comfirmed important aspects of Bohr's model of the atom. Before leaving Germany in 1933 he had been Professor of Experimental Physics and Director of the Second Institute for Experimental Physics at the Georg-August University of Göttingen. He and Max Born, who was the director of the Institute of Theoretical Physics in Göttingen, had made the university a preeminent center of research in atomic physics. In 1938 he accepted a position at the University of Chicago. In early 1942 he became involved in the Manhattan Project and served as director of the Chemistry Division of the Metallurgical Laboratory at the University of Chicago, the laboratory where the first atomic reactor was constructed under the leadership of Enrico Fermi. In June 1945 he became the chair of a committee consisting of Leo Szilard, Donald J. Hughes, James J. Nickson, Eugene Rabinowitch, Glenn T. Seaborg, and Joyce C. Stearns that was to address the political and social problems connected with the use of atomic bombs. The famous Franck Report was the summary of the deliberations of the committee and a detailed statement of its recommendations. See Smith 1964.

53. Bohr's July 1944 Memorandum to Roosevelt can be found at http://www.atomicarchive.com/Docs/ManhattanProject/Bohrmemo.shtml (accessed September 2007). See also Cantelon 1991.

54. For insightful expositions of Bohr's intentions, his efforts, and his meetings with Roosevelt and Churchill and the influential figures around them, see Aaserud 1999 and Gowing 1964, 346–366.

55. Frankfurter tried to assuage Roosevelt's concern by talking to Stimson. See, for example, Baker 1984, 387.

56. Clark 1973, 645–650.

57. John Anderson was the chancellor of the Exchequer in Churchill's wartime cabinet. He became a good friend of Bohr. See Oppenheimer 1964b and Gowing 1964.

58. Tolman, an outstanding theoretical physicist, was the dean of the graduate school at Cal Tech. During World War II he served as the scientific adviser to James Conant on matters relating to the Manhattan Project.

59. Gowing 1964, 359.

60. I am indebted to Finn Aaserud for this information.

61. See Gowing 1964, 346–366, for a detailed account of Bohr's interview with Churchill. It is interesting to note that the Frankfurter papers at the Harvard Law School Archives contain no materials relating to Frankfurter's interactions with Bohr and Einstein during World War II.

62. In addition to serving as chancellor of the Exchequer, Anderson oversaw the Tube Alloys project, the British nuclear program. Sir Henry Dale, a physiologist and pharmacologist, was the president of the Royal Society and served on Anderson's Consultative Council for the Tube Alloys project. Lindemann (Lord Cherwell) was a physicist who had made Oxford an outstanding center in low temperature physics. He was a close friend of Churchill and became his science adviser during World War II. Field marshall Jan Smuts, the prominent South African and commonwealth statesman, played an important role in Churchill's cabinet during World War II. He was an old friend of Bohr's and, together with Henry Dale, had helped him obtain a meeting with Churchill to discuss the long-term problem of the atomic bomb. See Gowing 1964, chap. 13.

63. Peter Kapitza was a Russian physicist who had been one of the outstanding experimentalists in Ernest Rutherford's Cavendish Laboratory in Cambridge and was a good friend of Bohr's. He was not allowed to return to Great Britain following a visit to Russia in 1934. He had access to Stalin, who evidently respected his views. Abram Joffe was a prominent Soviet physicist who made important contributions to the development of solid state physics in the Soviet Union.

64. Clark 1973, 577.

65. Einstein to Otto Stern, December 26, 1945, EA 22-240.

66. Einstein to Roosevelt, March 25, 1945, EA 39-485. Szilard's letter can be found in Lanouette 1992, 261–262.

67. The U.S. Department of Energy Archive offers this note on "MAUD": Although many people assume MAUD is an acronym, it actually stems from a simple misunderstanding. Early in the war, while Bohr was still trapped in German-occupied Denmark, he sent a telegram to his old colleague Frisch. Bohr ended the telegram with instructions to pass his words along to "Cockroft

and Maud Ray Kent." "Maud," mistakenly thought to be a cryptic reference for something atomic, was chosen as a codename for the committee. Not until after the war was Maud Ray Kent identified as the former governess of Bohr's children. From http://www.mbe.doe.gov/me70/manhattan/maud.htm (accessed May 2006).

68. Both Otto Frisch and Rudolf Peiels left Germany after Hitler's assumption of power in Germany. Frisch, together with his aunt, Lise Meitner, had given the explanation of the fission process in uranium induced by its bombardment by neutrons that had been observed by Hahn and Strassman in the winter of 1938. When World War II broke out in September 1939 they were both in England at Birmingham University. The Frisch-Peiels memorandum, written in March 1940, summarized their calculations concerning the possible construction of "super-bombs" based on a nuclear chain reaction in uranium. See Gowing 1964; Rhodes 1988. The memorandum is available at http://www.stanford.edu/class/history5n/FPmemo.pdf (accessed September 2007).

69. See, for example, *EoP* 350, 519. In September 1952 the editor of *Kaizo* asked him: "Why did you co-operate in the production of the atomic bomb although you were well aware of its tremendous destructive power?" His answer to the question was that "My participation in the production of the atomic bomb consisted of one single act: I signed a letter to President Roosevelt, in which I emphasized the necessity of conducting large-scale experimentation with regard to the feasibility of producing an atomic bomb. I was well aware of the dreadful danger which would threaten mankind were the experiments to prove successful."

70. Linus and Ava Helen Pauling Papers, Oregon State University Library, box 104, E: individual correspondence. Available on the Web at http://osulibrary.oregonstate.edu/specialcollections/coll/pauling/calendar/1954/11/16.html.

71. Einstein to von Laue, March 19, 1955, *EoP,* 620–621.

72. I thank Gerald Holton for pointing me to Heisenberg's remarks. See Holton 2005. It should be noted that in 1953 Einstein's *Mein Weltbild* had been published in Zurich by Europa Verlag.

73. It is quite possible that Heisenberg at the time was aware of the letter Einstein wrote to von Laue a few weeks before his death. Einstein to von Laue, March 19, 1955, *EoP,* 620–621.

74. See Rhodes 1988, chap. 19.

75. EA 39–488, 2.

76. Ibid., 352–353. See also Masters and Way's 1946 *One World or None,* which accurately and forcefully pointed out how atomic weapons had changed

the character of warfare, and described the steps that had to be taken to avoid their use in the future. The exposition was by the leading scientists who had been involved in the Manhattan Project.

77. See Smith 1965.

78. *Chicago Round Table,* August 12, 1945. Quoted in the pamphlet, *The First One Hundred Days of the Atomic Age.* Sydnor Walker edited the pamphlet in the fall of 1945 for the Woodrow Wilson Foundation (p. 14).

79. Einstein to Hutchins, September 10, 1945, *EoP* 337.

80. In October 1931 Einstein had written to Russell that Dr. Imre Revesz was soon traveling to England and asked him to grant him a brief interview (EA 33.157).

81. See, for example, Orville Prescott's review of it on June 13, 1945 in the *New York Times,* 21.

82. Reves to Einstein, August 24, 1945, EA 57–290.

83. Ibid. In fact, the reaction continued well into 1946. The December issue of *Reader's Digest,* an issue of some 10 million copies, carried the first of three installments of Reves's book. The same three extracts were published in the British, Spanish, Portuguese, Swedish, and Arabic editions of *Reader's Digest.* In addition, during the first half of 1946 *Reader's Digest* organized a discussion of *The Anatomy of Peace* in some 15,000 American clubs and discussion groups. Reves to Einstein, October 17, 1945, EA 57–299 and November 27, 1945, EA 57–300.

84. Reves to Einstein, August 24, 1945, EA 57–290.

85. Einstein to Reves, August 28, 1945, EA 57–292. In the fall of 1945 during Einstein's interview with Raymond Swing, Einstein commended Reves "whose book *The Anatomy of Peace*" is intelligent, brief, clear, and if I may use the abused term, dynamic on the topic of war and the need for world government." Before Einstein's endorsement, some 8,000 copies of the book had been sold. During the six months after Einstein had stated that the proper response to the atomic bomb was contained in Reves's *The Anatomy of Peace,* some 180,000 copies were sold in the United States. It was subsequently translated into 25 languages. Eventually, some 800,000 copies of the book were sold worldwide.

86. *EoP,* 340–341.

87. Reves, New York, December 19, 1945 to Winston S. Churchill, Gilbert 1997, 257.

88. Shortly after the death of Roosevelt, Truman on his way back from the San Francisco UN Charter meeting told a crowd in Kansas City: "It will be just as easy for nations to get along in a republic of the world as it is for you to get

along in the republic of the United States. Now when Kansas and Colorado have a quarrel over water in the Arkansas River they don't call out the National Guard in each state and go to war over it. They bring suit in the Supreme Court of the United States and abide by the decision. There isn't a reason in the world why we cannot do that internationally." Truman's statement was quoted in the *Open Letter to the American People* initiated by Justice Roberts and signed by Einstein. The entire open letter was reprinted on the cover of Emery Reves's *The Anatomy of Peace*.

89. Reves to Einstein, September 27, 1945, EA 57–293.00.

90. Einstein to Oppenheimer, September 29, 1945, EA 57–213. Indicative of how impressed Einstein was with Reves's book is the fact that on October 10, 1945 Einstein, together with Thomas Mann and eighteen other prominent men, signed a letter to the editor of the *New York Times* advocating "a Federal Constitution of the world, a working world-wide legal order, if we hope to prevent an atomic war" (*EoP*, 340–341). In that same letter, they collectively called attention to Reves's *The Anatomy of Peace* and urged people to read it, think about its conclusions, and discuss it with friends and neighbors privately and publicly.

91. Weber, "Politics as Vocation," in Weber 1946, 77–128.

92. "A man who believes in an ethic of responsibility takes account of precisely the average deficiencies of people; as Fichte has correctly said, he does not even have the right to presuppose their goodness and perfection. He does not feel in a position to burden others with the results of his own actions so far as he was able to foresee them; he will say: these results are ascribed to my action." Weber, "Politics as Vocation."

93. Stone 1946. See also Bernstein 1974.

94. A. H. Compton, "Swords into Plowshares," in Dibner 1959, 87–88.

95. However, nuclear weapons required and still require the creation of a huge infrastructure, namely, a nuclear industry, before the weapons can be produced, however cheaply or abundantly. Undoubtedly this was the reason Stalin and Beria insisted that the first Soviet weapon be an exact copy of the Alamogordo plutonium device, giving uranium enrichment (another huge infrastructure) a lower priority. See Holloway 1994.

96. See Sumner Wells's sharp dissent from the views expressed by Einstein, and Einstein's reply to these criticisms in the January 1946 *Atlantic Monthly*. See also *EoP*, 352–353.

97. William Golden at the time was attached to the office of Lewis Strauss at the AEC.

98. *Foreign Relations of the United States, 1947*, vol. 1 (Washington, D.C.: Government Printing Office, 1947), 487–489.

99. Einstein to von Schönaich, February 18, 1949.

100. The speech was delivered on April 27, 1948 and reported in the *New York Times* on April 29. See Nathan and Norden 1968.

101. Joe 1 was the U.S. moniker for the first Soviet nuclear test, which was announced by President Truman several weeks later.

102. Einstein to Muste, October 31, 1949, EA 60–630, 1.

103. In an addendum to the GAC's conclusions, Fermi and Rabi called the weapon genocidal and immoral. See York 1989.

104. With advice from Golden, Acheson, and Bradley, Truman's rejection of the GAC arguments was extremely swift. It came in late January 1950 on the heels of Klaus Fuchs's arrest.

105. For a brief biographical statement of this remarkable man, see http://www.ajmuste.org/ajmbio.htm.

106. Muste to Einstein, January 19, 1950, EA 60–630, 1, 2.

107. Einstein to Muste, January 23, 1950, EA 60–631, 1; *EoP*, 519–520.

108. Telegram, Muste to Einstein, January 30, 1950, EA 60–635, 1, 2.

109. Einstein to Muste, January 30, 1950. See also EA 60–636; *EoP* 520.

110. Muste to Einstein. September 29, 1950, EA 60–637 and October 9, 1950, EA 60–638. See also EA 60–639, EA 60–640, 1–4.

111. Einstein to Muste, October 11, 1950, EA 60–641.

112. The March 1, 1954 test, Bravo, developed three times the expected force (15 megatons instead of the expected 5). It was a "dry" lithium deuteride bomb whose weight made it deliverable by the rockets to which the United States was just committing itself to build. Further tests in the series, called "Castle," were indeed carried out. See Rhodes 1995.

113. Einstein to Muste, April 6, 1954, EA 60–651. The reference to the voice of angels is to be understood as implying slight mocking: Muste always had ministers joining him in his appeals—in this case Paul Scherer, "the country's foremost Lutheran preacher," and Howard Thurman, "negro poet and chaplain of Boston University."

114. EA 28–925, 1.

115. EA 28–920, 1.

116. *New York Times*, June 12, 1953. See also Jerome 2002, 241, 237–248.

117. Russell 1969, 18.

118. See Russell 1969. See also the numerous entries for Russell on Google. The above biographical material is based on http://www.san.beck.org/GPJ24-Russell,Muste.html#3 (accessed June 2005).

119. A copy of the Manifesto is reprinted in the Appendix as taken from http://www.pugwash.org/about/manifesto.htm (accessed May 2006).

120. Pugwash was the town in Nova Scotia, Canada, where Cyrus Eaton's estate, the site of the meeting, was located. See Rotblat 2001.

121. Consult the site of the Pugwash Organization. http://www.pugwash .org/site_index.htm, for further details about its conferences, workshops, and publications (accessed May 2006).

122. Joseph Rotblat was one of the very few scientists to leave Los Alamos; he left in December 1944, when it had become clear that Germany had been defeated. He was a Polish Jew who had emigrated to England before the war, and he was one of the sixteen British physicists to join the Los Alamos project in late 1943. At a private dinner in March 1944 at the home of James Chadwick, the head of the British Mission to Los Alamos, Groves had stated: "You realize, of course, that the whole purpose of this project is to subdue the Russians" (Rotblat 2002). Groves's statement shocked Rotblat. He had thought that the purpose of Los Alamos had been to cope with the German threat, but evidently its mission had already changed. Groves's declaration initiated Rotblat's doubts about the enterprise. Toward the end of 1944 he was informed by Chadwick, who was close to British intelligence and had been apprised of the findings of the Alsos mission (which searched Europe for signs of German atomic research), that Germany had stopped work on the bomb. He accordingly decided to end his participation in the project. Rotblat had believed "that if the US and Great Britain had developed the bomb, then even if Germany had it, we could have made the Germans give up using it. It was the idea of deterrence" (Rotblat 1995). Groves testified at the Oppenheimer hearings in 1954 that the views regarding the aim of the Los Alamos project expressed at the Chadwick dinner had shaped his policy regarding security matters at Los Alamos. "There was never from about 2 weeks from the time I took charge [of the atomic bomb] project any illusion on my part but that Russia was our enemy and the project was conducted on that basis. "In the Matter of J. Robert Oppenheimer," Transcript of Hearing Before Personnel Security Board, Washington, D.C., April 12, 1954 to May 6, 1954 (Washington, D.C., 1954), 173. Also quoted in Davis 1968, 151.

123. Leibniz in his *Monadology* and his essay on "Principles of Nature and of Grace, Founded on Reason. 1714" had introduced the notion of a "pre-established harmony" to account for the coherent views of the world of his monads—his windowless fundamental substances. See Leibniz 1934; Buchdahl 1969, 388–469.

124. Poincaré's list of useful principles included the two principles of thermodynamics, the principle of action and reaction (Newton's third law), the principle of relativity, the principle of conservation of mass, and the principle of least action. All the principles, except Carnot's, could be extracted from mechanics. Quoted in Darrigol 1995, 10–11.

125. In a letter to his friend Cornelius Lanczos, written on January 24, 1938, Einstein corroborated this: "I began with a skeptical empiricism more or less like that of Mach. But the problem of gravitation converted me into a believing rationalist, that is, into someone who searches for the only reliable source of Truth in mathematical simplicity" (EA 15.269.00).

126. The observed regularity that *all* objects near the surface of the earth fall with the same acceleration

$$g = \frac{GM_G}{R^2}\left(\frac{m_G}{m_r}\right)$$

where $m_G$ is the gravitational "charge" of an object—the mass that enters in Newtons law of universal gravitation for the attraction of two bodies of mass $m_{G1}$ and $m_{G2}$ separated by a distance $R$

$$F = G\frac{m_{G1}m_{G2}}{R^2}$$

and $m_I$ is the inertial mass of a body—which determines its response to a force $F$ and accelerates as determined by Newton's second law of motion: $F = m_I a$—implies that $m_G/m_I$ is a (universal) constant. For Newton, and every one before Einstein, this was an empirical fact, an *unexplained* regularity of nature, to be checked with ever greater accuracy.

That $m_G/m_I$ is a constant also implies that gravitation could be mimicked (locally) by transforming to an accelerated frame of reference, which in turn led Einstein to state his equivalence principle, which in 1907 he formulated as follows: "We have no reason to suppose that relatively accelerated reference systems can be distinguished from each other in any way. We shall therefore assume complete physical equivalence between the gravitational field and the corresponding acceleration of the reference system." Later in life, Einstein recalled, if his memory served him right, that he had "no serious doubt about the strict validity [of the law of the equality of inertial and gravitational masses] even without knowing the admirable experiments of Eötvös."

The road from this insight to general relativity was an arduous one. It first of all required Einstein to recognize the spatial and temporal components of his special relativistic description as elements of a *Minkowskian four-dimensional space-time*. The realization that the spatial geometry in an accelerated frame (such as a rotating frame) could not be Euclidian led him to a study of Riemannian geometry with his friend Marcel Grossman and a struggle to understand the meaning of general covariance. For a splendid, readily accessible exposition of Einstein's genesis of general relativity, see Einstein 1920 and

Eisennstaedt 2006. See also Schwinger 1986; Cao 1997; Stachel 2002; the proceedings of the conferences on the history of general relativity; and in particular Renn and Schemmel 2007.

127. First explicitly stated in Einstein's article in *The London Times*, November 28, 1919.

## 2. Albert Einstein and the Founding of Brandeis University

1. Einstein to Abba Eban, November 18, 1952, quoted in Gerald E. Tauber, "Einstein and Zionism," in French 1979.

2. Sayen 1985 gave a brief account of this story.

3. *Östjuden* were Jews coming from Czekoslovakia, Hungary, Roumania, Russia, Latvia, Lithuania. and Poland, They were less polished socially than German Jews and many of them adhered to orthodox Judaism. They were fleeing the poverty and dismal economic conditions that existed after World War I in the countries they came from. German Jews ostracized them and did not wish to be identified with them.

4. Beider 2000. In that speech he also reiterated his attitude toward Jewish nationalism: "Jewish nationalism is today a necessity because only through a consolidation of our national life can we eliminate those conflicts from which the Jews suffer today." But as he had done before, Einstein almost apologized for Jewish nationalism and expressed the following hope: "May the time soon come when this nationalism will have become so thoroughly a matter of course that it will no longer be necessary for us to give it a special emphasis."

5. Einstein to Paul Ehrenfest, April 12, 1926, EA 10–136.00.

6. See Batterson 2006 for the details of this story.

7. See the letter from Boris Young to Drioshua Loth Liebman, March 11, 1947, where Einstein is quoted as holding this view. Alpert Papers, Einstein Folder, RDFA.

8. See Batterson 2006, chap. 9.

9. Einstein to Lilienthal, July 9, 1946, EA 40–398.

10. Einstein to Boris Young, August 17, 1946, EA 40–203.

11. Newman 1923. For reactions, see the Stephen S. Wise Papers, American Jewish Historical Society, Waltham, Massachusetts, and New York, New York (STWP); in particular, Wise to Seiman, January 11, 1924, and Gross to Wise, March 25, 1930. At the recommendation of Ralph Lazrus, the chairman of the Board of Directors of the Albert Einstein Foundation, the fund-raising instrument for the embryonic Brandeis University, Sidney Hertzberg was commissioned in late 1946 to write a brochure to be sent to potential donors. It contains a brief history of the early efforts to found a Jewish-sponsored university. His

twenty-three-page manuscript can be found in the George Alpert Papers in RDFA. See also Sachar 1976, chap. 1.

12. Silver had been associate-counsel of the Senate's Committee on Banking and Currency.

13. Goldstein had been president of the Jewish National Fund and of the Zionist Organization of America. He played a significant role in the deliberations and negotiations that brought about a Jewish National state in Palestine. See Goldstein 1984; Lacqueur 1972.

14. Goldstein 1951; Einstein Papers, EA 40–378. The original copy of the letter is in the George Alpert Papers, Einstein Folder, RDFA.

15. As quoted in a letter from Boris Young, the director of the Einstein Foundation for Higher Learning, to Dr. Joshua Loth Liebman, March 11, 1947. See George Alpert Papers, Einstein Folder, RDFA.

16. The latter was organized and incorporated by late February 1946 with Goldstein as president. See Goldstein to Einstein, March 7, 1946, EA 40–375.

17. Einstein to Goldstein, March 4, 1946, EA 40–374. In his letter Einstein described Nathan as being "one of the most excellent persons I know concerning unselfishness of character and devotion to his work as scholar and researcher and to the Jewish cause. He is also experienced in question of organization and administration."

18. Otto Nathan and Helen Dukas, Einstein's longtime secretary, became the executors of Einstein's estate. In 1955, after Nathan had become the executor of the Einstein estate, he was refused a passport under the McCarran Act, which justified such action by the State Department if the trip abroad "might be for the purpose of advancing the Communist movement." Nathan filed suit to get his passport and at the same time filed an affidavit that he never was a member of the Communist Party. In May 1957, Nathan was ruled in contempt of Congress and indicted for refusing to answer questions posed by the House un-American Activities Committee as to whether he was a member of the Communist Party. He refused to answer, pleading the First Amendment, and challenged the committee's jurisdiction.

19. *Boston Traveler,* April 5, 1946.

20. Einstein to Goldstein, April 16, 1946, EA 40–376. Nathan was interested in becoming involved in the selection of the faculty and possibly in obtaining either a faculty or an administrative position. He had contacted Goldstein and had spoken to him by telephone. See Goldstein's response to Einstein's letter in which he indicates to him that it is somewhat premature to worry about faculty appointments. Goldstein to Einstein, April 19, 1946, EA 40–377.

21. Stephen S. Wise was the spiritual leader of the Free Synagogue in New York, whose creed melded a liberal and progressive religious outlook with

a traditionally oriented Judaism. He was at that time one of the most influential Reform rabbis in the United States.

22. A copy of the letter of invitation to the reception, essentially the draft in which Goldstein had outlined his vision and plans for the university, was included in his letter to Wise. Goldstein to Wise, April 29, 1946, Box 49, Folder 9, RDFA.

23. There is an extensive correspondence between Wise and Einstein in the Wise Papers; see Boxes 36 and 108, STWP. Similarly, there is a significant correspondence between Wise and Nathan dating back to the 1930s. See Box 78, STWP. Nathan, as Einstein's protégé and confidant, saw Wise socially on a fairly regular basis.

24. It was at Otto Nathan's suggestion that Goldstein's list of "endorsers and sponsors" was enlarged to include "prominent Scientists and Scholars," in particular Paul Klapper, president of Queens College; Karl Compton, president of MIT; and Alvin Johnson, one of the founders of the New School for Social Research in New York, and from 1922 until 1945 its director. In the 1930s and 1940s Johnson helped many European scholars, who had been dismissed from their posts in Germany or were fleeing from the Nazi occupation of their homeland, find positions in the United States. He created a "University in Exile" at the New School to accommodate them. See Einstein to Goldstein, May 14, 1946, EA 40–386 and Goldstein to Einstein, May 16, 1946, EA 40–387.

25. Wise to Nathan, May 16, 1946. See Box 78, Folder 18, STWP.

26. Thus, in late December 1945 Wise had written Goldstein to let him know "how fully I appreciate your understanding and wisdom, as well as your personal friendship which saved, perhaps averted, a really nasty situation [when dealing with the State Department]." And he concluded his letter: "With warmest regards, dear Israel." Wise to Goldstein, December 29, 1944, Box 49, Folder 9, STWP.

27. Quoted in Laqueur 1972, 577. Laqueur gives a succinct account of this momentous Congress, which came on the heels of the bombing of the King David Hotel by the Irgun. Weizmann viewed such acts of terrorism as a "cancer in the body politic of the yishuv." See also Goldstein 1984, vol. 1, 192–210, and 201 in particular.

28. Wise to Nathan, June 6, 1946, EA 40–388.

29. Goldstein to Wise, June 25, 1946, Box 49, Folder 9, STWP.

30. Einstein to Goldstein, July 1, 1946, EA 40–391. Behind the scenes Nathan had made his position clear to Einstein. See Nathan to Einstein, June 28, 1946, EA 40–390, 1.

31. Einstein to Wise, June 26, 1946, EA 35–263.

32. Wise to Einstein, June 28, 1946, Box 36, Folder 6, STWP.

33. Einstein to Wise, June 29, 1946, EA 35–265.

34. Goldstein to Einstein, July 20, 1946, EA 40–395.

35. Einstein to Lilienthal, July 9, 1946, EA 40–398.

36. Einstein to Boris Young, August 17, 1946, EA 40–403.

37. Einstein to Jackson, August 19, 1946, EA 40–402.

38. Einstein to Frankfurter, August 20, 1946, EA-404. In a handwritten addendum to the letter, Einstein indicated that this was an *"official"* letter that he had signed but not written—but that it is "wirklich richtig" (truly correct).

39. See Boris Young to Helen Dukas, August 29, 1946, EA-40.404, 9.

40. Abram Leon Sachar was the son of immigrant parents. He was born in New York City in 1899 and raised in St. Louis. He obtained a B.A. and an M.A. in history from Washington University and went to England to earn a Ph.D. from Cambridge University in 1923. Upon his return he joined the history faculty of the University of Illinois from 1923 to 1948, and became a well-known and respected historian of modern times. His *History of the Jews* went through five editions and twenty-six printings. He was one of the organizers, and later the director, of the B'nai Brith Hillel Foundation, the foundation responsible for establishing Hillel Houses for Jewish students on the campuses of many American universities. After World War II, he became very active in making it possible for gifted young people who had been in concentration camps to attend American colleges.

41. Einstein to Goldstein, September 2, 1946, EA 40–407. Einstein sent a copy of the letter to Stephen Wise on September 4, 1946, and a copy of it can be found in the Wise Papers, STWP. Einstein sent copies of his letter to Goldstein to the members of the Board and the advisory committee. In the letter Einstein sent to Frankfurter on September 2, 1946, he indicated that the reasons for his severing his ties with the project were because "behind my back, Cardinal Spellman has been invited to deliver the invocation (benediction) at the dinner and . . . some other irregularities which seem important to me occurred"; EA 40–407. See Frankfurter to Einstein, September 6, 1946, for Frankfurter's reply, EA 40–412; see also Paul Klapper to Einstein, September 7, 1946, EA 40–414.

42. Goldstein's September 12, 1946 letter to Einstein is reprinted in Goldstein 1951. Not "the slightest tinge of a commitment" had been made in Goldstein's conversation with Sachar and had been so reported by Goldstein to Nathan. Goldstein's retelling of the story in his memoirs is a fair and even-handed account. See Goldstein 1984, 172–185.

43. Minutes of the meetings of the Board of Directors of the Albert Einstein Foundation for Higher Learning, Inc., RDFA.

44. Alpert had also indicated to Nathan that a new academic advisory committee would be set up and that the original agreements would be adhered to. Nathan to Einstein, EA 40–416.

45. Wise to Nathan, November 8, 1946, Box 78, Folder 18, STWP.

46. For example, Einstein to Lilienthal, October 10, 1946, EA 40–417.

47. Einstein to Sachar, October 23, 1946, EA 40–425. Einstein and Sachar knew each other since the 1930s. As head of the Hillel foundations, Sachar had been active in creating scholarships for refugee students to attend American colleges and universities and had corresponded with Einstein regarding these matters. See the Einstein-Sachar correspondence in the Einstein Papers.

48. On November 11, 1946, Nathan became a member of the Board. Minutes of the Board, Albert Einstein Foundation, RDFA.

49. Otto Nathan, November 9, 1946, EA 40–427. Evidently, the Education Board also asked Paul Sweezy, a distinguished Marxist economist who taught at Harvard, to formulate a proposal for the structure of Brandeis University, for in January 1947 he submitted an eighty-nine-page document entitled "A Plan for Brandeis University" (EA 40–461). The assumptions that underlay his plan were clearly stated at the outset: (1) that the heart and soul of a university is its faculty, and the ultimate source of its authority; (2) that a university is, or should be, a *community* of scholarship and learning. Neither standards nor incentives can be imposed from without; they must be evolved and maintained from within. Sweezy's plan is interesting because several of its features were adopted. Sweezy recommended that the institution become a first-rate small university and that it should start out as a college with a faculty of 100 and a student body of about 500. The initial disciplinary emphasis ought to be on the social sciences and the humanities. The faculty ought to be organized into schools and not into departments. Sweezy also recommended that special attention be paid to the problem of "attracting Negroes for both faculty and student body" and suggested that a certain number of scholarships be set aside "exclusively for Negro students." For a brief biography of Sweezy, see http://www.monthlyreview.org/1004jbf .htm (accessed January 2007).

50. Minutes of the meeting of the Board of Directors of the Albert Einstein Foundation, January 6, 1947. The entry in the minutes reported that "In his conference with these educators he discussed many facets of academic administration, such as life tenure of office, the tutorial system, the size of the student body, the age of the faculty and related subjects. A more complete report of the trip will be submitted to the Board in due course." Nathan's expenses were paid by the Albert Einstein Foundation. His involvement and expenses intensified

during the first four months of 1947. See the Otto Nathan 1946–1941 folder in the Alpert Papers, RDFA.

51. See Kramnick and Sheerman 1993 for a thorough account of Laski's life and works. See also Deane 1972 for an analysis of Laski's writing.

52. See the *Manchester Guardian* and the *New York Times* for December 3, 1946. Both newspapers carried extensive coverage of the trial. A full record of the trial is given in Laski 1947. Kramnick and Sheerman devote an entire chapter to it (1993, 516–543).

53. Nathan wrote to his friends, among them Ralph Lazrus and Stephen Wise, asking for contributions to help Laski. Wise sent him a check for $25, and Lazrus also contributed.

54. Thus in the spring of 1947, Nathan had become involved in finding a new dean for the veterinary school. Whether he had made a certain moral commitment on behalf of the university to a particular candidate became a contentious matter at the Board meeting of June 4, 1947.

55. Einstein to Lazrus, July 19, 1947, Alpert Papers, Einstein Folder, RDFA. In a letter to Judge Steinbrick written on February 2, 1949, Alpert gives a differing account: "In 1947 at a meeting in Professor Einstein's home at which, in addition to Professor Einstein, Mr. Lazrus, Professor Nathan and I were present, it was suggested by Professor Nathan that authority be given to Professor Einstein to select and engage a President of the University. My reaction to the suggestion was that this was a function of the Board which it could not delegate. Professor Nathan, however, insisted that the Board should have sufficient confidence in Professor Einstein to leave the selection entirely up to him. I then asked whether Professor Einstein had anyone in particular in mind and was informed that it was Professor Laski. I pointed out that I had heard of the libel suit brought by Professor Laski in England and of the outcome of the suit, and that I did not feel that a figure as controversial as he even though he be a learned person, should be selected as president of Brandeis University. Furthermore, while the University was Jewish sponsored, it was to be an American institution,—a contribution by the Jews of America to American education and that I believed it to be very desirable that its first president be an American. With this point of view, Mr. Lazrus and Professor Nathan did not agree. . . . It was left that a meeting of the Board be held the following week and the matter placed before the Board." Alpert to the Hon. Meier Steienbrink, February 2, 1949, Alpert Papers, RDFA.

56. Minutes of the Meetings of the Board of Directors of the Albert Einstein Foundation, p. 104, RDFA.

57. Boris Young, the director of the Albert Einstein Foundation, attended the

April 14 meeting "by invitation." On June 5, on the heels of Lazrus and Nathan's resignations from the Board, he submitted a report of that meeting to Julius Silver. It stated: Present: S. Ralph Lazrus, chairman, Otto Nathan, Milton Bluestein, Israel Rogosin. Boris Young by invitation. After Chairman inquired if a quorum was present, and upon being informed by Mr. Young that there was no quorum, he decided to hold informal discussion of matters pending for Board consideration. Major Abraham F. Wechsler joined the meeting shortly after this point. Throughout the discussion no resolutions or motions were offered or acted upon, nor was an acting secretary appointed to substitute for Dr. Dushkin, who was absent. Nathan who was shown that report by Young, crossed out the last sentence and wrote in pencil: "From this point on a quorum existed." Boris Young to Julius Silver, June 5, 1947. A report submitted relating to the meeting of April 14, 1947 at 4:00 P.M. in Suite 4 of Hotel Pierre. Minutes of the Meetings of the Board of Directors of the Albert Einstein Foundation, p. 104, RDFA.

58. Silver to Lazrus, May 12, 1947, Alpert Papers, Einstein Folder, RDFA. At the June 4, 1947 meeting of the Board, after the rift between Einstein, Lazrus, Nathan, and the Board had occurred, "Dr. Nathan desired the minutes to disclose that it was his understanding that at the meeting held on April 14, 1947 Professor Albert Einstein was authorized to choose a President for Brandeis University despite the contention by others that no quorum was present."

59. Einstein to Laski, April 16, 1947, EA 40–432.

60. Laski to Einstein, April 25, 1947, EA 40–435.

61. There is a further somewhat curious aspect to the story: In 1975 Granville Eastwood, a retired English trade union man, was preparing his biography of Harold Laski. He contacted Nathan, who allowed him to read his correspondence with Laski. Among the letters was a copy of Einstein's letter to Laski offering him the presidency of Brandeis. Eastwood wanted to make use of this material and asked Nathan for permission to use its content. In an October 1975 letter to Helen Dukas asking for her approval of the material Eastwood wanted to include in his book (see Eastwood 1977, 85–86), Nathan added that "for reasons unknown to me I carried Laski's reply to Einstein's letter in my wallet for many years. When I gave you Einstein's letters to me to have them photostated I do not recall whether I gave you also Laski's letter which was addressed to Einstein or whether that letter was lost when I lost my wallet last year. I say all that because Eastwood's draft might lead to inquiries from other people about your and my own files in regard to Brandeis University. There is nothing in it, from our point of view, that could not see the light of day but I do not quite know what we ought to do and would like to know how you feel." Nathan

to Helen Dukas, October 28, 1975, EA 40–436. Nathan had clearly given Laski's reply to Dukas, as it is to be found in the Einstein Papers.

62. Silver to Lazrus, May 12, 1947, Alpert Papers, RDFA. Again, in his letter of February 2, 1949 to Steinbrick, Alpert gives a differing account. In the letter Alpert claimed that immediately after the April 14, 1947 meeting a conference was arranged in New York with Lazrus, Nathan, and Silver. "At this conference there was a lengthy and at times, spirited discussion. To my objection that nothing should be done in the selection of a president until the Board authorized, Professor Nathan repeatedly stated that Professor Einstein did not propose to be a "rubber stamp for any Board of Directors." The conference ended in a tone that was anything but amiable. Immediately after this conference, Mr. Silver and I drove to Princeton. We invited Mr. Lazrus and Professor Nathan to go along but they refused, Professor Nathan stating that he didn't understand what right I had to make an appointment to call on Dr. Einstein without his, Nathan's, prior consent." The tenor of Silver's May 11, 1947 letter to Lazrus does not jibe with Alpert's account. The already strained relations between Alpert and Nathan must have turned acrimonious in late May 1947.

63. Max Grossman to Nathan, April 20, 1947, Alpert Papers, Otto Nathan Folder, RDFA.

64. Alpert to Nathan, March 24, 1947, Alpert Papers, Otto Nathan Folder, RDFA.

65. Einstein to Lazrus, May 16, 1947, Alpert Papers, Einstein Folder, RDFA.

66. In the typewritten minutes in the Farber Archives, the last part of the last sentence was crossed out and amended by Nathan to read "that Professor Einstein and Mr. Lazrus and he do not want to injure the University and wished the project well." Minutes of the Meetings of the Board of Directors of the Albert Einstein Foundation of May 19, 1947, 114–115, RDFA.

67. The name of the Foundation was changed to the Brandeis Foundation. The press statement that was released in early June stated that the change was made after conferring with Professor Einstein "because the Albert Einstein Foundation had accomplished its purpose, namely the actual establishment of the school." It also explained that maintaining the Albert Einstein Foundation designation would confuse contributors, and that the new name was descriptive of the its present function: the support of Brandeis University.

68. Einstein to Susan Brandeis Gilbert, June 20, 1947, Alpert Papers, Einstein Folder, RFDA.

69. See *Boston Sunday Herald,* June 29, 1947.

70. *New York Times,* June 23, 1947.

71. Zukerman 1947, 6.

72. Wise Papers, Box 36, Folder: Brandeis University, STWP.

73. *PM* printed the first paragraph on June 29, 1947.

74. EA 40–447.

75. Thus Boris Young, the executive director of the Brandeis Foundation (the new name of the Albert Einstein Foundation after July 1947), wrote to Evelyn Van Gelder on October 14, 1947, stating that Albert Einstein had opposed Spellman's invocation for political reasons. Also that "In the Spring of this year, 1947, Mr. Lazrus and Dr. Nathan decided (although this was outside their own sphere, for neither Lazrus nor Nathan served on the Board of Trustees of the University) to install as first president of the University one of the most controversial political and educational figures of our times." In fact, earlier that year, on January 7, 1947, Young in his letter to E. Saveth had listed the membership of the Board of Trustees of the university. The list names S. Ralph Lazrus as a member! Alpert Papers, Einstein Folder, RDFA.

76. See, for example, the *Boston Sunday Herald,* June 29, 1947, from which article the following is abstracted.

77. It should be noted that three other "New York" Board members resigned at the meeting at which Lazrus and Nathan tendered their resignation.

78. Incidentally, for a while Stephen Wise remained very hostile to the Brandeis enterprise. He strongly advised Sachar not to take its presidency. See Wise to Sachar, March 26, 1948, STWP. Parts of this letter were reprinted in Sachar 1976, 39. Wise eventually changed his mind, and one of his last public appearances was a visit to the Brandeis campus.

79. On May 17, 1949, Sachar wrote Einstein to inform him of developments at the newly founded university: "We are admitting a second class of freshmen numbering approximately 150, and curriculum and faculty will naturally expand. In developing the area of Biology we have had the cordial cooperation of Dr. Selman Waksman with whom we are clearing personnel, and of course we are delighted with his interest in this first Jewish sponsored university. We must now bring in a fine Physicist. Quite a number of applications and recommendations have come to us. One of the applicants is Dr. Nathan Rosen, Professor of Physics at the University of North Carolina." Rosen was a friend and former associate of Einstein, who had drafted the EPR paper. Einstein had earlier indicated that, given past events, he would not suggest the name of young physicists who might be approached to join Brandeis, but would give his evaluation of candidates who had applied on their own. In his answer Einstein indicated that he did not believe that Rosen possessed the requisite stature for the job.

80. Albert Einstein to Sidney Shalett, December 31, 1951, EA 40–454.

81. Sachar to Einstein, April 5, 1952, EA 61–213.

82. "To forget injuries suffered, is like throwing well earned money out the window."

83. Einstein to Sachar, March 30, 1952, EA 61–212.

84. Einstein to Chakravarty, December 2, 1952, EA 40–195, 2.

85. Einstein to Sachar, February 22, 1953, Sachar Papers, RDFA.

86. Einstein to Sachar, January 12, 1954, EA 61–214.

87. Einstein to Nathaniel L. Goldstein, May 9, 1953, EA 40–450.

88. Sachar to Goldstein, June 2, 1948. In Goldstein 1951, 113. But see the somewhat less generous account in Sachar 1976.

89. These quotations from Einstein can be found in Isaiah Berlin's essay "Einstein and Israel," in Berlin 1981.

90. The speech that Einstein wrote for the fund-raising dinner of March 20, 1947 can be interpreted as making this point (EA 40–431). The speech was not delivered, but its text was published in the *Jewish Advocate* in April 1947.

91. That high academic standards were likewise imposed on the students is evidenced by the fact that of the 100 freshmen admitted in 1948 many did not graduate in 1952. There were 104 students graduating in 1952—but this reflected the fact that after 1948 a sizable number of students were being admitted with advanced standing. Of the 104 B.A. and B.S. degrees conferred in 1952, 23 were cum laude, 10 were magna cum laude, and 1 was summa cum laude.

## 3. J. Robert Oppenheimer

1. A remark Oppenheimer had made to John Edsall, his close friend at the time (Smith and Weiner 1980, 93).

2. Kennan first saw Oppenheimer in the fall of 1946 when Oppenheimer came to deliver a lecture at the National War College. In his *Memoirs* Kennan described that encounter as follows: "He shuffled diffidently and almost apologetically out to the podium: a frail, stooped figure in a heavy brown tweed suit with trousers that were baggy and too long, big feet that turned outward, and a small head and face that caused him, at times, to look strangely like a young student. He then proceeded to speak for nearly an hour, without the use of notes—but to speak with such startling lucidity and such scrupulous subtlety and precision of expression that when he had finished, no one dared to ask a question—everyone was sure that somehow or other he answered every possible point. [But] curiously enough, no one could remember exactly what he had said. The fascination exerted by his personality, the virtuosity of the performance, and the extreme subtlety of expression had actually interfered with the receptivity of the audience to the substance of what he was saying. This was to dog him

throughout his life whenever it fell to him to address any other than a scientifically specialized audience" (Kennan 1972, 18).

3. For a much fuller, insightful account of Oppenheimer's early years, see
Bird and Sherwin 2005.

4. There is an eighty-year rule at Harvard University that prohibits access
to students' records. All records are sealed for eighty years, and only
students/alumni have access to their own records. In the case of death, access
can only be given to the closest family members. In order to access a file, a nota-
rized request along with proof of relationship to the alumnus and a copy of the
death notice must be submitted in order to view the records. A list of courses
taken—without information about grades received—can only be obtained from
a student's record card and thus is unavailable before eighty years have elapsed.

5. As a freshman, instead of the usual Introductory Chemistry course, Op-
penheimer took the more advanced Organic Chemistry (Chem 2) and Qualitative
Analysis (Chem 3). James Bryant Conant was his instructor in the Experimental
Organic Chemistry course (Chem 22) that he took as a sophomore. He also took
Richards's Chemistry 8, Elementary Theoretical and Physical Chemistry including the Historical Development of Chemistry, as a sophomore, and Chemistry 6,
Physical Chemistry, a course primarily for graduates, as a junior and complained
that both were not mathematical enough. Interestingly, he only took the introductory Mathematics C course, Analytic Geometry—Introduction to Calculus, as
a freshman (Mathematics C was a prerequisite for all the numbered mathematics
courses and was the introduction to the calculus for students who had taken
trigonometry in high school), and the more advanced Math 5a, Differential and
Integral Calculus, as a sophomore. He took Math 9, a course on probability, during his sophomore year, for which he obtained the grade of B. He was evidently
dissatisfied with his freshman chemistry courses because during 1923–1924, his
sophomore year, he took three physics courses: Physics 6a, Heat and Elementary
Thermodynamics, with Edwin Kemble; Physics 6b, Advanced Thermodynamics,
with Percy Bridgman; and Physics 16a, Quantum Theory with Applications to Series Spectra, Atomic Structure, and the Kinetic Theory of Gases, with Kemble. To
satisfy the language requirement he took French B, French Grammar, Translation and Composition as a freshman, and French 6, General View of French Literature, Reading of Texts, as a sophomore. To fulfill his other requirements as a
freshman, he took English A, Rhetoric and English Composition, and obtained a
B in the course, and Philosophy A, History of Philosophy, in which he earned an
A; and as a sophomore he took Philosophy 9c, Theory of Knowledge—a study of
judgment, inference, truth and allied topics, taught by a Dr. Eaton, who gave him
an A. During his third year he took two chemistry courses to fulfill the degree

requirement as a Chemistry major: Chemistry 6, Physical Chemistry, "primarily for graduates," with Theodore Richards, the Nobel Prize–winning chair of the Chemistry department, and Chemistry 14c, Colloid Chemistry, with Jones. Except for the seminar on Metaphysics, Philosophy 20h, Philosophic Presuppositions of Science, with Alfred North Whitehead (for which he earned a B) and History 12, The History of England from 1688 to the Present Time, all the other courses he took during that year were either mathematics or physics courses: Mathematics 8, the graduate course in Dynamics, with Oliver Kellogg; Mathematics 10a, Introduction to the Theory of Potential Functions and Laplace's Equation, also with Kellogg; Mathematics 10b, The Analytic Theory of Heat and Problems in Elastic Vibrations, with Garrett Birkhoff; and Physics 9, Mathematical Theory of Electricity and Magnetism, with Bridgman.

6. Recall Oppenheimer's March 1932 letter to his brother Frank regarding discipline (Smith and Weiner 1980, 155–156).

7. For details, see Smith and Weiner 1980; Schweber 2000; and particularly Bird and Sherwin 2005.

8. See in this connection the interviews of Alice Kimball Smith with Bernheim. A. K. Smith Papers, MIT Archives. Pais in his autobiography indicates that already in the early 1950s he was convinced that a "strong, latent homosexuality was an important ingredient in Robert's emotional makeup" (Pais 1997, 241). In Oppenheimer's FBI files, the following piece of extremely derogatory information was obtained during an interview with (name deleted) of the University of California at Berkeley: "It was further stated by (name deleted) that it was common knowledge on campus that prior to Oppenheimer's marriage he was possessed [sic] with homosexual tendencies and at the time was having an affair with Harvey Hall, . . . a mathematics student at the University, who was an individual of homosexual tendencies and at the time was living with Robert Oppenheimer" (V. P. Kay to Mr. Ladd, November 10, 1947, J. R. Oppenheimer, FBI security file [microform]. Wilmington, Del.: Scholarly Resources, Reel 1).

9. The quoted passage is to be found in the first volume of *Du Coté de chez Swann*. It comes at the end of Proust's depiction of the frustration of Mademoiselle Vinteuil, a lesbian and sadist, for being incapable of gratifying her lust for evil. It is ironical that Oppenheimer would recite the Proustian passage to Chevalier. It is Oppenheimer's disclosure that Chevalier had approached him to see whether he would be willing to share atomic secrets with Eltenton, to be transmitted to the Soviet Union, that wrecked Chevalier's life and contributed to the savage attack on Oppenheimer at his clearance trial in 1954. The changing character of the stories that Oppenheimer told the security officers at the

Radiation Laboratory and at Los Alamos have some similarities with his telling
of the Blackett incident.

10. See Oppenheimer 1984, 173, for his recollection of these researches and
Merzbacher 2002, for a history of the explanation of quantum mechanical
tunneling.

11. See also Serber 1998.

12. See Barkan 1993.

13. The contributions of Oppenheimer and that of his students to particle
physics during the 1930s were detailed in Serber 1983, 206–221.

14. He had contracted dysentery when visiting his grandfather in Bohemia as
a result of exploring a cave to collect interesting geological specimens.

15. See Schweber 1994 for the postwar developments in renormalization theory, which corroborated this.

16. In his letter to general Kenneth D. Nichols, appealing the revocation of his
security clearance, Oppenheimer adumbrated his relationship with Jean Tatlock
as follows: "In the spring of 1936, I had been introduced by friends to Jean Tatlock, the daughter of a noted professor of English at the university [of California, Berkeley]; and in the autumn, I began to court her, and we grew close to each
other. We were at least twice close enough to marriage to think of ourselves as
engaged. Between 1939 and her death in 1944 I saw her very rarely. She told me
about her Communist Party memberships; they were on again, off again affairs,
and never seemed to provide for her what she was seeking. I do not believe that
her interests were really political. She loved this country and its people and its
life. She was, as it turned out, a friend of many fellow travelers and Communists,
with a number of whom I was later to become acquainted" (Oppenheimer 1970,
8). Tatlock suffered from severe depression and committed suicide in July 1944.

17. The role played by Lise Meitner in this discovery, unacknowledged by
them, is beautifully narrated in Sime 1996.

18. For the best account of the development of atomic weapons, see Rhodes
1988. For Frisch and Peierls's contributions, see Gowing 1964. For Oppenheimer, see Bird and Sherwin 2005.

19. In his presentation to the Gray Board that reviewed his security clearance
in 1954, Oppenheimer noted that "Los Alamos was a remarkable community,
inspired by a high sense of mission, of duty and of destiny, coherent, dedicated
and remarkably selfless. There was plenty in the life of Los Alamos to cause irri-
tation . . . but I have never known a group more understanding and more devoted to a common purpose, more willing to lay aside personal convenience and
prestige, more understanding in the role that they were playing in their country's
history. Time and time again we had in the technical work almost paralyzing

crises. Time and again the laboratory drew itself together and faced new problems and got on with the job. We worked by night and by day; and in the end the
many jobs were done" (Oppenheimer 1970, 14).

In the Reith lectures Oppenheimer delivered over the BBC in 1953, he recalled
Los Alamos in more personal terms, speaking of his experiences of "the power
and the comfort in even bleak undertakings, of common, concerted, co-operative
life." "Each one of us knows how much he has been transcended by the group of
which he is a part; each one of us has felt the solace of other men's knowledge to
stay his own ignorance, of other men's wisdom to stay his folly, of other men's
courage to answer his doubts and weakness" (Oppenheimer 1989, 71).

20. Jornada del Muerto is variously translated from the Spanish as either
"Journey of Death" or "Trail of the Dead Man."

21. Incidentally, that answer is most often given by the members of orchestras themselves.

22. The acronym CERN originally stood for *Conseil Européen pour la Recherche
Nucléaire* (European Council for Nuclear Research), the provisional council responsible for setting up the nuclear high energy laboratory that was established
in 1952 in Geneva by eleven European nations. The acronym was retained for
the new laboratory after the provisional council was dissolved, even though in
1954 the name changed to the current *Organisation Européenne pour la Recherche
Nucléaire* (European Organization for Nuclear Research).

23. Stimson chaired the committee. It consisted of Ralph Bard, the assistant
secretary of the Navy, Vannavar Bush, William Clayton, Karl T. Compton,
James Conant, and George Harrison, Stimson's scientific aide. After Roosevelt's
death, James Byrnes was added to the committee.

24. As chairman of the Scientific Panel and a member of the Target Committee, which was to give detailed recommendations on such matters as the height
of the detonation, psychological factors in target selections, and radiological effect, Oppenheimer considered himself a personal consultant to Stimson on the
use of the bomb.

25. For a detailed account of the rest of the deliberations of the Interim
Committee on May 31, see Bird and Sherwin 2005, 293–297.

26. Oppenheimer to Stimson, August 17, 1945; Smith and Weiner 1980,
293–294.

27. It was Stimson who drafted the text of Truman's address. For a statement of the draft and Truman's final version see http://www.trumanlibrary
.org/whistlestop/study_collections/bomb/large/documents/fulltext.php?full
-textid=20.

28. For a detailed history of the May-Johmson bill, see Smith 1965.

29. The other signatories included Frank Aydelotte, Norman Cousins, Allan Nevins, G. Bromley Oxnam, Alexander Sachs, Henry DeWolf Smythe, Raymond Swing, Harold Urey, and the members of the executive committee of the Oak Ridge, Manhattan Project, and Chicago Atomic Scientists.

30. EA 57–10.

31. John McCloy was assistant secretary of war during World War II. As the energies of seventy-seven-year-old Secretary of War Henry Stimson declined, McCloy's became ever more involved in policy matters regarding both the conduct of the war and the planning for the postwar peace. McCloy took part in many of the most important governmental meetings in 1945 on how to obtain Japan's surrender. He was present in the discussions at the meetings of Secretary of War Stimson, Secretary of the Navy James Forrestal, and Acting Secretary of State Joseph Grew, later known as the Committee of 3.

McCloy later recalled having stated at the June 18, 1945 meeting that President Truman convened with the commanders of the U.S. military at which he approved an invasion of mainland Japan, scheduled for November 1, 1945, that "we ought to have our heads examined if we don't explore some other method by which we can terminate this war than just by another conventional attack and landing" (Herbert Feis Papers, container 79, Len Giovannitti and Fred Freed interviews, Library of Congress). When Truman asked McCloy to indicate what he had in mind, McCloy offered the following ideas for obtaining surrender by diplomatic methods: "Some communication to the Japanese government which would spell out the terms that we would settle for—there would be a surrender: I wouldn't use again the term 'unconditional surrender,' but it would be a surrender that would mean that we would get all the important things that we were fighting for . . . if we could accomplish our objectives without further bloodshed, there was no reason why we shouldn't attempt to do it" (Herbert Feis Papers, container 79, Len Giovannitti and Fred Freed interviews, Library of Congress).

McCloy proposed that the United States remind Japan of its great military superiority, and that it would "permit Japan to continue to exist as a nation . . . , that [it] would permit them to choose their own form of government, including the retention of the Mikado, but only on the basis of a constitutional monarchy" (Herbert Feis Papers, container 79, Len Giovannitti and Fred Freed interviews, Library of Congress).

In addition, McCloy raised the question of "whether we oughtn't to tell them that we had the bomb and that we would drop the bomb." He also remembered that "as soon as I mentioned the word 'bomb'—the atomic bomb—even in that select circle . . . it was sort of a shock. You didn't mention the bomb out loud. . . . Well, there was a sort of a [gasp] back at that" (Herbert Feis Papers, container 79, Len Giovannitti).

Truman asked McCloy to present his ideas to the State Department for consideration. However, James Byrnes, who would soon become secretary of state, rejected McCloy's ideas. So the Committee of 3 took up the issue of how to convince Japan to surrender. They appointed McCloy to a committee that wrote the surrender proposal, which became known as the Potsdam Proclamation because it was announced during the Potsdam Conference in July 1945.

The Potsdam Proclamation as given to Truman by Stimson included a sentence that Stimson told Truman "would substantially add to the chances of acceptance" of the surrender terms by Japan. The sentence stated: "This may include a constitutional monarchy under the present dynasty if it be shown to the complete satisfaction of the world that such a government will never again aspire to aggression" (U.S. Department of State, *Foreign Relations of the U.S., The Conference of Berlin; The Potsdam Conference, 1945,* vol. 1 (Washington, D.C.: U.S. Government Printing Office, 1960), 892–894). However, Truman removed that sentence from the Proclamation (Len Giovannitti and Fred Freed interviews, Library of Congress). McCloy later commented that "everyone was so intent on winning the war by military means that the introduction of political considerations was almost accidental" (McCloy 1953, 42).

To his dying days McCloy believed that "we missed the opportunity of effecting a Japanese surrender, completely satisfactory to us, without the necessity of dropping the bombs." The use of nuclear weapons on Japan "was not given the thoroughness of consideration and the depth of thought that the president of the United States was entitled to have before a decision of this importance was taken" (McCloy quoted in Reston 1991, 500). From Doug Long at http://www.doug-long.com/mccloy.htm (accessed May 2005); see also Bird 1992.

32. See Badash 1995.

33. Oppenheimer prefaced his assessment of the failure of the Acheson-Lillienthal-Baruch plan in a 1948 *Foreign Affairs* article with the following parable: One day in a clearing in the forest, Confucius came upon a woman in deep mourning, wracked by sorrow. He learned that her son had just been eaten by a tiger; that her husband had been eaten by the same tiger a year earlier and that the year before that her father had been eaten by the tiger. When he failed to console her and help her restore her composure Confucius said to her "This would not seem to be a very salutary neighborhood. Why don't you leave it? The woman wrung her hands and said: "I know, I know; but you see the government is so excellent" (Oppenheimer 1948, 239).

34. On November 17, 1953, William Borden, who until the summer of 1953 had been the executive director of the congressional Joint Committee on Atomic Energy and in this capacity had had access to Oppenheimer's security file, sent a letter to J. Edgar Hoover, the director of the F.B.I., the purpose of which was to state that "based upon years of study of the available classified evidence, that more probably than not J. Robert Oppenheimer is an agent of the Soviet Union" (http://en.wikisource.org/wiki/Letter_from_William_L._Borden_to_J._Edgar_Hoover,_November_7,_1953, accessed May 2007). Hoover forwarded the letter to President Eisenhower, who on December 3, 1953 revoked Oppenheimer's clearance and ordered that a "blank wall" be erected between Oppenheimer and atomic secrets. Oppenheimer appealed the decision. The hearing before the Gray Board in April 1954 failed to reinstate the clearance.

35. See McMillan 2005 for a definitive account of the story.

36. Oppenheimer to J. B. Conant, October 21, 1949 (Hershberg 1993).

37. The full text of the report can be found at http://www.atomicarchive.com/Docs/Hydrogen/GACReport.shtml.

38. Callon 2005.

39. Its author, Charles Murphy, had been engaged by Lewis Strauss, who had become the new chairman of the AEC, to write it. Pricilla McMillan's *The Ruin of Oppenheimer* (2005) is the most thorough and insightful account of the American H-bomb story and of the formation of the Livermore Laboratory.

40. See Schweber 2000 for the details of Oppenheimer's behavior with respect to Rossi, Peters, Peters, and Lomanitz. See also Bird and Sherwin 2005; Wang 1999.

41. The address is reprinted in Oppenheimer 1955b, *The Open Mind.*

42. The *Bhagavad Gita,* whose title can be translated as the "Song of the Lord," is a masterpiece and monument of Hinduism. One expert calls it "the most important single text for the Hindu religion.

43. See Heilbron's (2005) article, "Oppenheimer's Guru" in Carson and Hollinger 2005.

44. See also his letter to Frank of June 4, 1934 (Smith and Weiner 1980, 180).

45. Ryder 1929, viii.

46. Ibid., ix.

47. See for example the photograph by John Bartlett of one such bas-relief entitled "Great armies stir," which can be seen at www.songsouponsea.com/Promenade/Still2A5.html (accessed January 2007).

48. Quoted in Bird and Sherwin 2005, 99. See also Heilbron 2005.

49. Oppenheimer, *Tradition and Discovery*, a lecture delivered in 1960, in Oppenheimer 1984, 106. In another lecture from this same period he asserted: "There are no synapses, no gaps, no gulfs in the picture of the various orders and parts of nature; but they are not derivable from one another. They are branches, . . . each with its own order, each with its own concern, each with its own vocabulary" (Oppenheimer 1960, 15). The metaphor of a tree is invoked because it is a growing thing; it is a tree not a temple (ibid., 17).

## 4. J. Robert Oppenheimer and American Pragmatism

1. See Chapter 3 for three such crises in Oppenheimer's life.

2. In order to teach students to live a moral life and to do so without grounding ethics on denominational religion, Adler had turned to Kant. Adler accepted the Kantian view that education must be concerned with the intellectual mastery of nature, with the glorification of life in art, and with its consecration in morality. In the moral sphere, Adler's point of departure was Kant's Categorical Imperative—to treat every person as an end and not as a means. The evolution of the way ethics was taught at the school mirrored Adler's interpretation and his subsequent reformulations of the Kantian imperative and of the ethics of practical reason. It may well be that later in life Oppenheimer found unattractive the transcendental Kantian basis of ethics and that Oppenheimer's "dislike" for the school that Rabi attributed to him may have stemmed from this fact. It should be noted, however, that in February 1949 Oppenheimer gracefully accepted the school's invitation to be the speaker at the installation of Richard Boyd Ballou as director of the Ethical Culture School, and he addressed its student body on another occasion some years later.

3. In a lecture delivered in 1962, Oppenheimer noted that "in those high undertakings when man derives strength from and insight from public excellence, we have been impoverished. We hunger for nobility, the rare words and acts that harmonize simplicity with truth." Oppenheimer, "On Science and Culture," in Oppenheimer 1984, 137.

4. Recall the section on the Acheson-Lilienthal plan in Chapter 3, in particular the entry that Lilienthal made on July 24, 1946 in his journal after they had spent a good deal of the previous night discussing the consequences of Bernard Baruch's reformulation of the plan.

5. See Day 2001 and Thorpe 2005, 2006.

6. J. Robert Oppenheimer, the William James Lectures, 1957, Oppenheimer Papers, Library of Congress, Box 259 (JROWJ). A photocopy of these is available in the Harvard University Archives but does not include Oppenheimer's preparatory notes.

7. For a historical perspective on complementarity, see Holton 1970 and Bohr 1999.

8. Einstein and Infeld, when discussing the wave-particle duality of photons, characterized the issue thus: "But what is light really? Is it a wave or a shower of photons? There seems no likelihood for forming a consistent description of the phenomena of light by a choice of only one of the two languages. It seems as though we must use sometimes the one theory and sometimes the other, while at times we may use either. We are faced with a new kind of difficulty. We have two contradictory pictures of reality; separately neither of them fully explains the phenomena of light, but together they do" (Einstein and Infeld 1942, 262–263).

9. Oppenheimer had addressed some of these topics in earlier lectures. See Day 2001. In his article, Day emphasizes Bohr's influence on Oppenheimer, and the persistence of Oppenheimer's attachment to complementarity and of his attempts to explain it to the public at large.

10. JROWJ, Lecture 8.

11. Einstein 1918.

12. The proceedings of the conference were published in the winter 1958 issue of *Daedalus* (*Proceedings of the American Academy of Arts and Sciences* 87[1]). It was later reprinted as Holton 1971. Oppenheimer's contribution was entitled "The Growth of Science and the Structure of Culture" (Oppenheimer 1958a).

13. The Institute for Advanced Study, Report of the Director 1948–1953, Box 233, JROLC. Note Oppenheimer's emphasis on method. He believed that the different sciences belonged to different realms, with method, the scientific method, providing a unifying nexus.

14. Ibid.

15. Notes on the Director's Fund, Box 233, JROLC. From 1948 until 1954, "twenty men" were brought to the Institute with the support of the Director's Fund. The invited members were usually selected with advice from members of the schools. But in some cases an advisory committee was appointed to help select the members. This was the case in psychology, whose invited members were: Edward Chase Tolman, 1952; David Levy, 1951–1953; Jean Piaget 1954; and Hans Wallach, 1954–1955. The other invited members were:

in biology: A. Szent-Gyorgyi, 1950; George Wald, 1954
in philosophy: Jean de Menasce, 1951 and 1953; Morton White, 1953–1954
in the history of science: Chaucey D. Leake, 1950 and 1952; Henry Guerlac,
    1954–1955; Alexandre Koyré, 1955–1956

in intellectual history: Perry Miller, 1953–1954

in literature, literary history, and criticism: Erich Auerbach, 1949–1950; Kenneth Burke, 1948–1949; Amiya Chakrabarty, 1950–1951; Ernst Robert Curtius, 1949–1950; Thomas Stearns Eliot, 1949; Francis Ferguson, 1948–1949; Richard P. Blackmur, 1950–1951

in law: Max Radin, 1949–1950; John Lord O'Brian, 1949–1950; Edward S. Greenbaum, 1949–1950; John Palfrey, 1950–1954; Mark deWolfe Howe, 1955–1956

in contemporary history: George Kennan, 1950–1952, 1953–1955; Herbert Feis, 1951–1953

16. Notes on the Director's Fund, Box 233, JROLC.

17. Kroeber had practiced psychoanalysis for two years in San Francisco after World War I. He was a structuralist, for whom culture was "superorganic." He believed that a culture's structures and the laws governing its evolution were independent of the vagaries of individual behavior. In 1944 he published *Configurations of Culture Growth,* in which he raised the question as to why a "clustering" of genius occur at certain times and places and why certain periods seem to be richer in "genius" than others.

18. The Institute for Advanced Study, Report of the Director 1948–1953, Box 233, JROLC, 16.

19. Ibid., 16–17.

20. Tolman's Berkeley lecture was amplified and given in an International Congress address and was subsequently published as Tolman 1948.

21. Oppenheimer had first met Ruth Tolman in the spring of 1928 during his first visit to Cal Tech. She was the wife of Richard Tolman, a well-known and distinguished professor of mathematical physics and physical chemistry there, who became the dean of its graduate school. She had completed her graduate studies in psychology at Berkeley in the early 1930s and became an outstanding clinical psychologist and a perceptive therapist. Soon after Oppenheimer accepted a position at Cal Tech, he and the Tolmans became very good friends, and Ruth and Robert became very close in the late 1930s after his relationship with Jean Tatlock had broken up. During World War II Richard Tolman was Conant's deputy at the Office of Scientific Research and Development, with responsibility for matters relating to atomic weaponry, and was stationed in Washington. After he became director of Los Alamos, whenever Oppenheimer was in Washington he stayed with the Tolmans. By then his friendship with Ruth Tolman was an intimate one, a deeply meaningful and cherished relationship that lasted until her death in 1956. For further details on the relationship,

see Bird and Sherwin 2005, 363–364. See the correspondence between Ruth Tolman and Robert and Kitty Oppenheimer during and after the war in the Ruth Tolman folder, Box 72, JROLC.

22. Oppenheimer first encountered Jerome Bruner, then an up-and-coming young psychologist, during the war on one of his stays at the Tolmans. In 1941, shortly after obtaining his Ph.D. from Harvard, Bruner had taken a job with the Foreign Broadcast Monitoring Service in Washington, D.C., to monitor German, Italian, and Japanese broadcasts and to report his findings to the State and War departments. After Pearl Harbor he accepted a position in the Program Surveys Division of the Bureau of Agricultural Economics. The Bureau was a component of the Office of Facts and Figures, an office that Roosevelt had set up to study such issues as the state of American morale, the public response to the draft and to various war appeals, and its reaction to the absorption of women into war work. It was there that Bruner met Ruth Tolman, who held a similar position in the Department of Agriculture. The work of the Program Surveys Division of the Bureau of Agricultural Economics is described in an article that Ruth Tolman wrote with Rensis Likert, the person in charge of the Office of Facts and Figures. Somewhat later, Bruner accepted a job in Princeton as associate director of the Office of Public Opinion Research that was headed by Hadley Cantril. His job was to assess the public understanding and support of US foreign policy and to report weekly to the State Department his analysis of the extensive interviews being carried out. As he and his family had moved to Princeton, he would stay with the Tolmans in Washington on his Thursday trips to Washington.

In his autobiography, Bruner described his first impression of Oppenheimer as follows: "Brilliant, discursive in his interests, lavishly intolerant, ready to pursue any topic anywhere, extraordinarily lovable.... We talked about almost anything, but psychology and the philosophy of physics were irresistible. We became close friends" (Bruner 1983, 44).

After the war, Bruner returned to Harvard as a faculty member, performing groundbreaking research in thinking and quickly rising in the ranks. In 1946 the Harvard department of psychology split. One branch combined with sociology and social anthropology to found the department of social relations; those that remained in the department, under the influence of Skinner, Boring, and S. S. Stevens, focused on operant conditioning and psychophysics. Together with Gordon Allport, Henry Murray, Samuel Stouffer, Clyde Kluckhorn, Pitirim Sorokin, and Talcott Parsons, Bruner joined the department of social relations, but kept close ties with the "old department" of psychology, in particular with Boring and Stevens. Together with his colleagues, Jacqueline Goodnow and

George Austin, Bruner shifted the emphasis in investigations of thinking to questions about the strategies, hypothesis, and goals that a subject uses in performing a task, leaving the questions of how these were learned in abeyance. As a result of their work, and that of Allen Newell and Herbert Simon, the analysis of performance took center stage, with learning at times looked upon as a form of problem solving. Bruner was a "wholist" and in experiments with his colleagues Leo Postman and Elliot McGinnies demonstrated that the perception of external stimuli is not independent of the internal context: "attitudes, values, expectancies and psychodynamic defenses all impinge upon perception." Bruner's approach to thinking resonated with von Neumann's thinking about computers and his metaphors about their relation to brain and mind. It is therefore not surprising that Oppenheimer would call on Bruner. See Bruner 1983, chap. 4.

23. Thus in 1950, with Harry Grayson, Ruth Tolman wrote a paper on "A Semantic Study of Concepts of Clinical Psychologists and Psychiatrists" (Tolman 1950) and in a wide-ranging address as the president of the Western Psychological Association in June 1953 critically analyzed the relation between theory and practice in psychology. Recalling Kant, she argued that the two should not be nor become separate enterprises but should always be in close communication for "Theory without the findings of practice, . . . abstraction without observation, become at the least impoverished and sterile, and often misleading and distorting. . . . Practice without theory remains an art or a technique and never becomes a science . . . [i.e., never] able to understand or predict on a sound and valid basis" (Tolman 1953).

24. This book review is one of the few documents that Oppenheimer actually wrote out. A typewritten copy is available in Box 23, JROLC. It is dated November 23, 1956. In Oppenheimer's words, the book was concerned with "the discovery and creation of order in man's cognitive life," studied a part of this great theme, "and exemplifies it" (Oppenheimer 1958b). In the preface of their book, the authors characterized it as "an effort to deal with one of the simplest and most ubiquitous phenomenon [sic] of cognition: categorizing or conceptualizing. . . . The spirit of the inquiry is descriptive. We have not sought 'explanation' in terms of learning theory, information theory, or personality theory. We have sought to describe and in small measure to explain what happens when an intelligent human being seeks to sort the environment into significant classes of events so that he may end by treating discriminately different things as equivalents" (Bruner et al. 1956). The book described the experiments they and others had carried out under simple, controllable situations to investigate how categories and concepts are learned on the basis of experiences under circumstances that are wholly adequate to teach them, and under circumstances that are only

partially adequate to do so. The authors suggested that their findings and conclusions "are applicable to any phenomenon where an organism is faced with
the task of identifying and placing events into classes on the basis of using criterial cues and ignoring others." Furthermore, they stressed the analogy between the learning problems they studied in their experiments and the
problems a scientist faces when planning his research program and later when
trying to decide between alternative concepts and theories that are seemingly
compatible with the evidence. In a pregnant phrase, which resonated with Oppenheimer, Bruner et al. stated: "We have found it more meaningful to regard a
concept as a network of sign-significate inferences by which one goes beyond a
set of observed criterial properties exhibited by an object or event to class identity of the object or event in question, and thence to additional inferences about
other unobserved properties of the object or event."

25. Clark was born in New York City in 1882, heir to a banking and railroad
fortune. He was educated at Harvard. In 1909, together with his law school classmates Francis W. Bird and Elihu Root Jr., Clark set up a small law office in New
York that within a few years grew into one of the largest and most prestigious law
firms in the United States, particularly noted for the large number of young
lawyers it trained who then went on to influential positions. In 1931 Clark was
elected to the seven-man corporation that governs Harvard University. During
World War II, he served as an assistant secretary of war under Stimson and spent
part of his time theorizing about ways and means of organizing a world at peace
under law. Already in 1939 with the outbreak of World War II, he had written a
pamphlet entitled *A Federation of Free Peoples.* In 1944 Stimson told him to "go
home and prevent World War III." The results were researches into the feasibility
of a world government. He criticized the structure of the United Nations Organization as outlined at the Dumbarton Oaks and San Francisco Conferences, and
in October 1945, he and former Justice Owen Roberts organized and hosted the
Dublin Conference to consider these matters. For the next several years Clark
served as an elder statesman and unofficial leader of the United World Federal-
ists. He wrote books and lectured extensively. His collaboration with Harvard Law
School Professor Louis B. Sohn began during this period and culminated in 1958
with a magisterial treatise, *World Peace through World Law.* After 1945, civil rights
also occupied Clark's time and energies, and because of his connections with Harvard he became a spokesman for academic freedom. Thus in 1949 Conant asked
him to explain to a wealthy alumnus why Harvard could not discharge professors
who held unpopular political views. In the 1950s he became an outspoken critic
of Senator Joseph McCarthy. His close friend Felix Frankfurter said of Clark: "He
is that rare thing in America, a man of independence, financially and politically,

who devotes himself as hard to public affairs as a private citizen as he would were he in public office." The historian Elting E. Morison characterized Clark as one who "appeared, in critical or confusing times, as a lobby for particular impulses of the national conscience." See J. Garry Clifford at http://www.harvardsquare library.org/unitarians/clark_grenville.html for a brief biography of Clark (accessed February 2005).

26. Oppenheimer to Glenville Clark, May 17, 1949, Box 233, Legal Studies, JROLC.

27. Ibid.

28. Ibid.

29. Box 233, Literary Studies, JROLC.

30. Ibid.

31. See, for example, his lecture on "Analogy in Science," which he delivered at the sixty-third annual meeting of the American Psychological Association on September 4, 1955, in which he gives proof of his extensive reading of James and Peirce (Oppenheimer 1956a, 127).

32. Bridgman 1927. The essence of the "operational" attitude, according to Bridgman, was that "the meanings of one's terms are to be found by an analysis of the operations one performs in applying the term in concrete situations or in verifying the truths of the statements or in finding the answers to questions" (Bridgman 1950, Preface, v).

33. In his letters to his brother Frank, Oppenheimer gave advice on what literature books to read—but I did not find any references to philosophical works. See Smith and Weiner 1980.

34. Wyzanski to Oppenheimer, June 2, 1949; Oppenheimer to Wyzanski, May 31, 1948, Box 122, JROLC. Box 122 is labeled "Harvard College Board of Overseers, Cambridge, Mass."

35. Wyzanski to Oppenheimer, March 7, 1949, and Oral Report of the Committee to Visit the Department of Philosophy, March 14, 1949, Box 122, JROLC.

36. David Bailey, secretary of the Board of Overseers of Harvard College, to Oppenheimer, June 28, 1949, Box 122, JROLC.

37. In addition, all changes in educational policy that affect the requirements of Harvard degrees must be approved by the Overseers and adopted by the Corporation. The Corporation, whose legal name is "President and Fellows of Harvard College," consists of the president, the treasurer, and five fellows, at the time appointed for life, and is the other central governing board of Harvard University. In general, "control of finances and executive management of the University vests in the Corporation," but subject to the consent of the Board of Overseers in many important matters. See Board of Overseers, Box 122, JROLC.

38. See Oppenheimer to Bailey, August 10, 1949; Bailey to Oppenheimer, August 17, 1949; and Oppenheimer to Bailey, September 9, 1949, Box 122, JROLC. The following year, 1950–1951, Oppenheimer also became a member of the Visiting Committee for Mathematics. In 1952–1953, membership on the Mathematics Committee was dropped. In 1953–1954, he was chairman of the Visiting Committee for Physics and Chemistry and also chairman of the Physical Sciences Coordinating Committee. After retiring as Overseer, Oppenheimer continued to serve as a lay member of the Philosophy and Physics Visiting Committees. See Bailey to Oppenheimer, May 4, 1955 and June 27, 1955. He continued serving as a lay member of the philosophy department until 1964. Oppenheimer to Friendly, March 1964, Box 122 and Box 124, JROLC.

39. As chairman, Oppenheimer had to arrange the annual meeting with the department and to host a dinner with it after a day of consulting and conferring with it. Expenses of the meal were paid for by the Board of Overseers so that no undue burden fell on the chairman. However, "a five dollar limit ha[d] been set per capita for such entertainment, and any expenses over this amount ha[d] to be met by the chairman in person." But Bailey assured Oppenheimer "that unless [the chairman] goes out of his way to encourage undue consumption of alcoholic liquors, the five dollars is likely to prove more than ample." See Bailey to Oppenheimer, August 17, 1949, Box 122, JROLC.

40. For example, he recommended that courses in Indian philosophy and the philosophy of law be added to the offerings by the department, and the department did so. See White to Oppenheimer, April 26, 1955, Box 122, JROLC.

41. The members of the Visiting Committee for Philosophy in 1949–1950 were Edwin De T. Bechtel, Edwin Canham, Harold F. Cherniss, Charles F. Curtis, Robert Cutler, Charles W. Gilkey, Walter Lippmann, Walter Louchheim Jr., Arthur O. Lovejoy, Jacques Maritain, Henry T. Moore, Ralph Barton Perry, and Charles E. Wyzanski. The composition of the committee at the end of Oppenheimer's chairmanship was the same except that the membership of Edwin De T. Bechtel, Robert Cutler, Charles W. Gilkey, Arthur O. Lovejoy, Jacques Maritain, and Charles E. Wyzanski had expired and they were replaced by Chadbourne Gilpatrick and Lucien Price.

42. Draft of the Interim Report of the Committee to visit the Department of Philosophy, December 13, 1950, Box 122, JROLC.

43. That situation changed when Morton White assumed the chairmanship of the department in 1955.

44. Report of the Committee to Visit the Department of Philosophy, January 1955, Box 122, JROLC.

45. Oppenheimer to Bailey, November 18, 1952, Box 122, JROLC.

46. Incidentally, the more traditional wing of the department was so attracted to Oppenheimer's proclivities in that direction that in 1952 they convinced the department to suggest to Conant that he appoint Oppenheimer as a university professor. This position is a highly prestigious appointment made at the discretion of the president. It entails no departmental affiliation and complete freedom concerning what courses to teach and when. Conant did sound out Oppenheimer, but he declined (White 1999, 138).

47. Morton White corroborated this in a telephone interview on April 11, 2004.

48. Pierre Duhem (1861–1916) was an emminent French physicist and historian and philosopher of science. As a physicist he is well known for his researches in thermodynamics, and is best known for his writings on the indeterminacy of experimental criteria and on scientific development in the Middle Ages.

49. Bohr had a semipermanent visiting member status at the Institute for Advanced Study. He was in residence there during the spring semester of 1948, 1950, and 1958 and the fall semester of 1954. Pauli was a visiting member of the School of Natural Sciences during the academic year 1949–1950 and the spring semesters of 1954 and 1956.

50. David Favroholdt in his "General Introduction" to volume 10, *Complementarity beyond Physics,* of *Collected Works* (Bohr 1999) stresses this point.

51. See Bohr 1999, which contains all his writings on the subject.

52. Oppenheimer had many admirers in the psychology department (e.g., Bruner and Boring), and the same was true in the philosophy department. See endnote 46.

53. The nature of the appointment made national news. Thus an editorial on February 25, 1957 in the *Harvard Crimson* took issue with the view that Raymond Moley, an analyst writing in *Newsweek,* had stated that Harvard should not have given honor to a man whose discretion had been challenged. He had also observed that evidently Harvard was "more concerned with repairing damaged careers than in the more prosaic task of pursuing and disseminating the truth." *The Harvard Crimson* on March 26, 1957 carried an article with the headline "Alumni Group to Probe Selection of Oppenheimer as Lecturer." The article reported that Col. Archibald Roosevelt, the only surviving son of Theodore Roosevelt, had become the chairman of the eight-member Veritas Committee and had sent a letter to the Harvard alumni asking for funds and help in determining why the William James invitation was extended to Oppenheimer.

54. On April 13, 1957, a few days after the initial William James lecture was delivered, *The Harvard Crimson* published some of the letters it had received over the controversy. The letters were overwhelmingly in support of the appoint-

ment. A recent graduate, Walter B. Raushenbush (class of 1950), concluded his letter to the Veritas Committee with the statement: "It is the dirty and discredited banner of McCarthyism which you are trying to raise anew, clothed in the honored respectability of the Harvard University motto. What you are doing seems to me bad for Harvard and bad for our country. I don't like it. Count me out." Whereas the *Harvard Crimson* strongly supported the appointment of Oppenheimer, *The Radcliffe News,* in an editorial on April 12, 1957, opposed it, stating: "Why should a top university place such high esteem and favor a man regarded by many as an unloyal American."

55. "First Oppenheimer Harvard Lecture Packs 'Em in Despite Early Protest," *Boston Globe,* April 9, 1957. The *New York Times* also covered the first lecture and headlined its report with "Harvard Cordial to Oppenheimer." Although there were no disturbances, the Atheneum, a Harvard undergraduate society that objected to Oppenheimer's appointment, organized a debate on the topic "Should Oppenheimer Be James Philosophy Lecturer at Harvard."

56. The first page of the transcripts gives the title of the first lecture as "A Pluralistic Universe," but the section that gives the content of the first lecture is called "A Pluralistic Multiverse." That was clearly the title that Oppenheimer had intended.

57. The transcript of this lecture is missing.

58. Reported in Bird and Sherwin 2005, 561.

59. A photocopy of the transcripts is on deposit in the Harvard Archives.

60. JROWJ, Lecture 1.

61. Ibid.

62. JROWJ, Lecture 7.

63. Ibid.

64. JROWJ, Lectures 1 and 2.

65. JROWJ, Lecture 4.

66. Thus in Lecture 4 Oppenheimer talked about mathematics as Wittgensteinian games with rules.

67. Its content was delivered on several different occasions later on. See, for example, the version in Bohr 1963, with the title "The Unity of Human Knowledge," 17–22.

68. Even though this is taken from a later essay, Bohr had spoken on the subject earlier.

69. Favroholdt in Bohr 1999, xlix.

70. I found useful the presentation of Dewey's changing philosophical views in Morton White's *Toward Reunion in Philosophy* (1956) and in White 1959. The changing character of the economics of science is reflected in the changing views

of Dewey with regard to science. Until World War II it was corporate America, principally through the foundations that had been established by Rockefeller, Carnegie, and their likes and in the laboratories attached to the large industrial corporations (e.g., GE, AT&T, Westinghouse, U.S. Steel, and Dupont), that supported scientific research. After World War II it was the government and the armed forces that did so—and on an unprecedented scale. See also Wilson 1990.

71. In an earlier lecture, Oppenheimer explained at length the sense in which he used the adjective "unified."

72. Like Dewey in his *Reconstruction in Philosophy*, Oppenheimer examined the relation between science and technology and characterized the latter in terms of "control."

73. All the subsequent quotations are from the first Oppenheimer lecture.

74. Oppenheimer in his 1956 AAAS talk had also spoken of a "reticular unity." A few months later he came across the epigram at the beginning of this chapter.

75. See, in particular, Dewey 1929.

76. "Visible nature is all plasticity and indifference, a multiverse, as one might call it, and not a universe" (James 1895, 10).

77. As noted earlier, Oppenheimer had written a lengthy review of the book in the *Sewanee Review* in 1958. See Oppenheimer 1958b.

78. See, for example, Heisenberg 1955a and 1974.

79. Throughout the lectures "objective" was synonymous with "unambiguously communicable." Until further notice, the quoted Oppenheimer remarks are from this second lecture.

80. Until further notice, the quotations are from Lecture 6.

81. The initial part of Lecture 7 was devoted to a further analysis of historical inquiries that recalled in detail the discussions of the 1949 Institute for Advanced Study history conference, and also his discussions with Morton White regarding macro theories of history, of which he said "nothing really follows from them; at least nothing that works out."

82. For example, that of a behaviorist, or someone interested in learning theory, personality theory, or psychotherapy.

83. See Chapter 5 for Oppenheimer's views on unification in the early 1960s.


## 5. Einstein, Oppenheimer, and the Extension of Physics

1. Wigner had made a similar statement at the Princeton Einstein Centennial Celebration in 1979 and there indicated that Dirac had had the same impression (Wigner 1980, 474).

2. See, in particular, Holton's essay, "Einstein and the Goal of Science," for an overview of the efforts to unify the description of physical phenomena at the end of the nineteenth century (Holton 1996, especially 161–164).

3. See Tilman Sauer's (2007, 2008) very informative "Einstein's Unified Field Theory Program."

4. In his review paper on the theory of relativity in 1907, Einstein asked: "Is it conceivable that the principle of relativity also holds for reference systems that accelerate with respect to each other?" By considering the next simplest situation from the case of two inertial observers moving with constant velocity relative to one another, namely, the case of constant acceleration, Einstein arrived at a deep insight. Suppose there were several bodies of different composition—for example, a lead one, an iron one, and a bag of feathers—all at rest relative to an inertial observer. To the observer in the accelerated frame, all these objects will appear to be moving with constant acceleration in a direction opposite to the one in which the observer is moving. But that is exactly the motion of masses in a constant gravitational field. Galileo had demonstrated this in his alleged experiment dropping bodies of different composition from the tower of Pisa. Newton's explanation recognized that near the surface of the earth the gravitational force between an object of mass $m_G$ and the earth, whose mass is $M_G$, is given by

$$F = G\frac{m_G\, M_G}{R^2}$$

where $R$ is the radius of the earth. This force on the object is what is meant by the weight of the object near the surface of the earth. Note that the masses that enter this equation are the gravitational masses of the body and that of the earth, that is, the measure of their property, which is responsible for their attracting one another. It can be thought of as their gravitational "charge," in the same way as the electrical charge of a body is a measure of its property to attract other charges. By Newton's second law of motion, $F = m_I a$, an object of (inertial) mass $m_I$ under the action of this force will move with an acceleration $a$. Note that it is the inertial mass of a body—the measure of its property on how it responds to forces acting upon it—which enters in the law of motion. The fact that all objects near the surface of the earth fall with the same acceleration

$$g = \frac{GM_G}{R^2}\left(\frac{m_G}{m_I}\right)$$

implies that $\left(\dfrac{m_G}{m_I}\right)$ is a constant, the same for all objects, and can be taken to be 1. As noted in Chapter 1, note 126, for Newton, and everyone before Einstein, this was an empirical fact, an *unexplained* regularity of nature, to be checked with ever greater accuracy. Einstein explained that regularity with his equivalence principle.

5. Einstein thus became very interested in Patrick Blackett's geomagnetic experiments after World War II. See Nye 2004.

6. See Sauer 2007 for an overview of the various approaches Einstein explored to formulate models of unified field theories.

7. The session took place in the MIT Armory on April 8, 1961. The records of the conference are available in the MIT Archives. I thank the archivist of the MIT Library for permission to quote from the transcripts of the session and Carl and Michelle Feynman for permission to quote from Richard Feynman's presentation at this session.

8. Unless otherwise indicated, all the quotations are from the presentations of the members of the panel at the session on "The Future of the Physical Sciences," MIT Centennial Celebration, MIT Archives.

9. MIT Centennial Celebration, 1961, MIT Archives; see also Yang ([1983] 2005), 319.

10. Here I am quoting Feynman's statement as he formulated it in Feynman 1963, 3–9.

11. The American high energy community had "sensed the apparent existence of some misunderstanding of the objectives of high energy physics, not only among the general public but also among the scientific community as a whole" (Yuan 1965, Preface). It was therefore felt that some statements addressed to both the scientific community and the general public regarding the aims of high energy physics were in order. Luke Yuan, at the Brookhaven National Laboratory, was asked to solicit statements from some thirty theoretical physicists, which together would "provide a comprehensive basis for a better understanding of the fundamental importance and great depth of high energy physics." The statements were gathered in a volume, *Nature of Matter: Purposes of High Energy Physics,* with Luke Yuan (1965) as its editor.

12. For a concise view of Kant on unification, see Morrison 2000.

13. I return to these themes in Chapter 6.

14. For a very accessible and insightful history and exposition of the theory of relativity, see Einsenstaedt 2006.

15. The increase in the size of the community and the ever expanding areas of interest to theorists makes the list of the principal contributors to the advances during this period much larger. The names Schwinger, Feynman, Dyson, Gell-Mann, T. D. Lee, Yang, Chew, Low, Goldberger, Feinberg, Weinberg, Glashow, Coleman, Bell, Goldstone, Nambu, Veltman, and Wilson (in high energy) and

Abrikosov, Anderson, Baym, de Gennes, Fisher, Kadanoff, Kohn, Luttinger, Martin, Pines, and Schrieffer (in condensed matter) are names that come immediately to mind.

16. This change has been chronicled in part by Mirowski and Sent (2002).

17. This division is particularly noticeable in Great Britain and in France.

18. See in particular Lepage 1989, 1997.

19. Just as physics has been transformed, so has chemistry. Undoubtedly, the biological and medical sciences have been most deeply affected by internal developments: Crick and Watson, genetic codes, recombinant technologies, DNA-sequencing, genome projects, bioinformatics. And it is in the biological sciences that the entrepreneurial aspects of the university are most visible.

20. The committee consisted of all the senior theorists holding professorial positions in the various Swiss universities at the time: A. Mercier, K. Bleuler, M. Fierz, W. Heitler, A. Houriet, F. G. Houtermans, R. Jost, D. Rivier, E. C. G. Stueckelberg, W. Scherrer, and M. Schüber.

21. Pauli, "Opening Talk," in Mercier and Kervaire 1956, 27.

22. See his opening remarks on p. 351 of the proceedings of that conference, which were published in the *Reviews of Modern Physics* 29 (1957): 351–546.

23. Komar, Misner, Regge, R. F. Baierlein, and D. H. Sharp are some of the Wheeler students who worked on GR for their Ph.D. thesis.

24. R. Penfield, R. Schiller, H. Zatzkis, and J. L. Anderson were among Bergmann's students doing their Ph.D. with him during the 1950s.

25. Andrzej Trautman was a student of Leopold Infeld.

26. The major ones are the following: R. V. Pound and G. A. Rebka Jr., "Gravitational Red-Shift in Nuclear Resonance," *Physical Review Letters* 3(9) (1959): 439–441; R. V. Pound and G. A. Rebka Jr., "Apparent Weight of Photons," *Physical Review Letters* 4(7) (1960): 337–341; R. H. Dicke, "New Research on Old Gravitation: Are the Observed Physical Constants Independent of the Position, Epoch, and Velocity of the Laboratory?" *Science* 129 (1959): 621–624; L. I. Schiff, "On Experimental Tests of the General Theory of Relativity," *American Journal of Physics* 28 (1960): 340–343; I. I. Shapiro, "Fourth Test of General Relativity," *Physical Review Letters* 13(26) (1964): 789–791; I. I. Shapiro, M. E. Ash, R. P. Ingalls, W. B. Smith, D. B. Campbell, R. B. Dyce, R. F. Jurgens, and G. H. Pettengill, "Fourth Test of General Relativity: New Radar Result," *Physical Review Letters* 26 (1971): 1132–1135.

27. C. H. Brans and R. H Dicke, "Mach's Principle and a Relativistic Theory of Gravitation," *Physical Review* 124 (1961): 925–935. (I have not entered these articles in the Bibliography so as to indicate their clustering.)

28. In his paper Utiyama considered field systems that are invariant under a certain group of transformations depending on $n$ parameters. He also formulated a general rule for introducing a new field that possessed a definite type of interaction with the original fields by postulating the invariance of the enlarged system under a wider group derived by replacing the parameters of the original group by a set of arbitrary functions. The transformation properties of this new field under the wider group was determined from the invariance postulate. Utiyama then derived the equations that the new fields obeyed, as well as the conservation laws that were a consequence of the invariance. Utiyama illustrated this approach with the electromagnetic, the gravitational, and the Yang-Mills fields as examples.

29. See Chrétien et al. 1969.

30. See section VI, "Relativity and Corpuscles," in Einstein's essay "Physics and Reality" in Einstein 1954, 290–324.

31. See Gross 2008.

## 6. Einstein, Oppenheimer, and the Meaning of Community

1. Writing to Robert's brother, Frank, their father Julius, who after his wife's death in October 1931 was staying with Robert, commented that "he [Robert] is always busy and has had two short talks with Einstein" (Smith and Weiner 1980, 153).

2. Already in 1920 he had described himself as "a person without roots anywhere . . . [who has] journeyed to and fro continuously—a stranger everywhere." Einstein to Born, March 3, 1920, in Einstein 1971, 26.

3. See Castagnetti et al. 1997; Castagnetti and Goenner 2004.

4. As Oppenheimer was to put it in his UNESCO speech of 1965.

5. For Wigner's recollection of the Kaiser Wilhelm colloquia, see Wigner 1982. Einstein missed the Third Solvay Congress in 1921 in order to go to the United States with Chaim Weizmann to raise money for the support of the Zionist cause and the establishment of the Hebrew University.

6. See in particular Lanczos 1974, 9.

7. See Einstein 1949a, 79 and 81; Infeld and Plebański 1960.

8. See Pais 1982, chap. 29, "Einstein's Collaborators."

9. Einstein to Otto Juliusburger, September 28, 1937, EA 38–163.00. Juliusburger was still in Berlin at the time.

10. J. R. Oppenheimer, Special Radio Program celebrating the 60th birthday of Albert Einstein, March 16, 1939. Oppenheimer's remarks were reported in *Science* 89(2311) (1939), 335–336. In 1960 Oppenheimer attributed the spectac-

ular success of the scientific tradition to "the methods and cooperative character of the undertaking . . . [which] have proven to be golden, one of the great events of human history" (Oppenheimer 1960, 8).

11. The one text on general relativity written during the 1940s does not refer to Oppenheimer's work. That book was written by Peter Bergmann (1942), a close associate of Einstein. See also Karl Hufbauer's essay in Carson and Hollinger 2005.

12. Oppenheimer, when declining the invitation, stated that he was "unprepared to make a public statement at this time on Atomic Energy with any confidence that the results will lead in the direction for which we all hope." Quoted in Bird and Sherwin 2005, 381–382.

13. For a brief history of the Institute, see Schweber 1993 and Regis 1987.

14. Freeman Dyson to his parents, October 4, 1948. I thank Freeman Dyson for permission to quote from this letter.

15. See Schrecker 1986; Wang 1999; Schweber 2000.

16. Oppenheimer did not contribute an article to the July 1949 *Reviews of Modern Physics* dedicated to Einstein's seventieth birthday, likely because of the pressure of other duties (his chairmanship of the General Advisory Committee of the Atomic Energy Commission, his membership on numerous governmental advisory committees, and the directorship of the Institute for Advanced Study) and because he was no longer doing any research in physics.

17. *New York Times*, June 12, 1953. Incidentally, in an editorial on June 13 the *Times* assailed the "forces of civil disobedience" as "illegal" and "unnatural." See Jerome 2002, 237, 241–248.

18. Sol Stein to Oppenheimer, March 5, 1954, JROLC, Box 32. The letter is quoted in Thorpe 2006. See endnote 24 for further details of the incident.

19. The convocation took place on October 3, 1954 at the Nassau Inn. On September 30 Oppenheimer told the secretary of the Technion Society in New York that he could not participate and would send a message.

20. JROLC, Box 256, Folder "Einstein's honorary degree, Technion 3/10/54." In his original letter dated February 18, 1954, Tulin stated that "Professor James Franck and Dr. Albert Einstein" were to be conferred degrees; in all the subsequent correspondence the order was Dr. Einstein and Professor Franck.

21. JROLC, Box 256, Folder "Einstein, Albert—Article for Princ. *Packet*."

22. Ibid.

23. Laurie Brown, who was a fellow at the Institute in 1953, invited Oppenheimer to Northwestern University in 1955 for a series of lectures. These lectures were presented shortly after Einstein's death. Brown recalls Oppenheimer

making these statements at a meeting with graduate students after the final lecture. L. Brown, personal communication, May 20, 2006.

24. The presence of Julian Huxley and Jean Piveteau is explained by the fact that the colloquium was to commemorate not only the tenth anniversary of the death of Einstein and the fiftieth anniversary of the theory of the general relativity but also the tenth anniversary of the death of Teilhard de Chardin. The colloquium was entitled "Science et Synthèse. Pour une connaissance de l'homme et de l'univers." Huxley and Piveteau talked about de Chardin.

25. JROLC, Box 235, Folder "UNESCO (Paris) Einstein, December 13, 1965 draft."

26. In the following unless otherwise noted, the quotations are from the original unedited transcription of Oppenheimer's talk. JROLC, Box 235, Folder "UNESCO (Paris) Einstein, December 13, 1965 draft."

27. Oppenheimer had told David Lilienthal a few years earlier that "men must have pity on myths. Life wouldn't be bearable if everything had to be based on provable propositions, on propositions that could be established as one does in physics" (Lilienthal 1969, Entry for January 31, 1959, 308).

28. Upon receiving the transcription of his address, Oppenheimer edited the statement "may very well prove wrong, no one but he would have done it for a long time" to read "is not well proved experimentally, no one but he would have done it for a long, long time," which is what appears in the version published in the *New York Review of Books* (Oppenheimer 1966).

29. Later edited to "it is this tradition which made him know that there had to be a field theory of gravitation."

30. This last sentence becomes edited to read "or is it something that we can actually study in nature by physical means."

31. Becomes edited to "but the laws of nature delimit the scope of observations."

32. Later edited to "no one could have been more ingenious in thinking up unexpected and clever examples."

33. "—the illusion—" added upon editing the transcription.

34. Frank Jotterand, "Oppenheimer," in the *Gazette de Lausanne,* December 18–19, 1965, 25.

35. Abraham Pais to Oppenheimer, December 20, 1965, JROLC, Box 285, Folder "UNESCO-Einstein Correspondence." See Pais 1997 for Pais's account of his long tenure at the Institute.

36. JROLC, Box 285, Folder "UNESCO (Paris) Einstein, December 13, 1965 draft."

37. Otto Nathan was at the conference. In a letter to Helen Dukas, dated December 16, 1965, he described Oppenheimer's talk. "Oppenheimer," Nathan wrote, "had played down Einstein's achievements and had assessed the last 25 years of his life as a "failure . . . in part because he didn't want to learn the new physics." Quoted by Diana Barkan and Kip Thorne in the Preface of the new edition of the Born-Einstein letters (Einstein 2005, xiii and note 18, xxx).

38. Oppenheimer to Helen Dukas, December 20, 1965, JROLC, Box 285, Folder "UNESCO-Einstein Correspondence."

39. Gerard Bonnot, "Oppenheimer parle d'Einstein," *L'Express,* December 20–26, 1965.

40. See Einstein 1987, 10–14.

41. See, for example, his article "Why Socialism," in Einstein 1954, 151–158.

42. Two years earlier Robert Oppenheimer had been the principal speaker at the installation of Richard Boyd Ballou as director of the Ethical Culture School in Riverdale.

43. Recall the statement of the faculty report recommending Einstein's appointment to the faculty of the University of Zurich in 1909: "[The] expressions of our colleague Kleiner, based on several years of personal contact, were all the more valuable for the committee as well as for the faculty as a whole since Herr Dr. Einstein is an Israelite and precisely to the Israelites among scholars are ascribed (in numerous cases not entirely without cause) all kinds of unpleasant peculiarities of character, such as intrusiveness, impudence, and a shopkeeper's mentality in the perception of their academic position. It should be said, however, that also among the Israelites there exist men who do not exhibit a trace of these disagreeable qualities and that it is not proper therefore, to disqualify a man only because he happens to be a Jew. . . . Therefore neither the committee nor the faculty as a wole considered it compatible with its dignity to adopt anti-Semitism as a matter of policy" (quoted in Pais 1982, 185–186).

44. EA 40-431-3.

45. Buber 1994, 459. Quoted in Skinner's introductory essay to chapter 1 of Scholem 2002, 10.

46. See Howard 1997, "A Peek behind the Veil of Maya," for a convincing presentation of Schopenhauer's extensive influence on Einstein.

47. See Hijiya 2000 for a detailed exposition of Oppenheimer's relation to the *Gita*.

48. "Books which had a fundamental philosophical or ideological outlook."

49. EA 39-19.

50. See Israel's *Radical Enlightenment* (2001) for an outstanding account of Spinoza's philosophy and of the role he played in shaping the Enlightenment.

51. See Jammer 1999, 42ff.

52. Incidentally, in 1920 Einstein composed a poem entitled "Zu Spinoza's Ethik," the first stanza of which read as follows:

> How do I love this man?
> More than I can express in words.
> Nonetheless I fear that he remains alone
> With his shining holy halo.

53. Lanouette 1992, 85.

54. Israel 2001, 231.

55. The first two stages being forms of religion brought about by fear and religion as justifying a moral code.

56. See, for example, Dauer 1969 and Safranski 1990.

57. These are sometimes encapsulated in a trinity: wisdom, uprightness, and self-concentration.

58. See also chapter xxxi of *The World as Will and Representation*, which is devoted entirely to a discussion of genius; there Schopenhauer states: "where the brain's power of forming representations has such a surplus that a pure, distinct, objective picture of the external world exhibits itself *without a purpose* as something useless for the will ... [t]is denoted by the name *genius*. ... Genius ... consists in an abnormal excess of intellect which can find its use only by being employed on the universal of existence. In this way it then applies itself to the service of the whole human race, just as does the normal intellect to that of the individual" (Schopenhauer 1966, vol. 2, 377).

59. Perception is Payne's translation of Schopenhauer's *Anschauung*. Schopenhauer uses *Anschauung* to describe what happens when the eye perceives an external object as the cause of the sensation on the retina (Schopenhauer 1966, vol. 1, viii).

60. See in particular Paty in Grene and Nails 1986, 267–302, Jammer 1999, and Holton 2005. To this should be added the acuity of Einstein's and Spinoza's political assessments. For example, Spinoza's analysis of Cromwell's regime in the *Tractacus Theologico-Politicus* and Einstein's analysis of militarism written shortly after the end of World War II.

61. Both Spinoza and the *Gita* formulate a strictly deterministic system.

62. Oppenheimer was first called "Opje" during his stay with Ehrenfest in Holland. In the 1930s he often signed his name "Opje." At Los Alamos the appellation Oppie became the accepted one.

63. See Morton White's essay, "The Philosopher and the Metropolis in America," in White 1973, 11–30.

64. See Lapp 1949, 162–164.

65. In Safranski 1990, 305–306.

66. See, in particular, Holton's essay, "Einstein's Scientific Program: The Formative Years," in Elkana and Holton 1982, 49–65.

67. Quoted in Holton 1998a, xxxi. See also Holton's discussion of the drive toward unification in Einstein in his essay "Einstein and the Goal of Science," in Holton 1996.

68. Quoted in Dukas and Hofmann 1979, 12.

69. Einstein to Robert S Marcus, of the World Jewish Congress, dated February 12, 1950, EA 60-424. Einstein's German draft is EA 60-425.

70. See Thorpe 2005.

71. JROWJ, Lecture 8: "The Hope for Order."

72. Recall the lectures at the April 1961 celebration of MIT's hundredth anniversary (see Chapter 5).

73. Oppenheimer, "Tradition and Discovery," a lecture delivered in 1960, in Oppenheimer 1984, 106. In another lecture from this same period he asserted: "There are no synapses, no gaps, no gulfs in the picture of the various orders and parts of nature; but they are not derivable from one another. They are branches, . . . each with its own order, each with its own concern, each with its own vocabulary" (Oppenheimer 1960, 15). The metaphor of a tree is invoked because it is a growing thing; it is a tree not a temple (ibid., 17).

74. Report of the Director, 1954, 26.

75. Einstein 1920, 1938; see also his essays in *I&O*.

## Some Concluding Remarks

1. See the introductions to Roth 1969a and Roth 1969b.

2. See Roth 1969a, 233–272.

# Bibliography

Aaserud, F. 1999. "The Scientist and the Statesmen: Niels Bohr's Political Crusade during World War II." *Historical Studies in the Physical and Biological Sciences* 30(1): 1–48.

Allmendinger, J., Hackman, J.R., and Lehman, E.V. 1994. *Life and Work in Symphony Orchestras: An Interim Report of Research Findings.* Cambridge, Mass.: Division of Research, Harvard Business School.

Alvarez, L. 1987. *Alvarez: Adventures of a Physicist.* New York: Basic Books.

Anderson, P. 1972. "More Is Different." *Science* 177(4047): 393–396.

Ashtekar, A., Cohen, R., Howard, D., Renn, J., Sarkar, S., and Shimony, A., eds. 2003. *Revisiting the Foundations of Relativistic Physics: Festschrift in Honor of John Stachel.* Dordrecht: Kluwer Academic Publishers.

Badash, L. 1995. *Scientists and the Development of Nuclear Weapons: From Fission to the Limited Test Ban Treaty, 1939–1963.* Atlantic Highlands, N.J.: Humanities Press.

Baker, L. 1984. *Brandeis and Frankfurter: A Dual Biography.* New York: Harper and Row.

Bargmann, V. 1980. "Working with Einstein." In *Some Strangeness in the Proportion: A Centennial Symposium to Celebrate the Achievements of Albert Einstein.* Edited by H. Woolf. Reading, Mass.: Addison-Wesley, 480–481.

Barkan, D.K. 1993. "The Witches' Sabbath: The First International Solvay Congress in Physics." *Science in Context* 6(1): 59–82.

Batterson, S. 2006. *Pursuit of Genius: Flexner, Einstein, and the Early Faculty of the Institute for Advanced Study*. Wellesley, Mass.: A. K. Peters.

Beider, J. 2000–2001. "Einstein in Singapore." *On the Page Magazine* no. 1 (Winter). http://www.onthepage.org/outsiders/einstein_in_singapore.htm (accessed June 2006).

Beller, M. 1993. "Einstein's and Bohr's Rhetoric of Complementarity." *Science in Context* 6: 241–256.

———. 1999. *Quantum Dialogue: The Making of a Revolution*. Chicago: University of Chicago Press.

Beller, M., Renn, J., and Cohen, R., eds. 1993. *Einstein in Context*. Special issue of *Science in Context*.

Ben-Menahem, Y. 1993. "Struggling with Causality: Einstein's Case." *Science in Context* 6: 291–310.

Bergmann, P. 1942. *Introduction to the Theory of Relativity*. New York: Prentice-Hall.

———. 1982. "The Quest for Unity." In *Albert Einstein: Historical and Cultural Perspectives*. Edited by G. Holton and J. Elkana. Princeton, N.J.: Princeton University Press, 27–38.

Berlin, I. 1981. *Personal Impressions*. Edited by H. Hardy. Introduction by N. Annan. New York: The Viking Press; London: Hogarth Press.

———. 1982. "Einstein and Israel." In *Albert Einstein: Historical and Cultural Perspectives*. Edited by G. Holton and J. Elkana. Princeton, N.J.: Princeton University Press, 281–292.

Bernstein, B. J. 1974. "The Quest for Security: American Foreign Policy and International Control of Atomic Energy, 1942–1946." *The Journal of American History* 60(4) (March): 1003–1044.

Bernstein, J. 2004. *Oppenheimer: Portrait of an Enigma*. Chicago: Ivan R. Dee.

Bethe, H. 1967. "J. Robert Oppenheimer: Where He Was There Was Always Life and Excitement." *Science* 155(3766) (March 3): 1080–1084. (Also reprinted in Bethe 1991, 221–230.)

———. 1968. "J. Robert Oppenheimer. April 22, 1904–February 18, 1967." *Biographical Memoirs of Fellows of the Royal Society* 14: 390–416. Reprinted in the *Biographical Memoirs of the National Academy of Sciences*. Available at http://www.nasonline.org/site/PageServer?pagename=MEMOIRS_O (accessed October 2007).

———. 1991. *The Road from Los Alamos*. New York: American Institute of Physics.

Bird, K. 1992. *The Chairman: John J. McCloy, the Making of the American Establishment*. New York: Simon and Schuster.

Bird, K., and Sherwin, M. 2005. *American Prometheus: The Triumph and Tragedy of J. Robert Oppenheimer*. New York: Alfred A. Knopf.

Bohr, N. 1935. "Can Quantum-Mechanical Description of Physical Reality Be Considered Complete?" *Physical Review* 48: 696–702.

———. 1950. "Open Letter to the United Nations." *Impact of Science on Society* 1(2): 68–72. Available at http://www.ambwashington.um.dk/en/menu/Information aboutDenmark/EducationandTraining/FamousDanishScientists/NielsBohr .htm (accessed September 2007).

———. 1958. *Atomic Physics and Human Knowledge.* Bungey, Suffolk: Richard Clay and Company.

———. 1963. *Essays 1958–1962 on Atomic Physics and Human Knowledge.* Bungey, Suffolk: Richard Clay and Company.

———. 1987. "Unity of knowledge." In *The Philosophical Writings of Niels Bohr: Volume II—Essays 1932–1957 on Atomic Physics and Human Knowledge.* Woodbridge, Conn.: Ox Bow Press.

———. 1999. *Collected Works: Complementarity beyond Physics.* Vol. 10. Edited by D. Favrholdt. Amsterdam: Elsevier.

Born, M., and Oppenheimer, J. R. 1927. "Zur Quantentheorie der Molekülen." *Annalen der Physik,* 4th ser. 84: 457–484.

Brans, C., and Dicke, R. H. 1961. "Mach's Principle and a Relativistic Theory of Gravitation." *Physical Review* 124: 925.

Bridgman, P. W. 1927. *The Logic of Modern Physics.* New York: Macmillan.

———. 1950. *Reflections of a Physicist.* New York: Philosophical Library.

Brown, A. P. 1997. *The Neutron and the Bomb: A Biography of Sir James Chadwick.* Oxford: Oxford University Press.

Brown, L. M., and Hoddeson, L. 1983. *The Birth of Particle Physics.* Cambridge: Cambridge University Press.

Bruner, J. 1983. *In Search of Mind: Essays in Autobiography.* New York: Harper and Row.

———. 2004. "The Psychology of Learning: A Short History." *Daedalus* (Winter): 13–20.

Bruner, J., Goodnow, J., and Austin, G. A. 1956. *A Study of Thinking.* New York: Wiley.

Buber, M. 1994. *Briefe I.* Munich: Beck Verlag.

Buchdahl, G. 1969. *Metaphysics and the Philosophy of Science. The Classical Origins: Descartes to Kant.* Cambridge, Mass.: MIT Press.

Callon, M. 2005. "Why Virtualism Paves the Way to Political Impotence: A Reply to Daniel Miller's Critique of *The Laws of the Markets.*" *Economic Sociology: European Electronic Newsletter* 6(2) (February): 3–20. Quoted in I. Hardie and D. MacKenzie, "Assembling an Economic Actor: The *Agencement* of a Hedge Fund." Paper presented at the workshop, "New Actors in a Financialised

Economy and Implications for Varieties of Capitalism," Institute of Commonwealth Studies, London, May 11–12, 2006. Available at http://www.sps.ed.ac.uk/staff/mackenzie.html (accessed April 2007).

Camus, A. 1955. *The Myth of Sisyphus and Other Essays*. Translated by J. O'Brien. New York: Alfred A. Knopf.

Cantelon, P. L., et al., eds. 1991. *The American Atom: A Documentary History of Nuclear Policies from the Discovery to the Present*. Philadelphia: University of Pennsylvania Press.

Cao, T. Y. 1997. *Conceptual Developments of 20th Century Field Theories*. Cambridge: Cambridge University Press.

Carson, C., and Hollinger, D., eds. 2005. *Reappraising Oppenheimer: Centennial Studies and Reflections*. Berkeley: University of California Press.

Cassidy, D. 1995. *Einstein and Our World*. Atlantic Highlands, N.J.: Humanities Press.

———. 2005. *J. Robert Oppenheimer and the American Century*. New York: Pi Press.

Castagnetti, G., and Goenner, H. 2004. "Directing a Kaiser-Wilhelm-Institute: Albert Einstein, Organizer of Science?" Planck-Institut für Wissenschaftsgeschichte, Preprint 260. Available at http://www.mpiwg-berlin.mpg.de/en/research/preprints.html.

Castagnetti, G., Goenner, H., Renn, J., Sauer, T., and Scheideler, B. 1997. "Foundations in Disarray: Essays on Einstein's Science and Politics in the Berlin Years." Max-Planck-Institut für Wissenschaftsgeschichte, Preprint 62. Available at http://www.mpiwg-berlin.mpg.de/en/research/preprints.html.

Cat, J. 1998. "The Physicists' Debate on Unification in Physics at the End of the 20th Century." *Historical Studies in the Physical Sciences* 28(2): 253–299.

Chevalier, H. 1965. *Oppenheimer: The Story of a Friendship*. New York: George Braziller.

Chrétien, M., Deser, S., and Goldstein, J., eds. 1969. *Astrophysics and General Relativity*. 2 vols. New York: Gordon and Breach.

Clark, R. W. 1973. *Einstein: The Life and Times*. London: Hodder and Stoughton.

Cohen, R. S., and Tauber, A. I., eds. 1998. *Philosophies of Nature: The Human Dimension*. Dordrecht: Kluwer Academic Publishers.

Cornwell, J., ed. 1995. *Nature's Imagination: The Frontier of Scientific Vision*. Introduction by F. Dyson. Oxford: Oxford University Press.

Corry, L. 1998. "The Influence of David Hilbert and Hermann Minkowski on Einstein's Views over the Interrelation between Physics and Mathematics." *Endeavour* 22(3): 97–99.

———. 2004. *David Hilbert and the Axionazition of Physics*. Dordrecht: Kluwer Academic Publishers.

Crease, R. P. 2005. "Oppenheimer and the Sense of the Tragic." In *Reappraising*

*Oppenheimer: Centennial Studies and Reflections*. Edited by C. Carson and D. Hollinger. Berkeley: University of California Press, 315–324.

Darrigol, O. 1995. "Henri Poincaré's Criticism of *Fin-de-Siècle* Electrodynamics." *Studies in History and Philosophy of Modern Physics* 26(1): 1–44.

Dauer, D. W. 1969. *Schopenhauer as Transmitter of Buddhist Ideas*. Berne: Herbert Lange & Company.

Davis, H. M. 1948. "The Man Who Built the A-Bomb." *New York Times Magazine*, April 18, 20ff.

Davis, N. P. 1968. *Lawrence and Oppenheimer*. New York: Simon and Schuster.

Day, M. A. 2001. "Oppenheimer on the Nature of Science." *Centaurus* 43: 73–112.

Deane, H. A. 1972. *The Political Ideas of Harold J. Laski*. Hamden, Conn.: Archon Books.

de Boer, J., Dal, E., and Ulfbeck, O., eds. 1986. *The Lesson of Quantum Theory: Neils Bohr Centenary Symposium Held 3–7 October, 1985 in Copenhagen, Denmark*. Amsterdam: North-Holland.

DeGrand, T., and Toussaint, D., eds. 1989. *From Actions to Answers: Proceedings of TASI'89*. Singapore: World Scientific.

Dennefink, D. 2005. "Einstein *versus* the *Physical Review*." *Physics Today* 58(9): 43–46.

Depew, M., and Obbink, D. 2000. *Matrices of Genre: Authors, Canons, and Society*. Cambridge, Mass.: Harvard University Press.

Deser, S., and Ford, K., eds. 1965. *Lectures on General Relativity*. New York: Prentice-Hall.

de-Shalit, A., Feshbach, H., and van Hove, L., eds. 1966. *Preludes in Theoretical Physics in Honor of V. F. Weisskopf*. Amsterdam: North-Holland.

Dewey, J. 1916. *Democracy and Education: An Introduction to the Philosophy of Education*. New York: Macmillan. (See also vol. 9 of Dewey 1980.)

———. 1920. *Reconstruction in Philosophy*. New York: Henry Holt and Company. (See also vol. 9 of Dewey 1980.)

———. 1927. *The Public and Its Problems*. New York: Henry Holt and Company.

———. 1929. *The Quest for Certainty*. New York: Milton Balch.

———. 1980. *The Middle Works, 1899–1924*. Vol. 9. Edited by J. Boynston. Carbondale: Southern Illinois University Press.

DeWitt, B. S. 1957. "Introductory Note" to papers from the conference on the role of gravity in physics held at the University of North Carolina, January 18–23. *Reviews of Modern Physics* 29: 351.

Dibner, B. 1959. *Science and the Technion*. New York: American Technion Society.

Dirac, P. A. M. 1971. *The Development of Quantum Theory*. J. Robert Oppenheimer Memorial Prize Acceptance Speech. New York: Gordon and Breach Science Publishers.

Doty, P. 1982. "Einstein and International Security." In *Albert Einstein: Historical and Cultural Perspectives*. Edited by G. Holton and J. Elkana. Princeton, N.J.: Princeton University Press, 347–368.

Dukas, H., and Hoffmann, B., eds. 1979. *Albert Einstein: The Human Side*. Princeton, N.J.: Princeton University Press.

Dyson, F. 1995. "The Scientist as Rebel." In *Nature's Imagination: The Frontier of Scientific Vision*. Edited by J. Cornwell. Introduction by Freeman Dyson. Oxford: Oxford University Press, 1–11.

Earman, J., and Norton, J. D., eds. 1997. *The Cosmos of Science: Essays of Exploration*. Pittsburgh: University of Pittsburgh Press.

Eastwood, G. G. 1977. *Harold Laski*. London: Mowbrays.

Eddington, A. 1924. *The Mathematical Theory of Relativity*. 2nd ed. Cambridge: Cambridge University Press.

Einstein, A. 1916. "Näherungsweise Integration der Feldgleichungen der Gravitation." *Königlich Preussische Akademie der Wissenschaften. (Berlin). Sitzungberichete* pt. 1: 688–696.

———. 1918. "Motiv der Forschens." In *Zu Max Plancks 60. Geburtstag: Ansprachen in der Deutschen Physicalischen Gesellschaft*. Karlsruhe: Müller. Reprinted in *Mein Weltbild* and its English translations in Einstein, *Ideas and Opinions*. Rev. ed. New translation and revisions by Sonja Bargmann. New York: Crown Publishers, 1954, 219–220.

———. [1920] 1961. *Relativity: The Special and the General Theory, a Popular Exposition*. 3rd ed. London: Methuen; New York: Wings Books.

———. 1931. *Cosmic Religion: With Other Opinions and Aphorisms*. New York: Covici-Friede.

———. 1934. *Essays in Science*. New York: Philosophical Library.

———. 1938. *The Evolution of Physics: The Growth of Ideas from Early Concepts to Relativity and Quanta*. New York: Simon and Schuster.

———. 1941. "The Common Language of Science." *Advancement of Science* 2(5): 109–110.

———. 1949a. "Autobiographical Notes." In *Albert Einstein: Philosopher-Scientist*. Edited and translated by P. A. Schilpp. Evanston, Ill.: The Library of Living Philosophers.

———. 1949b. *The World as I See It*. New York: Philosophical Library.

———. 1954. *Ideas and Opinions*. Rev. ed. New translation and revisions by S. Bargmann. New York: Crown Publishers.

——. 1955. *The Meaning of Relativity*. 5th ed. Princeton, N.J.: Princeton University Press.

——. 1956a. *Lettres à Maurice Solovine*. Paris: Gauthiers-Villars.

——. 1956b. *Out of My Later Years*. Rev. rpt. ed. New York: Citadel Press, 1995.

——. 1961. *Relativity: The Special and the General Theory*. New York: Wings Books.

——. 1971. *The Born-Einstein Letters: Correspondence between Albert Einstein and Max and Hedwig Born from 1916–1955*. Translated by I. Born. New York: Walker.

——. 1987. *The Collected Papers of Albert Einstein*. Multiple volumes. Edited by J. Stachel et al. Translations by A. Beck. Princeton, N.J.: Princeton University Press.

——. 1991. *The World as I See It*. New York: Carol Publishing Group.

——. 1999. *Einstein's Miraculous Year: Five Papers That Changed the Face of Physics*. Edited by J. Stachel et al. Princeton, N.J.: Princeton University Press.

——. 2005. *The Born–Einstein Letters: Friendship, Politics, and Physics in Uncertain Times: Correspondence between Albert Einstein and Max and Hedwig Born from 1916 to 1955 with Commentaries by Max Born. Translated by Irene Born; note on the new edition by Gustav Born . . . [et al.]*. New York: Macmillan.

——. 2007. *Einstein on Politics: His Private Thoughts and Public Stands on Nationalism, Zionism, War, Peace, and the Bomb*. Edited by D. E. Rowe and R. Schulmann. Princeton, N.J.: Princeton University Press.

Einstein, A., and Besso, M. 1972. *Correspondence 1903–1955*. Translation, notes, and introduction by P. Speziali. Paris: Hermann.

Einstein, A., and Infeld, L. 1942. *The Evolution of Physics*. New York: Simon and Schuster.

Einstein, A., and Marić, M. 1992. *The Love Letters*. Edited by J. Renn and R. Schulmann. Translated by S. Smith. Princeton, N.J.: Princeton University Press.

Einstein, A., Podolsky, B., and Rosen, N. 1935. "Can Quantum-Mechanical Description of Physical Reality Be Considered Complete?" *Physical Review* 47: 777–780.

Einstein, A., and Rosen, N. 1937. "On Gravitational Waves." *Journal of the Franklin Institute* 223: 43–54.

Einstein, A., Tolman, R. C., and Podolsky, B. 1931. "Knowledge of Past and Future in Quantum Mechanics." *Physical Review* 37: 780–781.

Eisenstaedt, J. 1993. "Dark Bodies and Black Holes. Magic Circles and Montgolfiers: Light from Newton to Einstein." *Science in Context* 6: 83–106.

——. 2006. *The Curious History of Relativity: How Einstein's Theory of Gravity Was Lost and Found Again*. Translated by A. Sangalli. Princeton, N.J.: Princeton University Press.

Eliot, T. S. 1963. *Collected Poems: 1909–1962*. London: Faber and Faber.

Elkana Y., and Holton, G., eds. 1982. *Albert Einstein: Historical and Cultural Perspectives. The Centennial Symposium in Jerusalem*. Princeton, N.J.: Princeton University Press.

Erikson, E. H. 1974. *Dimensions of a New Identity*. New York: W. W. Norton.

Feld, B. 1982. "Einstein and the Politics of Nuclear Weapons." In *Albert Einstein: Cultural Perspectives*. Edited by G. Holton and J. Elkana. Princeton, N.J.: Princeton University Press, 369–396.

Feld, B. T. 1979. "Einstein and the Politics of Nuclear Weapons." *Bulletin of the Atomic Scientists* 35(3): 5–16.

Feuer, L. 1974. *Einstein and the Generations of Science*. New York: Basic Books.

Feynman, R. P. 1963. *The Feynman Lectures*. Vol. 1. Reading, Mass.: Addison-Wesley.

——. 1964. "The Quantum Theory of Gravitation." *Acta Physica Polonica* 24: 697–722.

——. 1965. *The Character of Physical Law*. Cambridge, Mass.: MIT Press.

——. 1985. *QED: The Strange Theory of Light and Matter*. Princeton, N.J.: Princeton University Press.

Fine, A. 1986. *The Shaky Game: Einstein, Realism, and the Quantum Theory*. Chicago: University of Chicago Press.

——. 1993. "Einstein's Interpretations of the Quantum Theory." *Science in Context* 6: 257–274.

——. 1996. *The Shaky Game: Einstein, Realism, and the Quantum Theory*. 2nd ed. Chicago: University of Chicago Press.

Fölsing, A. 1997. *Albert Einstein: A Biography*. Translated from the German by E. Osers. New York: The Viking Press.

Forman, P. [1993] 2002. "Recent Science: Late Modern and Postmodern." In *Science Bought and Sold*. Edited by P. Mirowski and E.-M. Sent. Chicago: University of Chicago Press, 109–150.

——. 2007. "The Primacy of Science in Modernity, of Technology in Postmodernity, and of Ideology in the History of Technology." *History and Technology* 23(1/2): 1–152.

Foucault, M. 1975. *Surveiller et Punir: Naissance de la prison*. Paris: Editions Gallimard.

——. 1979. *Discipline and Punish: The Birth of the Prison*. Translated from the French by A. Sheridan. New York: Vintage Books.

——. 1980. "Truth and Power." In *Power/Knowledge: Selected Interviews and Other Writings 1972–1977*. Translated and edited by C. Gordon. Brighton, Sussex: Harvester Press, 109–133.

French, A. P. 1979. *Einstein: A Centenary Volume.* Cambridge, Mass.: Harvard University Press.

Friedman, M. 1992. *Kant and the Exact Sciences.* Cambridge, Mass.: Harvard University Press.

Friess, H. L. 1981. *Felix Adler and Ethical Culture: Memories and Studies.* New York: Columbia University Press.

Galison, E. 2003. *Einstein's Clocks, Poincaré's Maps: Empires of Time.* New York: W. W. Norton.

Galison, P., Holton G., and Schweber, S. S., eds. 2008. *Einstein for the 21st Century.* Princeton, N.J.: Princeton University Press.

Gardner, H. 1993. *Creating Minds: An Anatomy of Creativity Seen Through the Lives of Freud, Einstein, Picasso, Stravinsky, Eliot, Graham, and Gandhi.* New York: Basic Books.

Gilbert, M., ed. 1997. *Winston Churchill and Emery Reves: Correspondence, 1937–1964.* Austin: University of Texas Press.

Glazer, N. 1961. *The Social Basis of American Communism.* New York: Harcourt, Brace and World.

Golden, F. 1999. "Albert Einstein." *Time,* December 31, 62–66.

Goldstein, C., and Ritter, J. 2003. "The Varieties of Unity: Sounding Unified Field Theories 1920–1930." In *Revisiting the Foundations of Relativistic Physics: Festschrift in Honor of John Stachel.* Edited by A. Ashtekar et al. Dordrecht: Kluwer Academic Publishers, 93–149.

Goldstein, I. 1951. *Brandeis University, Chapter of Its Founding.* New York: Bloch Publishing Company.

———. 1984. *My World as a Jew: The Memoirs of Israel Goldstein.* New York: Herzl Press.

Goudsmit, S. A. 1989, 1947. *Alsos.* Los Angeles, Calif.: Tomash Publishers.

Gowing, M. 1964. *Britain and Atomic Energy 1939–1945.* London: Macmillan.

Graubard, S. R. 1999. "Forty Years On." *Daedalus.* Supplement to Vol. 128 of the *Proceedings of the American Academy of Arts and Sciences.*

Greenspan, N. T. 2005. *The End of the Certain World: The Life and Science of Max Born.* New York: Basic Books.

Grene, M., and Nails, D., eds. 1986. *Spinoza and the Sciences.* Dordrecht: D. Reidel Publishing Company.

Gross, D. 2008. "Einstein and the Quest for a Unified Theory." In *Einstein for the 21st Century.* Edited by P. Galison, G. Holton, and S. S. Schweber. Princeton, N.J.: Princeton University Press, 286–297.

Hackman J. R., ed. 1990. *Groups that Work (and Those that Don't): Creating Conditions for Effective Teamwork.* San Francisco: Jossey-Bass.

Hadamard, J. 1945. *An Essay on the Psychology of Invention in the Mathematical Field.* Princeton, N.J.: Princeton University Press.

Heilbron, J. L. 2005. "Oppenheimer's Guru." In *Reappraising Oppenheimer: Centennial Studies and Reflections.* Edited by C. Carson and D. Hollinger. Berkeley: University of California Press, 275–292.

Heilbron, J. L., and Seidel, R. W. 1989. *Lawrence and His Laboratory: A History of the Lawrence Berkeley Laboratory.* Berkeley: University of California Press.

Heisenberg, W. 1955a. *Das Naturbild der heutigen Physik.* Hamburg: Rowohlt.

———. 1955b. "The Scientific Work of Einstein." *Universitas* 10(9): 878–902. Reprinted in Heisenberg 1974.

———. 1958. *The Physicist's Conception of Nature.* New York: Harcourt Brace.

———. 1974. *Across the Frontiers.* New York: Harper & Row.

———. 1989. *Encounters with Einstein.* Princeton, N.J.: Princeton University Press.

Herbert, Z. 1991. "Spinoza's Bed." In *Still Life with a Bridle.* New York: The Ecco Press.

Herken, G. 2002. *Brotherhood of the Bomb: The Tangled Lives and Loyalties of Robert Oppenheimer, Ernest Lawrence, and Edward Teller.* New York: Henry Holt and Company.

Hershberg, J. 1993. *James B. Conant: Harvard to Hiroshima and the Making of the Nuclear Age.* New York: Alfred A. Knopf.

Hijiya, J. A. 2000. "The *Gita* of J. Robert Oppenheimer." *Proceedings of the American Philosophical Society* 144(2): 123–167.

Hoffmann, B., and Dukas, H. 1972. *Albert Einstein: Creator and Rebel.* New York: The Viking Press.

Holloway, D., 1994. *Stalin and the Bomb: The Soviet Union and Atomic Energy, 1939–1956.* New Haven, Conn.: Yale University Press.

Holton, G. 1952. *Introduction to Concepts and Theories in Physical Science.* Cambridge, Mass.: Addison-Wesley.

———. 1958a. "Perspective on the Issue 'Science and the Modern World.' " *Daedalus* (Winter): 3–7.

———. 1958b. *Science and the Modern Mind.* Boston: Beacon Press.

———. 1970. "The Roots of Complementarity." *Daedalus* 99: 1015–1055.

———, ed. 1971. *Science and the Modern Mind.* Freeport, N.Y.: Books for Libraries Press.

———. 1996. *Einstein, History, and Other Passions.* Reading, Mass.: Addison-Wesley.

———. 1998a. *The Advancement of Science, and Its Burdens.* Cambridge, Mass.: Harvard University Press.

———. 1998b. "Einstein and the Cultural Roots of Modern Science." *Daedalus* (Winter): 1–44.

———. 2005. *Victories and Vexations in Science.* Cambridge, Mass.: Harvard University Press.

Holton, G., and Elkana, J. 1982. *Albert Einstein: Historical and Cultural Perspectives.* The Centennial Symposium in Jerusalem. Princeton, N.J.: Princeton University Press.

Howard, D. A. 1993. " 'A Kind of Vessel in Which the Struggle for Eternal Truth Is Played Out': Albert Einstein and the Role of Personality in Science." In *The Natural History of Paradigms: Science and the Process of Intellectual Evolution.* Edited by J. H. Langdon and M. E. McGann. Indianapolis, Ind.: University of Indianapolis Press, 111–138.

———. 1997. "A Peek behind the Veil of Maya: Einstein, Schopenhauer, and the Historical Ground for the Individuation of Physical Systems." In *The Cosmos of Science: Essays of Exploration.* Edited by J. Earman and J. D. Norton. Pittsburgh: University of Pittsburgh Press, 87–152.

———. 2004. "Einstein's Philosophy of Science." In *The Stanford Encyclopedia of Philosophy.* Edited by E. N. Zalta. http://plato.stanford.edu/archives/spr2004/ entries/einstein-philscience/ (accessed June 2007).

Hufbauer, K. 2005. "J. Robert Oppenheimer's Path to Black Holes." In *Reappraising Oppenheimer: Centennial Studies and Reflections.* Edited by C. Carson and D. Hollinger. Berkeley: University of California Press, 36–48.

Hutchins, E. 1995. *Cognition in the Wild.* Cambridge, Mass.: MIT Press.

Infeld, L. 1941. *Quest: An Autobiography.* New York: Chelsea.

———. 1956. "On Equations of Motion in General Relativity Theory." In *Jubilee of Relativity Theory.* Edited by A. Mercier and M. Kervaire. Basel: Birkhäuser, 206–209.

———. 1980. *Quest: An Autobiography.* 2nd ed. New York: Chelsea.

Infeld, L., and Plebański, J. 1960. *Motion and Relativity.* Oxford: Pergamon Press.

Israel, J. I. 2001. *Radical Enlightenment: Philosophy and the Making of Modernity.* Oxford: Oxford University Press.

Jakobson, R. 1982. "Einstein on Language." In *Albert Einstein: Historical and Cultural Perspectives.* Edited by G. Holton and J. Elkana. Princeton, N.J.: Princeton University Press, 139–150.

James, W. 1895. "Is Life Worth Living?" *International Journal of Ethics* 6: 1–24.

———. 1907. *Pragmatism: A New Name for Some Old Ways of Thinking.* New York: Longmans, Green, and Company.

———. [1907] 1981. *Pragmatism.* Edited with an introduction by B. Kuklick. Indianapolis, Ind.: Hackett.

Jammer, M. 1999. *Einstein and Religion.* Princeton, N.J.: Princeton University Press.

Janssen, M., and Lehner, C., eds. 2008. *The Cambridge Companion to Einstein.* Cambridge: Cambridge University Press.

Jenkins, R. 2001. *Churchill*. London: Macmillan.

Jerome, F. 2002. *The Einstein File: J. Edgar Hoover's Secret War against the World's Most Famous Scientist*. New York: St. Martin's Press.

Jerome, F., and Taylor, R. 2005. *Einstein on Race and Racism*. New Brunswick, N.J.: Rutgers University Press.

Johnston, C. 1908. *Bhagavad-Gita*. Translated with an introduction and commentary by C. Johnston. New York: The Quarterly Book Department.

Kaiser, D. 2006. "Whose Mass Is It Anyway? Particle Cosmology and the Objects of Theory." *Social Studies of Science* 36: 533–564.

Kempton, M. 1994. *Rebellions, Perversities and Main Events*. New York: Times Books.

Kennan, G. F. 1967. "Oppenheimer." *Encounter* 28(4): 55–56.

———. 1972. *Memoirs 1950–1963*. Vol. 2. Boston: Little, Brown.

Klein, M. 1986. "Great Connections Come Alive: Bohr, Ehrenfest and Einstein." In The *Lesson of Quantum Theory: Neils Bohr Centenary Symposium held 3–7 October, 1985 in Copenhagen, Denmark*. Edited by J. de Boer, E. Dal, and O. Ulfbeck. Amsterdam: North-Holland, 325–341.

Kox, A. J. 1993. "Einstein and Lorentz: More than Just Good Colleagues." In *Einstein in Context: A Special Issue of Science in Context*. Edited by M. Beller, J. Renn, and R. Cohen, 43–53.

Kragh, H. 1999. *Quantum Generations*. Princeton, N.J.: Princeton University Press.

Kramnick, I., and Sheerman, B. 1993. *Harold Laski: A Life on the Left*. New York: Allen Lane, Penguin Press.

Krober, A. L. 1944. *Configurations of Culture Growth*. San Francisco: University of California Press.

Kuhn, T. S. 1963. "Interview with J. Robert Oppenheimer by Thomas S. Kuhn, November 18, 1963." *Archive for the History of Quantum Physics*. College Park, Md.: American Institute of Physics.

Kuklick, B. 1977. *The Rise of American Philosophy: Cambridge, Massachusetts 1860–1930*. New Haven, Conn.: Yale University Press.

Lanczos, C. 1974. *The Einstein Decade*. New York: Academic Press.

Langdon, J. H., and McGann, M. E., eds. 1993. *The Natural History of Paradigms: Science and the Process of Intellectual Evolution*. Indianapolis, Ind.: University of Indianapolis Press.

Lanouette, W. 1992. *Genius in the Shadows*. New York: Charles Scribner's Sons.

Lapp, R. 1949. *Must We Hide?* Cambridge, Mass.: Addison-Wesley.

Laqueur, W. 1972. *A History of Zionism*. New York: Holt, Rinehart, and Winston.

Laski, H. 1947. "Laski vs. Newark Advertiser Co., Ltd. & Parlby: Before Lord

Goddard, Lord Chief Justice of England and a Special Jury." *London Daily Express*.

Latour, B. 1993. *We Have Never Been Modern*. Translated by C. Porter. Cambridge, Mass.: Harvard University Press.

Laughlin, R. B., and Pines, D. 2000. "The Theory of Everything." *Proceedings of the National Academy of Sciences* 97(1): 28–31.

Laurikainen, K. V. 1988. *Beyond the Atom: The Philosophical Thought of Wolfgang Pauli*. Berlin: Springer Verlag.

Leibniz, G. W. 1934. *Philosophical Writings*. New York: E. P. Dutton.

Lepage, P. 1989. "What Is Renormalization?" In *From Actions to Answers: Proceedings of TASI'89*. Edited by T. DeGrand and D. Toussaint. Singapore: World Scientific.

———. 1997. "How to Renormalize the Schrödinger Equation." Lectures at the VIII Jorge André Swieca Summer School (Brazil, February 1997). arXiv:nucl -th/9706029 1(12) (June 12). Also available at arxiv.org/pdf/nucl-th9706029 (accessed June 2005).

Lilienthal, D. 1964. *The Journals of David E. Lilienthal: The Atomic Energy Years 1945–1950*. Vol. 2. New York: Harper and Row.

———. 1969. *The Journals of David E. Lilienthal: The Road to Change, 1955–1959—The Atomic Energy Years 1945–1950*. Vol. 4. New York: Harper and Row.

Maier, G. 1881. *Mehr Licht! Ein Wort zur "Judenfrage" an unsere christlichen Mitbürger*. Ulm: H. Kerler.

———. 1898. *Der Prozess Zola vor dem Schwurgerichte zu Paris im Februar 1898: Kritischer Bericht eines Augenzeugen*. Bamberg: Verlag der Handels-Druckerei

———. 1919. *Soziale Bewegungen und Theorien bis zur modernen Arbeiterbewegung*. Leipzig: B. G. Teubner.

Marshak, R. E., ed. 1966. *Perspectives on Modern Physics: Essays in Honor of H. A. Bethe*. New York: Interscience Publishers.

Masters, D., and Way, K., eds. 1946. *One World or None*. Foreword by N. Bohr. Introduction by A. H. Compton. New York: McGraw-Hill.

———, eds. 1972. *One World or None*. Freeport, N.Y.: Books for Libraries Press.

McCloy, J. 1953. *The Challenge to American Foreign Policy*. Cambridge, Mass.: Harvard University Press.

McMillan, P. 2005. *The Ruin of J. Robert Oppenheimer and the Birth of the Modern Arms Race*. New York: The Viking Press.

Menand, L. 2001. *The Metaphysical Club: The Story of Ideas in America*. New York: Farrar, Straus and Giroux.

Mercier, A., and Kervaire, M., eds. 1956. *Jubilee of Relativity Theory*. Basel: Birkhäuser.

Merzbacher, E. 2002. "The Early History of Tunneling." *Physics Today* 55(8): 44–49.

Michelmore, P. 1969. *The Swift Years: The Robert Oppenheimer Story.* New York: Dodd, Mead.

Miller, A. 1984. *Imagery in Scientific Thought: Creating 20th-Century Physics.* Boston: Birkhäuser.

———. 1996. *Insights of Genius: Imagery and Creativity in Science.* New York: Copernicus.

Mirowski, P., and Sent, E.-M., eds. 2002. *Science Bought and Sold.* Chicago: University of Chicago Press.

Misner, C. W. 1957. "Feynman Quantization of General Relativity." *Reviews of Modern Physics* 29: 497–509.

Monk, R. 1990. *Ludwig Wittgenstein: The Duty of Genius.* New York: The Free Press.

Morrison, M. 2000. *Unifying Scientific Theories: Physics Concepts and Mathematical Structures.* Cambridge: Cambridge University Press.

Most, G. 2000. "Generating Genres: The Idea of the Tragic." In *Matrices of Genre: Authors, Canons, and Society.* Edited by M. Depew and D. Obbink. Cambridge, Mass.: Harvard University Press, 15–36.

Nadler, S. 1999. *Spinoza: A Life.* Cambridge: Cambridge University Press.

———. 2004. *Spinoza's Heresy: Immortality and the Jewish Mind.* Oxford: Clarendon Press.

Nathan, O., and Norden, H., eds. 1968. *Einstein on Peace.* New York: Schocken Books.

Newman, L. I. 1923. *A Jewish University in America? With a Symposium of Opinions by Educators, Editors and Publicists, and a Bibliography on the Jewish Question in American Colleges.* New York: Bloch Publishing Company.

Nye, M. J. 2004. *Blackett: Physics, War, and Politics in the Twentieth Century.* Cambridge, Mass.: Harvard University Press.

Oppenheimer, J. R. 1939. "Celebration of the Sixtieth Birthday of Albert Einstein." *Science* 89: 335. See also folder "Einstein's 60th birthday—3/16/39," J. Robert Oppenheimer Papers, Manuscript Division, Library of Congress, Washington, D.C., Box 256.

———. 1941. "The Mesotron and the Quantum Theory of Fields." In University of Pennsylvania Bicentennial Conference, *Nuclear Physics, by Fermi [and others].* Philadelphia: University of Pennsylvania Press.

———. 1946. "The Atom Bomb as a Great Force for Peace." *New York Times Magazine,* June 9.

———. 1948. "International Control of Atomic Energy." *Foreign Affairs* 26(2): 239–253.

———. 1950. "The Age of Science: 1900–1950." *Scientific American* 183(3): 20–24.

———. 1953. "Atomic Weapons and American Policy." *Foreign Affairs* 31(4): 525–536.

——. 1954. *Science and the Common Understanding: The Reith Lectures.* New York: Oxford University Press.

——. 1955a. "Article on Einstein for *Princeton Packet.*" J. Robert Oppenheimer Papers, Manuscript Division, Library of Congress, Washington, D.C., Box 256.

——. 1955b. *The Open Mind.* New York: Simon and Schuster.

——. 1956a. "Analogy in Science." *The American Psychologist* 11(3): 127–136.

——. 1956b. "Einstein." *Reviews of Modern Physics* 28: 1–2.

——. 1958a. "The Growth of Science and the Structure of Culture." *Daedalus* (Winter): 67–76.

——. 1958b. "A Study of Thinking." Book Review. *The Sewanee Review* 46: 481–490.

——. 1959. "Talk to Undergraduates." *Kansas Teacher* (February): 18–20, 39–40.

——. 1960. *Some Reflections on Science and Culture.* Chapel Hill: University of North Carolina Press.

——. 1964a. *The Flying Trapeze: Three Crises for Physicists.* The Whidden Lectures for 1962. New York: Oxford University Press.

——. 1964b. "Neils Bohr and Nuclear Weapons." *New York Review of Books* 3 (December 17): 6–8.

——. 1965. "The Symmetries of Forces and States." In *Preludes in Theoretical Physics in Honor of V. F. Weisskopf.* Edited by A. de-Shalit, H. Feshbach, and L. van Hove. Amsterdam: North-Holland, 70–78.

——. 1966a. "On Albert Einstein." *New York Review of Books,* March 17, 4–5.

——. 1966b. "Perspectives on Modern Physics." In *Perspectives on Modern Physics: Essays in Honor of H. A. Bethe.* Edited by R. E. Marshak. New York: Interscience Publishers, 9–20.

——. 1967. *Oppenheimer.* New York: Charles Scribner's Sons.

——. 1970. *In the Matter of J. Robert Oppenheimer: Transcript of Hearing before Personnel Security Board and Texts of Principal Documents and Letters.* Foreword by P. M. Stern. Cambridge, Mass.: MIT Press.

——. 1980. *Letters and Recollections.* Edited by A. K. Smith and C. Weiner. Cambridge, Mass.: Harvard University Press.

——. 1984. *Uncommon Sense.* Edited by N. Metropolis, G. C. Rota, and D. Sharp. Basel: Birkhäuser.

——. 1989. *Atom and Void: Essays on Science and Community.* Princeton, N.J.: Princeton University Press. (This collection is a reprint of Oppenheimer 1954 and Oppenheimer 1964.)

Oppenheimer, J. R., and Snyder, H. 1939. "On Continued Gravitational Contraction." *Physical Review* 565: 455–459.

Oppenheimer, J. R., and Volkoff, G. M. 1939. "On Massive Neutron Cores." *Physical Review* 55: 374–381.

Oppenheimer, J. R., et al. 1965. *Comment vivre demain?* Textes des conférences et des entretiens organizes par les Rencontres internationales de Genève 1964. Neuchâtel: La Baconnière.

Pais, A. 1982. *"Subtle Is the Lord—": The Science and the Life of Albert Einstein.* New York: Oxford University Press.

———. 1991. *Neils Bohr's Times in Physics, Philosophy and Polity.* Oxford: Clarendon Press.

———. 1994. *Einstein Lived Here: Essays for the Layman.* New York: Oxford University Press.

———. 1997. *A Tale of Two Continents: The Life of a Physicist in a Turbulent World.* Princeton, N.J.: Princeton University Press.

———. 2000. *The Genius of Science: A Portrait Gallery of Twentieth Century Physicists.* Oxford: Oxford University Press.

———. 2006. *J. Robert Oppenheimer: A Life.* With supplemental material by R. P. Crease. New York: Oxford University Press.

Palevsky, M. 2000. *Atomic Fragments: A Daughter's Questions.* Berkeley: University of California Press.

Pauli, W. 1922. "Relativitätstheorie." In *Encyklopädie der Mathematischen Wissenschaften.* Vol. 5. Leipzig: B. G. Teubner, 621–858.

———. 1979–2005. *Scientific Correspondence with Bohr, Einstein, Heisenberg, among Others.* Edited by A. Hermann, K. van Meyenn, and V. F. Weisskopf. New York: Springer.

Paulsen, F. 1906. *The German Universities and University Study.* Translated by F. Thilly and W. W. Elwang. New York: C. Scribner's Sons.

Peierls, R. 1970. "J. Robert Oppenheimer." In *Dictionary of Scientific Biography.* Edited by C. C. Gillespie. New York: Charles Scribner's Sons.

Polenberg, R., ed. 2002. *In the Matter of J. Robert Oppenheimer: The Security Clearence Hearing.* Ithaca, N.Y.: Cornell University Press.

Price, D. K. 1967. "J. Robert Oppenheimer." *Science* 155(3766): 1061.

Quine, W. V. 1985. *The Time of My Life: An Autobiography.* Cambridge, Mass.: MIT Press.

Rabi, I., Serber, R., Weisskopf, V., Pais, A., and Seaborg. G. T. 1969. *Oppenheimer.* New York: Charles Scribner's Sons.

Regis, E. 1987. *Who Got Einstein's Office?* Reading, Mass.: Addison-Wesley.

Reiser, A. 1931. [pseudonym, Rudolf Kayser] *Albert Einstein: A Biographical Portrait.* London: Thoernton Butterworth Limited.

Renn, J. 1993. "A Comparative Study of Concept Formation in Physics." *Science in Context* 6: 311–344.

Renn, J., and Schemmel, M., eds. 2007. *The Genesis of General Relativity*. 4 vols. Dordrecht: Springer.

Reston, J. 1991. *Deadline: A Memoir*. New York: Random House.

Reves, E. 1942. *A Democratic Manifesto*. New York: Random House.

——. 1945. *The Anatomy of Peace*. New York: Harper and Brothers.

Rhodes, R. 1988. *The Making of the Atomic Bomb*. New York: Simon and Schuster.

——. 1995. *Dark Sun*. New York: Simon and Schuster.

Rieff, P. 1969. *On Intellectuals: Theoretical Studies, Case Studies*. Garden City, N.Y.: Doubleday.

Rotblat, J. 1967. *Pugwash: A History of the Conferences on Science and World Affairs*. Prague: Czechoslovak Academy of Sciences.

——. 1995. "Post-Hiroshima Campaigns by Scientists to Prevent the Future Use of Nuclear Weapons." http://www.spusa.org/pubs/speeches/Rotblatspeech. pdf (accessed June 2005).

——. 1996. "A War-Free World." http://www.cnn.com/EVENTS/1996/world. report.conference/news/08/rotblat.speech/index.html (accessed June 2005).

——. 2001. "The Early Days of Pugwash." *Physics Today* 54(6): 50–55.

——. 2002. "The Nuclear Issue after the Posture Review." http://www.waging peace.org/articles/2002/06/00_rotblat_nuclear-issue.htm (accessed June 2005).

Roth, J. K., ed. 1969a. *The Moral Philosophy of William James*. New York: Thomas Y. Crowell.

——, ed. 1969b. *Freedom of the Moral Life: The Ethics of William James*. Philadelphia: The Westminster Press.

Rummel, J. 1992. *Robert Oppenheimer: Dark Prince*. New York: Facts on File.

Russell, B. 1969. *Autobiography*. Vol. 3. London: George Allen and Unwin.

Russell, B., and Whitehead, A. N. 1910–1913. *Principia Mathematica*. Cambridge: Cambridge University Press.

Ryder, A. W. 1929. *The Bhagavad-Gita*. Chicago: University of Chicago Press.

Sachar, A. 1976. *Brandeis University: A Host at Last*. Boston: Little, Brown.

——. 1996. *Brandeis University: A Host at Last*. Rev. ed. Hanover, N.H: University Press of New England [for] Brandeis University Press.

Safranski, R. 1990. *Schopenhauer and the Wild Years of Philosophy*. Cambridge, Mass.: Harvard University Press.

Sauer, T. 2007. "Einstein's Unified Field Theory Program." http://philsci -archive.pitt.edu/archive/00003293/ (accessed June 2006).

——. 2008. "Einstein's Unified Field Theory Program." In *The Cambridge Companion to Einstein*. Edited by M. Jannsen and C. Lehner. Cambridge: Cambridge University Press.

Sayen, J. 1985. *Einstein in America: The Scientist's Conscience in the Age of Hitler and Hiroshima.* New York: Crown Publishers.

Scheideler, B. 2003. "The Scientist as Moral Authority: Albert Einstein between Elitism and Democracy, 1914–1933." *Historical Studies in the Physical and Biological Sciences* 32(2): 319–346.

Schilpp, P. A., ed. 1949. *Albert Einstein, Philosopher-Scientist.* Evanston, Ill.: Library of Living Philosophers.

Schlick, M. 1918. *Algemeine Erkentnislehre.* Berlin: J. Springer.

Scholem, G. 2002. *A Life in Letters, 1914–1982.* Edited and translated by A. D. Skinner. Cambridge, Mass.: Harvard University Press.

Schopenhauer, A. 1966. *The World as Will and Representation.* Translated from the German by E. F. J. Payne. Vols. 1 and 2. New York: Oxford University Press.

———. 1974. *Parerga and Paralipomena; Short Philosophical Essays.* Translated from the German by E. F. J. Payne. Vols. 1 and 2. Oxford: Clarendon Press.

Schrecker, E. 1986. *No Ivory Tower: McCarthyism and the Universities.* New York: Oxford University Press.

Schweber, S. S. 1988. "The Empiricist Temper Regnant: Theoretical Physics in the United States 1920–1950." *Historical Studies in the Physical Sciences* 17: 17–98.

———. 1994. *QED and the Men Who Made It.* Princeton, N.J.: Princeton University Press.

———. 2000. *In the Shadow of the Bomb.* Princeton, N.J.: Princeton University Press.

———. 2003. "J. Robert Oppenheimer: Proteus Unbound." *Science in Context* 16: 219–242.

———. 2005. "Intersections: Einstein and Oppenheimer." In *Reappraising Oppenheimer: Centennial Studies and Reflections.* Edited by C. Carson and D. Hollinger. Berkeley: University of California Press, 343–360.

———. 2006. "Einstein and Oppenheimer: Interactions and Intersections." *Science in Context* 19(4): 513–559.

Schwinger, J. 1962. "Non-Abelian Gauge Fields: Relativistic Invariance." *Physical Review* 127: 324–330.

———. 1963. "Quantized Gravitational Field." *Physical Review* 130: 1253–1258.

———. 1986. *Einstein's Legacy: The Unity of Space and Time.* New York: Scientific American Books.

Scott, J., and Keates, D., eds. 2001. *Schools of Thought: Twenty-five Years of Interpretive Social Science.* Princeton, N.J.: Princeton University Press.

Seelig, C. 1956. *Albert Einstein—A Documentary Bibliography.* London: Staples Press.

Serber, R. 1983. "Particle Physics in the 1930s: A View from Berkeley." In *The Birth of Particle Physics*. Edited by L. M. Brown and L. Hoddeson. Cambridge: Cambridge University Press, 206–221.

Serber, R., with Crease, R. P. 1998. *Peace and War: Reminiscences of a Life on the Frontiers of Science*. New York: Columbia University Press.

Shimony, A. 1998. "The Relationship between Physics and Philosophy." In *Philosophies of Nature: The Human Dimension*. Edited by R. S. Cohen and A. Tauber. Dordrecht: Kluwer Academic Publishers, 177–182.

Sime, R. L. 1996. *Lise Meitner: A Life in Physics*. Berkeley: University of California Press.

Slaughter, S., and Rhoades, G. 2002. "The Emergence of a Competitiveness and Development Policy Coalition and the Commercialization of Academic Science and Technology." In *Science Bought and Sold*. Edited by P. Mirowski and E. -M. Sent. Chicago: University of Chicago Press, 69–108.

Smith, A. K. 1965. *A Peril and a Hope: The Scientists' Movement in America, 1945–1947*. Chicago: University of Chicago Press.

Smith, A. K., and Weiner, C., eds. 1980. *Robert Oppenheimer: Letters and Recollections*. Cambridge, Mass.: Harvard University Press.

Smyth, H. de Wolf. 1989. *Atomic Bombs. Atomic Energy for Military Purposes: The Official Report on the Development of the Atomic Bomb under the Auspices of the United States Government, 1940–1945; with a new foreword by Philip Morrison and an essay by Henry DeWolf Smyth*. Stanford, Calif.: Stanford University Press.

Snow, C. P. 1959. *The Two Cultures and the Scientific Revolution: The Rede Lecture 1959*. Cambridge: Cambridge University Press.

———. 1963. *The Two Cultures: A Second Look*. Cambridge: Cambridge University Press.

Soames, M., ed. 1998. *Speaking for Themselves: The Personal Letters of Winston and Clementine Churchill*. New York: Doubleday.

Stachel, J. 1993. "The Other Einstein: Einstein contra Field Theory." *Science in Context* 6: 275–290.

———. 2002. *Einstein from "B" to "Z."* Boston: Birkhäuser.

Stern, A. 1945. "Interview with Einstein." *Contemporary Jewish Record* (June): 245–249.

Stone, I. F. 1946. "Atomic Pie in the Sky." *The Nation* 162 (April 6).

Strauss, L. 1965. *Spinoza's Critique of Religion*. New York: Schocken Books.

Szilard, L., and Zinn, W. H. 1939. "Instantaneous Emission of Fast Neutrons in the Interaction of Slow Neutrons with Uranium." *Physical Review* 35(8): 799–800.

Taylor, C. 2001. "Modernity and Identity." In *Schools of Thought: Twenty-five Years of Interpretive Social Science*. Edited by J. Scott and D. Keates. Princeton, N.J.: Princeton University Press, 139–153.

Telegdi, V. L. 2000. "Szilard as Inventor: Accelerators and More." *Physics Today* 53(10): 25–28.

Thorpe, C. 2005. "The Scientist in Mass Society: J. Robert Oppenheimer and the Postwar Liberal Imagination." In *Reappraising Oppenheimer: Centennial Studies and Reflections*. Edited by C. Carson and D. Hollinger. Berkeley: University of California Press, 293–314.

———. 2006. *Oppenheimer*. Chicago: University of Chicago Press.

Tolman, R. 1942. "Psychologists' Services in the Field of Agriculture." *Journal of Consulting Psychology* 6(2): 62–68.

———. 1948. "Cognitive Maps in Rats and Men." *Psychological Reviews* 55(4): 189–208.

———. 1950. "A Semantic Study of Concepts of Clinical Psychologists and Psychiatrists." *The Journal of Abnormal and Social Psychology* 45: 216–231.

———. 1953. "Virtue Rewarded and Vice Punished." *The American Psychologist* 8: 721–733.

United States Atomic Energy Commission. 1954. *In the Matter of J. Robert Oppenheimer*. Washington, D.C.: U.S. Goverment Printing Office.

Utiyama, R. 1956. "Invariant Theoretical Interpretation of Interaction." *Physical Review* 101: 1597–1607.

Vizgin, V. P. 1994. *Unified Field Theories in the First Third of the 20th Century*. Basel: Birkhäuser.

Wang, H. 1987. *Reflections on Kurt Gödel*. Cambridge, Mass.: MIT Press.

Wang, J. 1999. *American Science in an Age of Anxiety: Scientists, Anticommunism and the Cold War*. Chapel Hill: University of North Carolina Press.

Weber, M. 1946. *Essays in Sociology*. Translated and edited by H. H. Gerth and C. Wright Mills. New York: Oxford University Press.

Weil, S. 1952. *The Need for Roots*. London: Routledge & Kegan Paul.

Wells, H. G. 1914. *The World Set Free*. London: Macmillan.

Wertheimer, M. 1959. *Productive Thinking*. New York: Harper.

Weyl, H. 1922a. *Raum, Zeit, Materie: Vorlesungen über allgemeine Relativitätstheorie*. Berlin: J. Springer.

———. 1922b. *Space, Time, Matter*. London: Methuen.

Wheeler, J. A. 1980a. *Albert Einstein: His Strength and His Struggle*. Leeds: Leeds University Press.

———. 1980b. "Einstein." *Biographical Memoirs of the National Academy of Sciences* 51: 97–117.

White, M. 1949. *Social Thought in America: The Revolt against Formalism*. New York: The Viking Press.

———. 1955. *The Age of Analysis: Twentieth Century Philosophers*. Boston: Houghton Mifflin.

——. 1956. *Toward Reunion in Philosophy*. Cambridge, Mass.: Harvard University Press.

——. 1957. *Social Thought in America: The Revolt against Formalism*. Paperback edition with a new preface and an epilogue. Boston: The Beacon Press.

——. 1959. "Experiment and Necessity in Dewey's Philosophy." *Antioch Review* 19: 329–344.

——. 1973. *Pragmatism and the American Mind: Essays and Reviews in Philosophy and Intellectual History*. New York: Oxford University Press.

——. 1999. *A Philosopher's Story*. University Park: Pennsylvania State University Press.

Wigner, E. P. 1957. "Relativistic Invariance and Quantum Phenomena." *Reviews of Modern Physics* 29(3): 255–268.

——. 1980. "Thirty Years of Knowing Einstein." In *Some Strangeness in the Proportion: A Centennial Symposium to Celebrate the Achievements of Albet Einstein*. Edited by H. Woolf. Reading, Mass.: Addison-Wesley, 461–468.

——. 1983. "The Glorious Days of Physics." In *The Unity of the Fundamental Interactions*. Edited by A. Zichici. New York: Plenum Press, 765–774. Also reprinted in Wigner 1992, vol. 6, 610–625.

——. 1992. *Collected Works of E. P. Wigner*. Part B. *Philosophical and Reflections and Synthesis*. Edited by Jagdish Mehra. Berlin: Springer Verlag.

Will, C. M. 1986. *Was Einstein Right?* New York: Basic Books.

Williams, W. A. 1962. *The Tragedy of American Diplomacy*. New York: Dell Publishing Company.

Wilson, D. J. 1990. *Science, Community, and the Transformation of American Philosophy, 1860–1930*. Chicago: University of Chicago Press.

Wilson, R. R. 1996. "Hiroshima: The Scientists' Social and Political Reaction." *Proceedings of the American Philosophical Society* 140(3) (September): 350–357.

Wittgenstein, L. 1922. *Tractacus Logico-Philosophicus*. English-German. London: Kegan Paul, Trench, Trubner & Company.

Wolff, S. L. 2000. "Physicists in the 'Krieg der Geister': Wilhelm Wein's 'Proclamation'." *Historical Studies in the Physical Sciences* 33: 7–368.

Woolf, H., ed. 1980. *Some Strangeness in the Proportion: A Centennial Symposium to Celebrate the Achievements of Albert Einstein*. Reading, Mass.: Addison-Wesley.

Yang, C. N. [1983] 2005. *Selected Papers (1945–1980): With Commentary*. Singapore: World Publishing.

York, H. 1987. *Making Weapons, Talking Peace*. New York: Basic Books.

——. 1989. *The Advisors: Oppenheimer, Teller, and the Superbomb*. With a historical essay by H. A. Bethe. Stanford, Calif.: Stanford University Press.

Yourgrau, P. 1999. *Gödel Meets Einstein: Time Travel in the Gödel Universe.* Chicago: Open Court.

———. 2005. *A World without Time: The Forgotten Legacy of Gödel and Einstein.* New York: Basic Books.

Yuan, L. 1965. *Nature of Matter: Purposes of High Energy Physics.* Upton, N.Y.: Brookhaven National Laboratories.

Zichichi, A., ed. 1982. *The Unity of the Fundamental Interactions.* New York: Plenum Press.

Zukerman, W. 1947. "March of Jewish Events." *The American Hebrew,* July 4, 2–10.

# Acknowledgments

What appears between these covers is the result of lectures I gave during the 2005 Einstein celebrations and of my continued involvement with the life of Oppenheimer. In my undertaking I have had the benefit of the very helpful, constructive criticisms and assistance of friends and colleagues, in particular Finn Aaserud, Cathryn Carson, Leo Corry, Raine Daston, Paul Forman, Snait Gissis, Gerald Holton, Don Howard, Evelyn Fox Keller, Victor McElheny, Howard Schnitzer, John Stachel, and Skuli Sigurdsson. I am deeply indebted to them. Similarly, the encouragement and support of Michael Fisher, editor-in-chief at Harvard University Press, is gratefully acknowledged. I am also grateful to the anonymous reviewer of my manuscript whom I had "angered" with my preface for his thoughtful, incisive, and sharp but valuable criticisms. It is a better book because of them. It is also much more readable by virtue of the editorial assistance I received. John Donohue and Betty Pessagno of Westchester Book Services were responsible for the copy editing, and I thank them for their valuable help. I am also deeply appreciative of Anne Zarrella's efforts at Harvard University Press to obtain photographs for the book, and for her labors in trying to get hold of some that proved unobtainable.

Some of the sections of Chapter 1 are based on a paper delivered at the April 10–13, 2005, Israel Academy of Sciences Symposium, "Albert Einstein's Legacy—A One Hundred Years Perspective." Some of these materials are contained in the chapter titled "Einstein and Nuclear Weapons," my contribution to the volume *Einstein for the Twenty-First Century*, edited by Peter Galison, Gerald Holton, and Silvan S. Schweber, to be published by Princeton University Press. I thank Princeton University Press for permission for their use. I would like to thank Springer Science and Business Media for the use in Chapter 2 of materials in my article "Albert Einstein and the Founding of Brandeis University," which was my contribution to the festschrift in honor of John Stachel, *Revisiting the Foundations of Relativistic Physics*, edited by Ashtekar, Cohen, Howard, Renn, Sarkar, and Shimony (Kluwer Academic Publishers, 2003), vol. 234 of the Boston Studies in the Philosophy of Science (BSPH), pp. 615–640.

Some of the sections of Chapter 6 are based on materials that appeared in my article "Intersections: Einstein and Oppenheimer," in *Reappraising Oppenheimer*, edited by Cathryn Carson and David A. Hollinger (Berkeley: Office for History of Science and Technology, University of California, 2005). I thank Cathryn Carson, the editor of the *Berkeley Papers on the History of Science*, and the Office for History of Science and Technology, the publishers of the series, for permission to use those materials. Similarly, I thank the editors of *Science in Context*, Leo Corry, Aléxandre Métraux, and Jürgen Renn, and Cambridge University Press for permission to make use of my two articles on Oppenheimer that were published in *Science in Context*: "J. Robert Oppenheimer: Proteus Unbound," *Science in Context* 16 (2003): 219–242; and "Einstein and Oppenheimer: Interactions and Intersections," *Science in Context* 19 (2006): 513–559.

I would like to thank Saul Cohen for allowing me to read the manuscript of his autobiography, *At Brandeis*, and Art Reis for sharing with me his research on the early history of Brandeis and for allowing me to consult the materials in the Farber University Archives.

My thanks to the archivists at the Hebrew National Library, Chaya Becker and Barbara Wolf, for their help with the Einstein Papers and to Zeev Rozenkranz, the Bern Dibner curator of the Albert Einstein Archives, for permission to quote from them; to Abigail A. Schooman, the archivist at the American Jewish Historical Society, for permission to quote from the Wise Papers; and to Eliot Wilczek and Lisa C. Long, the archivists in Brandeis University's Farber University Library, for their courteous and helpful assistance and for permission to quote from the Alpert and Sachar Papers, from the minutes of the Board of Trustees of Brandeis University, and of the Albert Einstein and Brandeis Foundations.

It is traditional for authors to take responsibility for all the mistakes that are still between the covers of their books. Needless to say, I do so. I am very conscious of the fact that much in the book is an amalgamation, distillation, and synthesis of materials I have read and heard elsewhere—in lectures, in seminars, and in workshops—and that I could not acknowledge this specifically. Being a member of the history of science community, and in particular of the history of science community in the Greater Boston area, being an associate of the department of the history of science at Harvard is what made this book possible. I do not know how to thank them.

# Index